OPTICAL INTERFEROMETRY

Edited by **Alexander A. Banishev, Mithun Bhowmick** and **Jue Wang**

Optical Interferometry
http://dx.doi.org/10.5772/63683
Edited by Alexander A. Banishev, Mithun Bhowmick and Jue Wang

Contributors

Dahi Ghareab Abdelsalam Ibrahim, Baoli Yao, Daodang Wang, Rongguang Liang, Rogerio Dionisio, Alexander P. Vladimirov, Alexey A. Bakharev, Feng Gao, Lutang Wang, Nian Fang, Sanichiro Yoshida, David Didie, Jong-Sung Kim, Ik-Keun Park, Fei Yang, Dan Xu, Haiwen Cai, Ronghui Qu, qui Tran-Cong-Miyata, Dan-Thuy Van-Pham, Hideyuki Nakanishi, Tomohisa Norisuye, Aurelien Bruyant, Julien Vaillant, Yi Huang, Yunlong Zhu, Tzu-Heng Wu, Abeer Al Mohtar, Jesus Muñoz Maciel, Francisco J. Casillas, Francisco G. Peña-Lecona, Victor Manuel Duran Ramirez, Miguel Mora-González

Published by InTech

Janeza Trdine 9, 51000 Rijeka, Croatia

Publishing Process Manager Martina Usljebrka
Technical Editor SPi Global
Cover InTech Design team

First published February, 2017
Printed in Croatia
Legal deposit, Croatia: National and University Library in Zagreb

Additional hard copies can be obtained from orders@intechopen.com

Optical Interferometry, Edited by Alexander A. Banishev, Mithun Bhowmick and Jue Wang
p. cm.
Print ISBN 978-953-51-2955-4
Online ISBN 978-953-51-2956-1

For EU product safety concerns:
IN TECH d.o.o., Prolaz Marije Krucifikse Kozulić 3, 51000 Rijeka, Croatia,
info@intechopen.com or visit our website at intechopen.com.

Meet the editor

Dr. Alexander Banishev is a research associate at the University of Illinois, Urbana-Champaign (UIUC) (USA). He received his PhD degree in Physics at Lomonosov Moscow State University (Russia) in 2008. Before joining the University of Illinois, he was working at the University of California, Riverside. His primary research interests include optical and laser physics, light interaction with nano- and molecular materials, and optical properties of materials under extreme conditions and precise measurements. He reviews articles for *Optical Society of America* journals.

Dr. Mithun Bhowmick completed his PhD degree in Physics from Virginia Tech in 2012 and currently is working as a research associate at the University of Illinois, Urbana-Champaign (UIUC). Before Joining the UIUC, he worked at Nazarbayev University and National Autonomous University of Mexico as full-time and visiting professors, respectively. His research interests are optical properties of semiconductors, many body effects in scattering mechanisms of carriers/spins, and energetic liquids under extreme conditions. He is a recipient of Jamie Dunn Award, Philip Morris Fellowship, and Ministry of Education and Science grant from the Republic of Kazakhstan for his studies on semiconductor heterostructures. He is a member of American Physical Society and Optical Society of America and reviews articles for international journals on materials science/applied physics (*Physical Review B*, *Journal of Applied Physics*, etc.).

Dr. Jue Wang is a research associate at the University of Illinois, Urbana-Champaign, USA. He received his PhD degree in Geosciences at Princeton University in 2014. Before joining the University of Illinois, he was studying equation of state and structure of materials at high pressures and high temperatures at Princeton University. His research interests include physical and chemical properties of materials at extreme conditions. He reviews articles for *Scientific Reports* and *Optics Letters*.

Contents

Preface IX

Chapter 1 **Digital Processing Techniques for Fringe Analysis 1**
Jesús Muñoz Maciel, Francisco Javier Casillas Rodríguez, Miguel Mora González, Francisco Gerardo Peña Lecona and Víctor Manuel Duran Ramírez

Chapter 2 **Applications of Mach-Zehnder Interferometry to Studies on Local Deformation of Polymers Under Photocuring 25**
Dan-Thuy Van-Pham, Minh-Tri Nguyen, Hideyuki Nakanishi, Tomohisa Norisuye and Qui Tran-Cong-Miyata

Chapter 3 **Interferometry for Online/In-Process Surface Inspection 41**
Feng Gao

Chapter 4 **Application of Optical Interferometry for Characterization of Thin-Film Adhesion 61**
Sanichiro Yoshida, David R. Didie, Jong-Sung Kim and Ik-Keun Park

Chapter 5 **Interferometry and its Applications in Surface Metrology 81**
Dahi Ghareab Abdelsalam and Baoli Yao

Chapter 6 **Dynamic Speckle Interferometry of Thin Biological Objects: Theory, Experiments, and Practical Perspectives 103**
Alexander P. Vladimirov and Alexey A. Bakharev

Chapter 7 **Applications of Fiber-Optic Interferometry Technology in Sensor Fields 143**
Lutang Wang and Nian Fang

Chapter 8 **Interferometry Applications in All-Optical Communications Networks 167**
Rogerio Pais Dionisio

Chapter 9 **Point Diffraction Interferometry 187**
Daodang Wang and Rongguang Liang

Chapter 10 **Interferometry Using Generalized Lock-in Amplifier (G-LIA): A Versatile Approach for Phase-Sensitive Sensing and Imaging 211**
Aurélien Bruyant, Julien Vaillant, Tzu-Heng Wu, Yunlong Zhu, Yi Huang and Abeer Al Mohtar

Chapter 11 **120° Phase Difference Interference Technology Based on 3 × 3 Coupler and its Application in Laser Noise Measurement 233**
Yang Fei, Xu Dan, Cai Haiwen and Qu Ronghui

Preface

The aim of this book is to present readers with the latest progress in optical interferometry. It has descriptions of some of the state-of-the-art interferometry techniques and their applications and implications for future research. At present, optical methods of measurements are the most sensitive techniques of noncontact investigations, and at the same time, they are fast as well as accurate which increases reproducibility of observed results. In recent years, the importance of optical interferometry methods for research has dramatically increased, and applications range from precise surface testing to finding extrasolar planets.

The book has a wide variety of topics in interferometry including detailed descriptions of novel apparatuses and methods, as well as application interferometry for studying biological objects, surface qualities, materials characterization, and optical testing. The book consists of eleven chapters. Chapter 1 reviews the techniques for interferogram processing and interpretation. Chapter 2 presents in situ study of local polymer deformation using Mach-Zehnder interferometer. Chapter 3 covers two approaches based on optical interferometry for accurate surface inspection. Chapter 4 discusses the application of optical interferometry for characterization of thin-film systems. Chapter 5 contains fundamentals of interferometry and its application in investigation of surface shapes. Chapter 6 describes the theory and experimental basis of dynamic speckle interferometry and its application in the investigation of biological objects. . Chapter 7 includes basics of fiber-optic interferometry techniques and their applications in sensor fields. Chapter 8 deals with applications of interferometric techniques in all-optical communications networks. Chapter 9 discusses point diffraction interferometer systems for wavefront and surface testing. Chapter 10 covers the method based on generalized lock-in amplifier for extracting phase and amplitude when arbitrary phase modulations are used. The examples of practical implementation of the method are described. Chapter 11 describes 120-degree phase difference technology and its application in laser phase and frequency noise measurement. The book will be interesting for scholars who are exploring theory and experiment in developing advanced interferometric techniques and their applications.

We would like to thank all the contributors who invested time and shared their expertise in determining the scientific merit of the chapters submitted to this book. Our special acknowledgment to the Publishing Process Manager, Ms. Martina Usljebrka, and other staffs of InTech publishing whose constant support and great efforts made this book both possible and successful.

Alexander Banishev
University of Illinois at Urbana-Champaign, USA

Mithun Bhowmick
University of Illinois at Urbana-Champaign, USA

Jue Wang
University of Illinois at Urbana-Champaign, USA

Chapter 1

Digital Processing Techniques for Fringe Analysis

Jesús Muñoz Maciel,
Francisco Javier Casillas Rodríguez,
Miguel Mora González,
Francisco Gerardo Peña Lecona and
Víctor Manuel Duran Ramírez

Additional information is available at the end of the chapter

http://dx.doi.org/10.5772/66474

Abstract

Digital image processing techniques are needed in order to recover the object information encoded in fringe patterns generated in a determined interferometric setup. Main fringe analysis techniques are reviewed in order to give the reader the most fundamental insights for the interpretation of interferograms. Phase shifted, open fringe, lateral shear and other types of interferograms make use of specific procedures to correctly retrieving the searched phase. Here, algorithms are described and tested in numerical simulations and real data.

Keywords: fringe analysis, phase recovery, phase unwrapping

1. Introduction

Interferometric techniques are widely used for measuring a wide range of physical variables including refraction index, deformations, temperature gradients, etc. Typically, an interferometer is used to generate one or several interferometric fringe patterns that contain the information of the physical variable that is being measured. Those images must be interpreted in order to recover the parameters that are encoded in the fringe patterns generated by the interferometric setup. Thus, fringe analysis methods deal with the problem of a three-dimensional reconstruction (the object information) from a two-dimensional intensity patterns (interferograms) acquired by a CCD camera. Digital interferometry became extensively used since the development of lasers and CCD devices. In those years, however, the resulting interferograms had to be interpreted visually, and only qualitative results were often achieved. Visual interpretation of an interferogram with only straight or circular fringes is not difficult, but things become more complicated for a fringe pattern that combines several regions with circular, straight and crossed fringes of varying density. Rapidly, it was recognized the need of automatic methods for fringe analysis. The first great advance arises with the development of the phase-shifting techniques. With those procedures, a set of interferograms is acquired with a phase shift among them. The phase shifts are usually introduced by a piezoelectric transducer moving the reference mirror in such way that the phase difference between two consecutive interferograms is a constant term. With phase-shifting techniques, it is possible to isolate the sine and cosine of the phase allowing the calculation of the wrapped phase distribution and consequently the continuous phase with an unwrapping algorithm. Another great success came with the method proposed by Takeda (also referred as the Fourier method) performing a band-pass filtering in the Fourier domain. The method of Takeda works only with interferograms that contain open fringes (patterns that consist in nearly straight fringes). In order to generate such interferograms, the reference beam (e.g., in a two arm interferometer) is tilted introducing a large carrier function to the phase. The Fourier transform of these interferograms is composed of three lobules, one at the center that corresponds to the background term and two lobules located symmetrically respect to the origin. One of this lobules and the one that is located at the origin are filtered out. The remaining spectrum is transformed back to the spatial domain from which the so-called wrapped phase can be calculated. A final step is to apply a phase unwrapping technique to recover the continuous phase. Interferometric measurements and fringe analysis techniques are a growing and fast-changing field of research. Through this chapter, we will review the most known procedures.

2. Interferogram acquisition

The wave nature of light can be studied theoretically by a homogeneous partial differential equation of second order, which satisfies the superposition principle:

$$\nabla U(x,y,z,t) = \frac{1}{c^2}\frac{\partial U(x,y,z,t)}{\partial t}. \tag{1}$$

If two waves of the same frequency are superimposed on a point in space, they excite oscillations in the same direction:

$$U_1(t) = A_1 \cos(\omega t + \varphi_1), \tag{2}$$

and

$$U_2(t) = A_2 \cos(\omega t + \varphi_2). \tag{3}$$

In the preceding equations and the subsequent ones in this section, we will drop the spatial dependence for displaying purposes. The amplitude of the resulting oscillation at that point is determined by the equation:

$$A^2 = {A_1}^2 + {A_2}^2 + 2A_1A_2\cos(\delta), \tag{4}$$

where

$$\delta = \varphi_1 - \varphi_2. \tag{5}$$

If the phase difference, δ, of the oscillations excited by waves remains constant with time, these waves are coherent. In the case of noncoherent waves, δ varies continuosly, taking any values with equal probability, so the average value of $cos(\delta)$ is zero. Therefore,

$$A^2 = A_1^2 + A_2^2. \tag{6}$$

So, we can conclude that the intensity observed in the superposed point by noncoherent waves equals the sum of the intensities, which create each separately. However, if the difference $\delta = \varphi_1 - \varphi_2$ is constant, the $cos(\delta)$ will also have a constant value over time, but own for each point in space, so that:

$$I = I_1 + I_2 + 2\sqrt{I_1 I_2}\cos(\delta) \tag{7}$$

where $I_1 = A_1^2$ and $I_2 = A_2^2$. This superposition at a point in space results in a different sum of the intensities of the separate components intensity [1]. This phenomenon is known as interference. The light and dark areas that observed on screen placed in the region of interference are called interference fringes, and the fringes are intensity which changes from minimum to maximum, which together form a pattern commonly called interference pattern or interferogram, as can be seen in **Figure 1**.

The interference of two or more electromagnetic waves can be usually achieved in two ways: by division of the wave front and by division of the amplitude. A mechanism used for the division of the wave front is, for example, the Young's experiment or double slit that is showed in **Figure 2**.

A mechanism used for dividing the amplitude of the wave is, for example, the Michelson interferometer that is shown in **Figure 3**.

Both mechanisms produce a pattern of light and dark intensities in the plane of interference fringes. The image resulting from the interference is known as interferogram. When the interferogram is captured by a recording medium, that is, a photographic film or a CCD

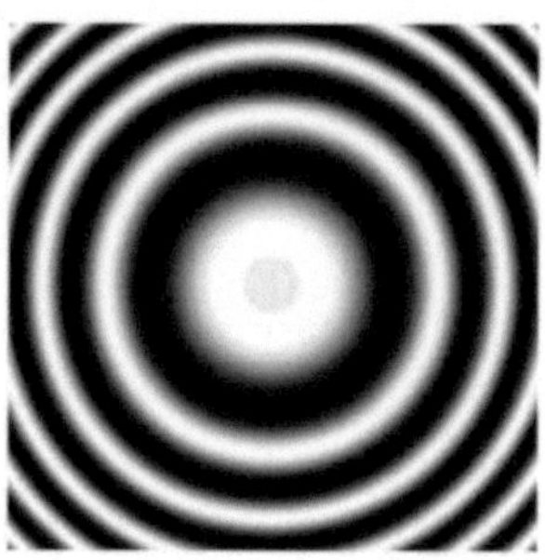

Figure 1. Interference pattern or interferogram showing circular fringes.

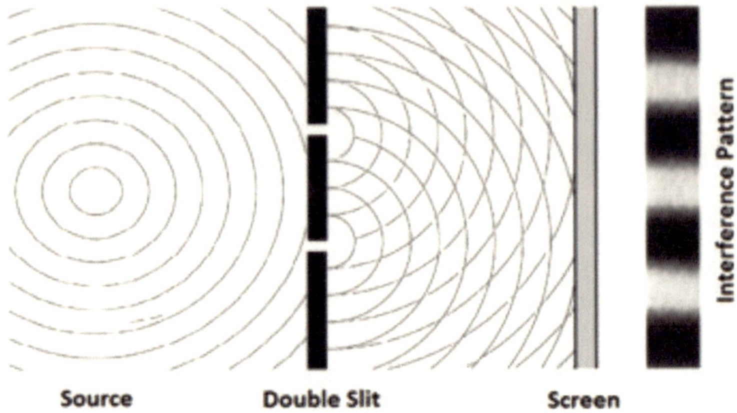

Figure 2. Young's experiment, where a double slit produces two wave fronts that interfere on a screen.

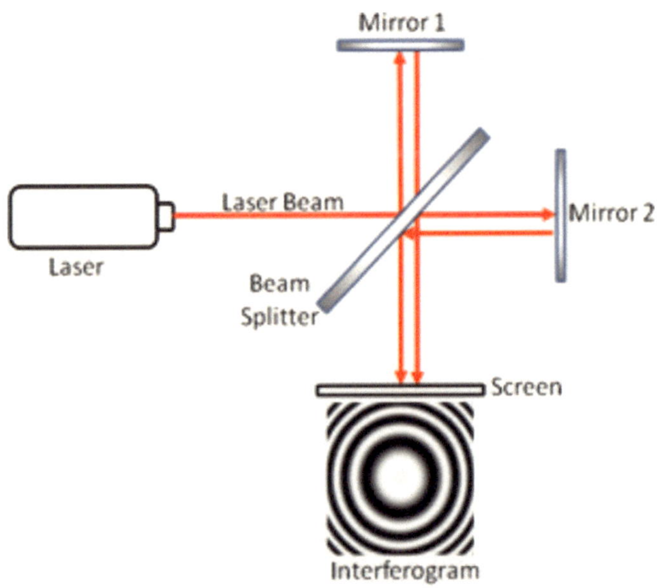

Figure 3. Michelson interferometer.

camera, the process commonly involves some optical system, which introduces imperfections with respect to the ideal image. Such imperfections are known as aberrations. Aberrations can be classified as chromatic and monochromatic. Chromatic aberrations are present to illuminate the object with white light or polychromatic light, that is, light with different wavelengths. These aberrations are the only ones that can be predicted by the theory of the first order, which states that an optical system consisting of lenses has different focal lengths for different wavelengths. These variations are related to the change of refractive index with respect to wavelength causing that both the position and the image size are different for each wavelength. Monochromatic aberrations occur when the object is illuminated with monochromatic light (i.e., light of a single wavelength), and the reflected or transmitted light is registered by a recording medium. This type of aberration causes that the captured image of a punctual object is no longer a point, but a blurred point. Monochromatic aberrations can be calculated roughly in the third-order theory, using the first two terms of the expansion in power series of the $sin\theta$ and $cos\theta$. Another alternative to calculate more accurately is to make the exact trigonometric trace rays through the system, where the deviation of the rays is calculated. These aberrations were studied in detail first by Ludwig von Seidel, hence often called Seidel aberrations, and are sphericity, coma, astigmatism, field curvature and distortion. These aberrations have a distinctive fringe distribution that appears frequently in the testing of a variety of objects with an interferometer. As the interferogram acquisition is only the capture of the interference phenomenon or interference pattern in a recording medium, it is convenient to study correcting aberrations problems. From these patterns, it can be known as the aberration coefficients using Zernike polynomials. Zernike polynomials have been successfully used in the recognition of patterns and image processing. Additionally, these have been used in astronomy to describe wavefront aberrations due to atmospheric turbulence and to describe wavefront aberrations in the human eye. This is because of the Seidel aberrations are related to the Zernike polynomials. Due to this relationship, polynomials are used to describe wavefront aberrations in order to calculate the aberration coefficients of a set of interferograms generated by an adaptive lens. These coefficients enable to describe the behavior of the aberrations present in the lens.

3. Interferogram filtering

In order to enhance the interferogram images and reduce the noisy caused by external factors at the interference phenomenon in an interferometer (interference produced by two or more controlled wave fronts), low-pass techniques are used as preprocessing filtering step. Interferogram smoothing and denoising are the principal purposes for the application of low-pass filters in interferometry. Filtering techniques can be grouped in spatial or frequency domains, where the spatial filtering is directly applied to the interferogram image, pixel to pixel, while the frequency filtering is usually performed in the Fourier domain. Band-pass and band-stop are some filtering techniques that can be applied in the Fourier domain, and these filters are used to attenuate some frequencies of specific noise.

3.1. Spatial filtering

The mathematical entity applied in spatial filtering is the convolution operation, also known as windowing [2], written as follows:

$$f(x,y)*g(x,y) = \sum_{m=-\frac{M+1}{2}}^{\frac{M+1}{2}} \sum_{n=-\frac{N+1}{2}}^{\frac{N+1}{2}} f(x-m,y-n)g(m,n), \tag{8}$$

where f, g, (x, y), (m, n) and $M \times N$ are the image to be filtered, the convolution mask, the original image coordinates, the coordinates where the convolution is performed and the size of the convolution mask, respectively. The kind of convolution filter is determined by the chosen convolution filter (averaging, Gaussian, quadratic, triangular, trigonometric, etc.). These convolution functions are presented in a mathematical form as follows: [3]

$$g(m,n) = \frac{1}{\sum_{m=1}^{M}\sum_{n=1}^{N}\omega_{m,n}}\begin{bmatrix}\omega_{1,1} & \cdots & \omega_{1,N}\\ \vdots & \ddots & \vdots \\ \omega_{M,1} & \cdots & \omega_{M,N}\end{bmatrix}, \tag{9}$$

where

$$\omega_{m,n} = \begin{cases} Ae^{(-B[m^2+n^2])}, & \text{Gaussian} \\ A-B(m^2+n^2), & \text{quadratic} \\ \frac{A}{2}+\frac{A}{4}[\cos(Bm)+\cos(Bn)], & \text{trigonometric} \\ A, & \text{average} \end{cases}. \tag{10}$$

Here A, B and ω are the amplitude, the width factor and the weight of the spatial filter function, respectively. The study about the magnitude spectrums of Eq. (10) (low-pass masks) was reported in Ref. [2], where the Gaussian and quadratic masks delivered the best results. Low frequencies were conserved by these filters, while the high frequencies were attenuated. However, the filtering results may vary depending on the interferogram under process, as can be seen in **Figure 4**, where the results of the filter on a ronchigram (a particular kind of interferogram) can be seen in **Figure 4a**. The four kinds of masks presented in Eq. (10) were employed to generate the filtered fringe patterns shown in **Figure 4b**. The parameters used by the filters were $A = 1$ and $B = 0.1$, with mask sizes of 3 × 3, 5 × 5 and 7 × 7 pixels.

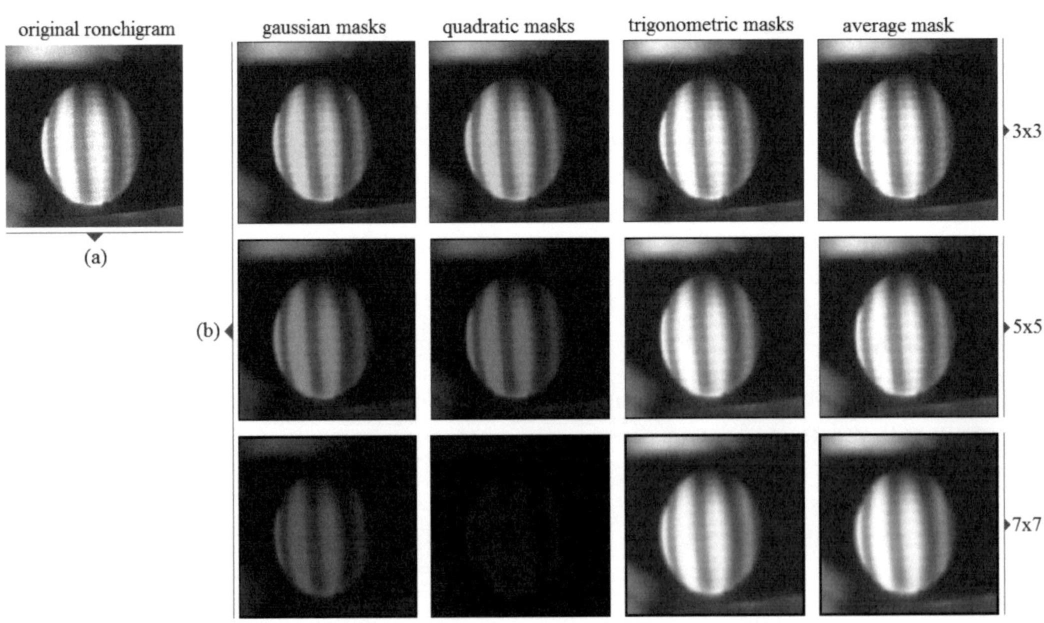

Figure 4. Spatial low-pass filters: (a) original ronchigram, (b) convolution image filtering with Gaussian, quadratic, trigonometric and average masks, as well as 3 × 3, 5 × 5 and 7 × 7 convolution mask sizes.

http://dx.doi.org/10.5772/66474

3.2. Frequency filtering

Frequency filtering is usually performed in the Fourier domain. The Fourier transform represents the change from spatial to frequency domain. Eq. (11) and Eq. (12) represent a pair of discrete Fourier transforms in two dimensions [2]

$$F(u,v) = \mathcal{F}\{f(x,y)\} = \sum_{x=1}^{U}\sum_{y=1}^{V} f(x,y)e^{-2\pi i\left(\frac{ux}{U}+\frac{vy}{V}\right)}, \tag{11}$$

and

$$f(x,y) = \mathcal{F}^{-1}\{F(u,v)\} = \sum_{u=1}^{U}\sum_{v=1}^{V} F(u,v)e^{2\pi i\left(\frac{ux}{U}+\frac{vy}{V}\right)} \tag{12}$$

where (u, v), $U \times V$, $\mathcal{F}$ and $\mathcal{F}^{-1}$ are the frequency coordinates, the image size in pixels, the Fourier transform and the inverse Fourier transform operators, respectively. A significant

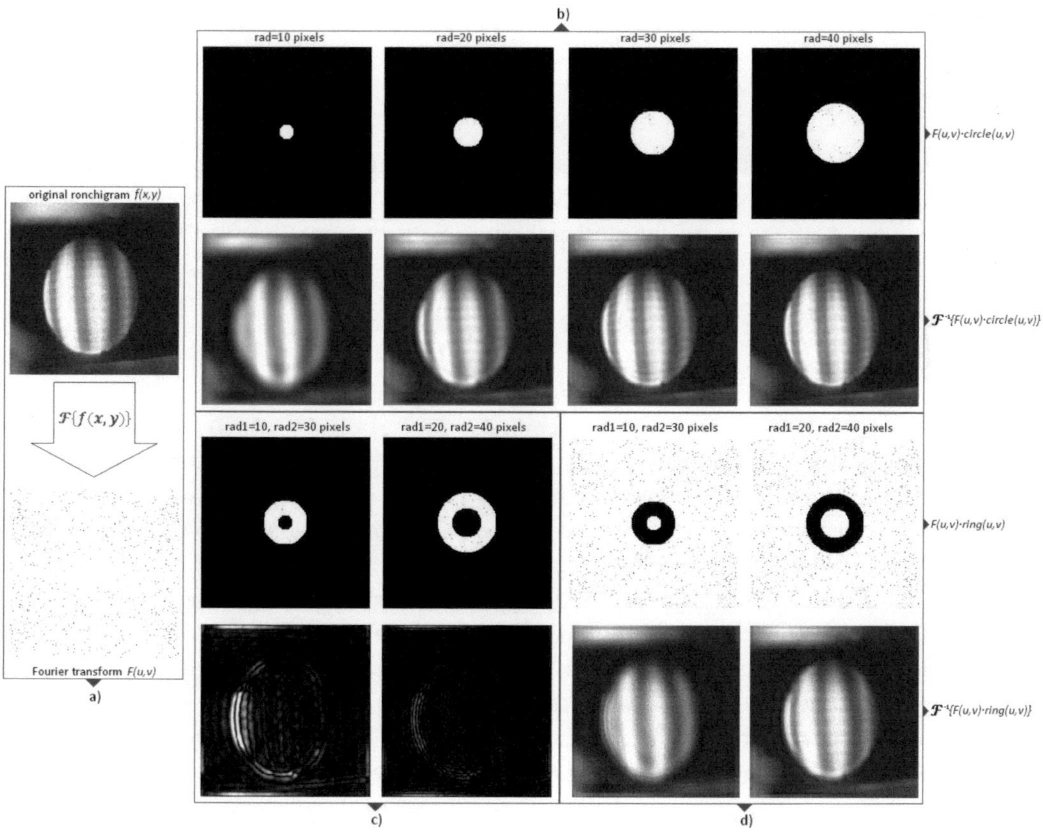

Figure 5. Frequency filtering: (a) Fourier transform of the original ronchigram, (b) low-pass filtering with different radius of binary circle mask, (c) band-pass filtering with two size of binary ring mask, and (d) band-stop filtering with two size of binary ring mask.

attribute obtained from the Fourier transform is that it gives the frequencies content of the image. Due to this property, frequency filter design is a very straight forward task for image processing. Low frequencies are located into the Fourier domain around the central coordinates; as frequencies gradually increase, they spread out from the center in a radial form. This characteristic is ideal for frequency filtering (low-pass, high-pass, band-pass and band-stop) [3]. The frequency filtering development consists in the multiplication between the Fourier transform with some kind convolution function. The kind of convolution mask will determine the class of filtering to be performed. The following is a summary of the most usual filtering masks: a white centered circle for low-pass; a black centered circle for high-pass; a white centered ring for band-pass and a black centered ring for band-stop.

In **Figure 5**, it is showed the masks described above along with the results of the filtering process. Low-pass filtering masks and results are presented in **Figure 5b**, band-pass filters and filtered images are seen in **Figure 5c** and finally, band-stop filters and filtering results can be appreciated in **Figure 5d**. The kind of filter or the size of geometrical mask depends directly in the image and in its noise content. There is no ideal filter; the kind of applied filter to process an image is dependent of the characteristics that are pretended to enhance or eliminate.

4. Phase-shifting interferometry

Interference fringe patterns obtained by means of interferometric techniques can be evaluated by using digital image processing techniques for the estimation of phase map distributions. The intensity of an image in an interference fringe pattern, according with Eq. (7), can be represented by Eq. (13)

$$I(x, y) = a(x, y) + b(x, y)cos[\phi(x, y)] \tag{13}$$

where $\phi(x, y)$ is the phase difference between the reference and testing wave fronts that interfere, $a(x, y) = I_1(x, y) + I_2(x, y)$ is the mean intensity and $b(x, y) = 2(I_1I_2)^{1/2}$ is the intensity modulation registered by each pixel (x, y) of a CCD camera. When analyzing a single image for the estimation of the relative phase difference in applied metrology, the sign of the phase cannot be assessed because of the argument in the sinusoidal function of the modulation term in Eq. (13) results in an identical interferogram for $\pm\phi(x, y)$ values. In order to overcome this difficulty, multiple interferograms can be sequentially registered introducing a known small phase change amount which is linear in time; this method is well known as phase shifting [4]. In practical phase shifting, a piezoelectric device PZT is included in the interferometric system to produce the phase-shifted interferograms. In general, when a voltage signal is used to polarize a PZT actuator, this electrical signal is converted directly into linear displacement motion. Then, phase-displaced intensity images can be represented by

$$I(x, y) = a(x, y) + b(x, y)cos[\phi(x, y) + \theta]. \tag{14}$$

Since there are three unknown terms in the representation of the interference intensity equation, then the measurement of at least three interferograms at known phase shifts is needed to determine the relative phase difference. However, one of the simplest modes to determinate the phase considers the use of four interferograms equally spaced by $\theta = \pi/2$, obtaining:

$$I_1 = a + b\cos(\phi), \tag{15}$$

$$I_2 = a\text{–}bsin(\phi), \tag{16}$$

$$I_3 = a\text{–}bcos(\phi), \tag{17}$$

and

$$I_4 = a + bsin(\phi). \tag{18}$$

Rearranging the simultaneous equation system from the above formulas, in which the spatial dependence (x, y) was not included, the phase distribution from intensity images can be calculated with:

$$tan(\phi) = \frac{I_4\text{–}I_2}{I_1\text{–}I_3}. \tag{19}$$

The phase difference estimated from the four equations is determined in the range between $-\pi$ and π when using the arctangent function, hence producing a wrapped phase distribution. In order to analyze an interference pattern by the phase-shifting techniques, a Twyman-Green interferometer is considered for the testing of optical elements; the basic arrangement is shown in **Figure 6**. A laser diode is used in the interferometer as illumination source, and then the emerging laser beam is collimated by a lens to obtain a plane wave front that propagates throughout a 50:50 beam splitter to produce two wave fronts of same amplitude. One beam is deviated to the reference mirror M1, and the second beam, to the mirror M2 under test. Next, the beams are reflected back toward the beam splitter, and part of the intensity overlaps on the

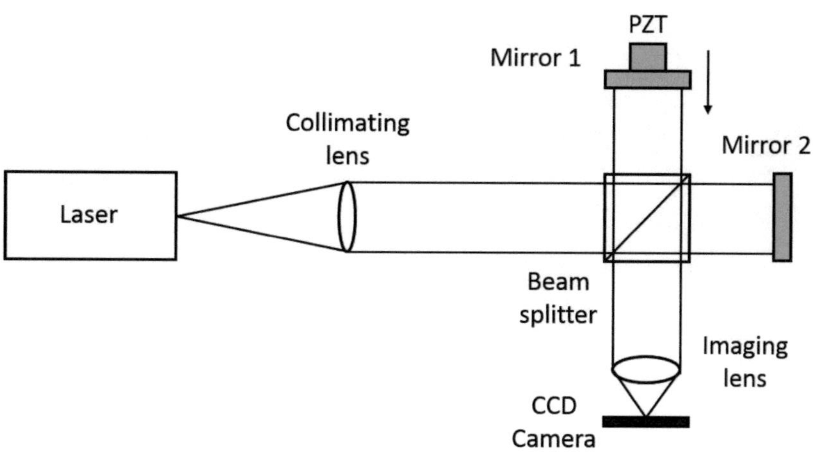

Figure 6. Twyman-Green interferometer for the testing of optical elements.

observation plane, where a CCD camera sensor is placed to register an image of the resulting interferogram that is then stored in a PC for subsequent processing.

In **Figure 7**, a set of four phase-shifted interferograms with phase shifts of $\pi/2$ is shown. A first interferogram with $\theta = 0$ is seen in **Figure 7a**, and subsequent interferograms with $\theta = \pi/2$, $\theta = \pi$ and $\theta = 3\pi/2$ are observed in **Figure 7b–d**, respectively. The wrapped phase is showed in **Figure 7e**, and the unwrapped phase related with the shape of the optical element being tested is shown in **Figure 7f**. A three-dimensional representation of the unwrapped phase seen in **Figure 7f** is presented in **Figure 8**.

In applied phase-shifting interferometry, there are concerns about the presence of errors that may affect the accurate phase extraction from phase-shifted interferograms. A typical systematical source of error introduced PZT arises when there is miscalibration of the phase-shifting actuator, causing detuning in the phase extraction process. The inclusion of phase-shifting algorithms with more than three or four interferograms can be implemented in order to reduce this systematic error [5].

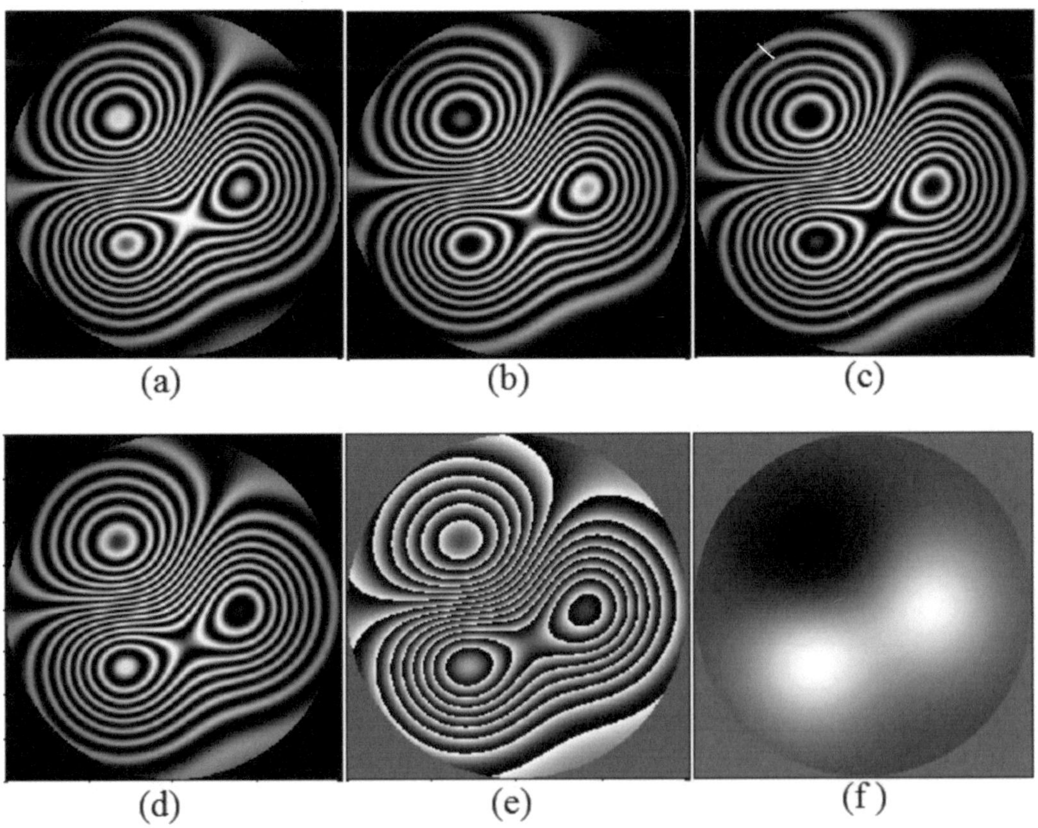

Figure 7. Four-step phase-shifted interferograms from (a) to (d), the wrapped phase (e) and the unwrapped phase (f).

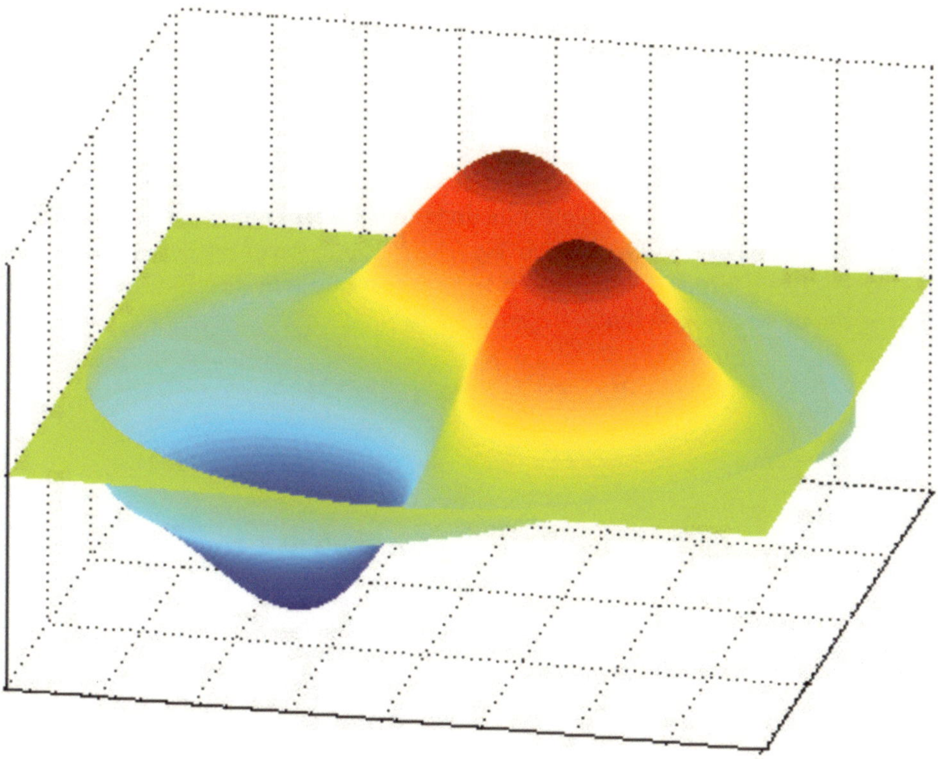

Figure 8. The unwrapped continuous phase obtained from a set of four phase-shifted fringe patterns and an unwrapping method.

5. The Fourier method

The method of Takeda or the Fourier method was developed in 1982 [6]. Unlike phase-shifting methods, see Ref. [7], a single interferogram with open fringes can be analyzed. This is useful when the object under study changes dynamically or environmental disturbances (vibrations or air turbulence) do not allow the use of phase-shifting methods unless special, and often, very expensive hardware is used to acquire several images simultaneously. However, in general, the accuracy and the dynamical range of the phase that can be measured are reduced. The Fourier method makes use of the fast Fourier transform technique to separate, in the frequency domain, the background and phase terms of the interferogram. Employing complex notation, an interferogram can be written as follows:

$$I(x,y) = a(x,y) + \frac{1}{2}b(x,y)e^{i[\phi(x,y)+\varphi(x,y)]} + \frac{1}{2}b(x,y)e^{-i[\phi(x,y)+\varphi(x,y)]}, \tag{20}$$

where $a(x, y)$ and $b(x, y)$ are the background intensity and the modulation term, respectively. The phase term is denoted by $\phi(x, y)$ and finally, the symbol $\varphi(x, y)$ denotes the lineal carrier function or tilt that is introduced usually by tilting the reference mirror in a two arm interferometer. The Fourier transform of the above expression can be written as:

$$\mathcal{F}\{I(x,y)\} = \tilde{I}(u,v) = \delta(u,v) + E(u+\alpha, v+\beta) + E^{*}(u-\alpha, v-\beta), \tag{21}$$

where (u, v) is the coordinates in the frequency domain, $\delta(u, v)$ is a delta function and $E(u+\alpha, v+\beta)$ and $E^{*}(u\text{-}\alpha, v\text{-}\beta)$ are complex conjugate functions that correspond to the transforms of the second and third terms of Eq. (20), respectively. The introduction of the linear carrier function, $\varphi(x, y) = \alpha x + \beta y$, shifts the terms $E(u, v)$ and $E^{*}(u, v)$ in opposite directions in the frequency spectrum as can be seen noting that:

$$E(u,v) = \mathcal{F}\left\{\frac{1}{2}b(x,y)e^{i\phi(x,y)}\right\}, \tag{22}$$

and

$$E(u+\alpha, v+\beta) = \mathcal{F}\left\{\frac{1}{2}b(x,y)e^{i\phi(x,y)}e^{i\varphi(x,y)}\right\}, \tag{23}$$

similarly

$$E^{*}(u,v) = \mathcal{F}\left\{\frac{1}{2}b(x,y)e^{-i\phi(x,y)}\right\}, \tag{24}$$

and

$$E^{*}(u-\alpha, v-\beta) = \mathcal{F}\left\{\frac{1}{2}b(x,y)e^{-i\phi(x,y)}e^{-i\varphi(x,y)}\right\}. \tag{25}$$

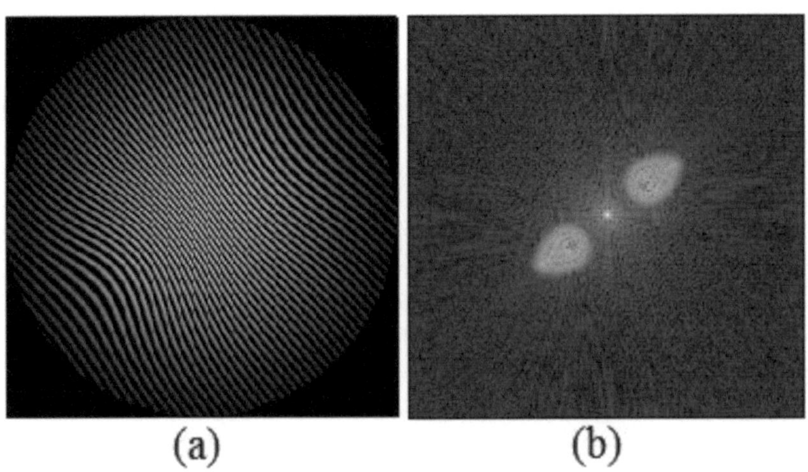

Figure 9. Interferogram with open fringes (a) and its Fourier spectrum (logarithm of the module) (b).

The separation of the terms in the Fourier spectrum due to the introduction of a linear carrier can be observed in **Figure 9**. An open-fringe interferogram and its Fourier spectrum can be seen in **Figure 9a** and **b**. It can be noted that the three terms of Eq. (21) are clearly separated. The central peak corresponds to the delta function $\delta(u, v)$ while the adjacent lobules correspond to $E(u + \alpha, v + \beta)$ and $E^*(u - \alpha, v - \beta)$. The interferogram was constructed as follows:

$$a(x,y) = 120e^{-(x^2+y^2)} \tag{26}$$

$$b(x,y) = 124e^{-(x^2+y^2)}, \tag{27}$$

$$\phi(x,y) = 2\pi\left\{4e^{-4[(x-0.2)^2+(y-0.3)^2]}-5e^{-6[(x+0.2)^2+(y+0.3)^2]}\right\}, \tag{28}$$

and

$$\varphi(x,y) = 2\pi(16x + 20y) \tag{29}$$

where x and y vary from 1 to −1 along the vertical and horizontal directions and the interferogram was multiplied by a circular function of radius one.

In order to recover the phase, we need to isolate one of the lateral lobules of the Fourier domain. To this end, we employ a band-pass filter. The filtered spectrum is then transformed back to the spatial domain to obtain $I_f(x, y)$ and the wrapped phase is found with the arctangent function of the ratio of the imaginary and real parts of $I_f(x, y)$ as shown in **Figure 10**. **Figure 10a** shows the band-pass filter $H(u, v)$, the filtered spectrum can be observed in **Figure 10b** and finally the wrapped phase is shown in **Figure 10c**.

The band-pass filter has the following form:

$$H(u,v) = e^{-3000\left[(u-0.2)^2+(v-0.2)^2\right]^2}, \tag{30}$$

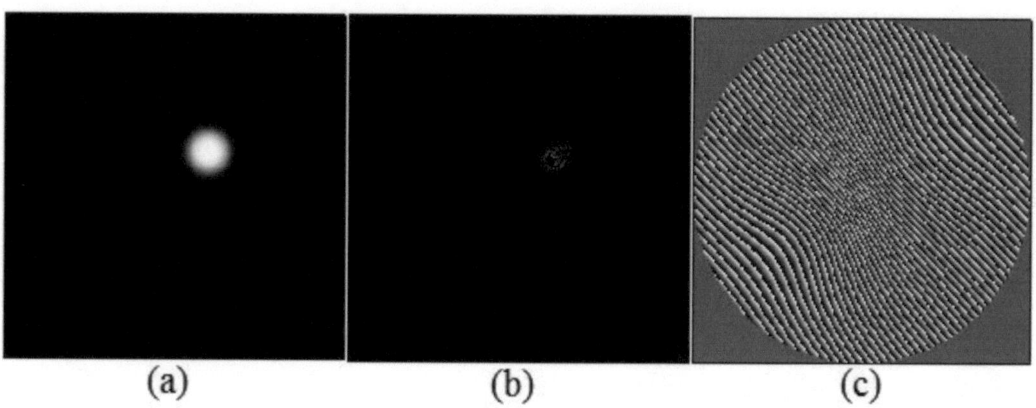

Figure 10. Band-pass filtering process of the Fourier spectrum seen in **Figure 9**. Pass-band filter (a), filtered spectrum (b) and recovered wrapped phase (c).

where u and v vary from 1 to −1 in vertical and horizontal directions, respectively. The filtered spectrum becomes,

$$\tilde{I}_f(u,v) = \tilde{I}(u,v)H(u,v). \tag{31}$$

Transforming back to spatial domain we obtain:

$$I_f(x,y) = \mathcal{F}^{-1}\left\{\tilde{I}_f(u,v)\right\}. \tag{32}$$

The wrapped phase is found by:

$$\psi_w(x,y) = atan2\left\{imag[I_f(x,y)], real[I_f(x,y)]\right\} \tag{33}$$

where the *atan2()* function accepts two arguments corresponding to the sine and cosine and returns the results modulo 2π. This wrapped phase, however, is not the desired one because it contains the introduced tilt that is not part of the object information. The wrapped desired phase is found with:

$$\phi_w(x,y) = atan2\left\{sin[\psi_w(x,y)-\varphi(x,y)], cos[\psi_w(x,y)-\varphi(x,y)]\right\}. \tag{34}$$

The final step is to apply an unwrapping method to obtain the continuous phase related with the object under study. This last procedures are shown in **Figure 11a** and **b** where the wrapped phase $\varphi_w(x, y)$ and the unwrapped reconstructed phase $\varphi_r(x, y)$ can be appreciated, respectively.

The Fourier method is not the unique procedure for phase retrieval from one interferogram with open fringes. Besides the Fourier approach there are other procedures in the spatial domain including phase locked loop [8] and spatial carrier phase-shifting methods [9, 10].

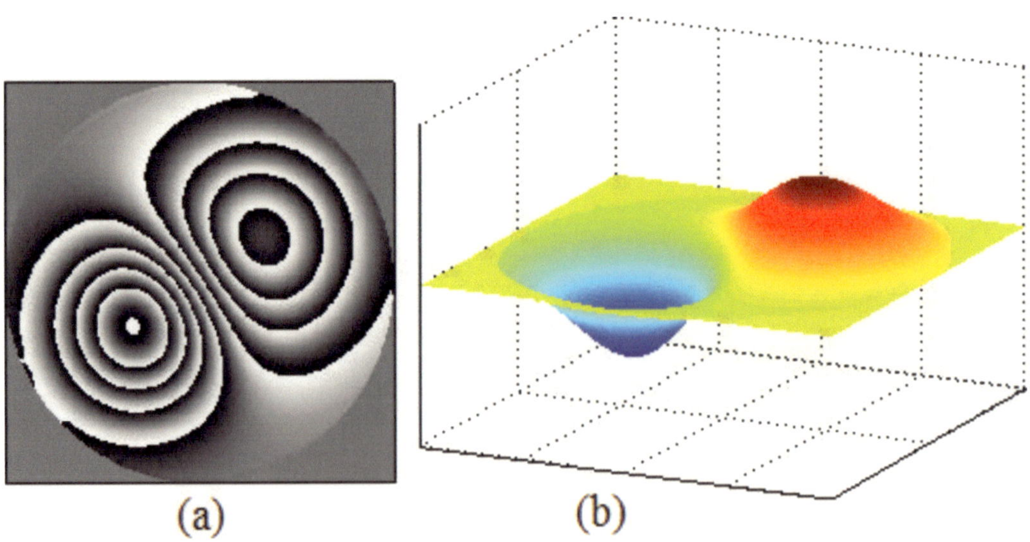

Figure 11. Phase reconstruction results. Wrapped phase (a) and unwrapped continuous phase (b).

6. Phase unwrapping

Phase unwrapping is a common step to finally find a continuous phase for several fringe analysis techniques such as phase shifting, Fourier, the phase synchronous and others methods that use the arctangent function of the ratio of the sine and cosine of the phase to obtain a wrapped phase. In its simplest form, phase unwrapping consists in adding or subtracting 2π terms to the pixel being unwrapped if a difference greater that π is found with a previous pixel already unwrapped [11]. The phase unwrapping problem in one dimension can be observed in **Figure 12**. The wrapped phase found with the arctangent function is seen in **Figure 12a** and the unwrapped continuous phase is showed in **Figure 12b**.

The described procedure works well only for wrapped phases with no inconsistencies and low noise levels, however, delivers wrong results when dealing with noisy wrapped phases or those obtained from interferograms with broken or unconnected fringes. A more consistent approach is achieved with the least squares phase unwrapping method [12]. The least square technique integrates the discretized laplacian of the phase. To this end, the laplacian of the phase is calculated as follows:

$$L_{i,j} = \phi^x_{i+1,j} - \phi^x_{i,j} + \phi^y_{i,j+1} - \phi^y_{i,j}, \tag{35}$$

where

$$\phi^x_{i,j} = atan2\left[\sin\left(\phi^w_{i,j} - \phi^w_{i-1,j}\right), cos\left(\phi^w_{i,j} - \phi^w_{i-1,j}\right)\right] p_{i,j} p_{i-1,j}, \tag{36}$$

and

$$\phi^y_{i,j} = atan2\left[\sin\left(\phi^w_{i,j} - \phi^w_{i,j-1}\right), cos\left(\phi^w_{i,j} - \phi^w_{i,j-1}\right)\right] p_{i,j} p_{i,j-1}. \tag{37}$$

In the last equations, we have used pixel subscript notation in order to limit the extension of the equations. The pupil function $p_{i,\,j}$ is defined as equal to one where we have valid data and zero otherwise. One may note that if $p_{i,\,j} = p_{i+1,\,j} = p_{i-1,\,j} = p_{i,\,j+1} = p_{i,\,j-1} = 1$, then

$$L_{i,j} = -4\phi_{i,j} + \phi_{i+1,j} + \phi_{i-1,j} + \phi_{i,j+1} + \phi_{i,j-1}. \tag{38}$$

Solving for the phase in the above equation, we obtain that the unwrapping problem under the least square approach consist in the resolution of a linear system of equations, as:

$$\phi_{i,j} = \frac{\phi_{i+1,j} + \phi_{i-1,j} + \phi_{i,j+1} + \phi_{i,j-1} - L_{i,j}}{4}. \tag{39}$$

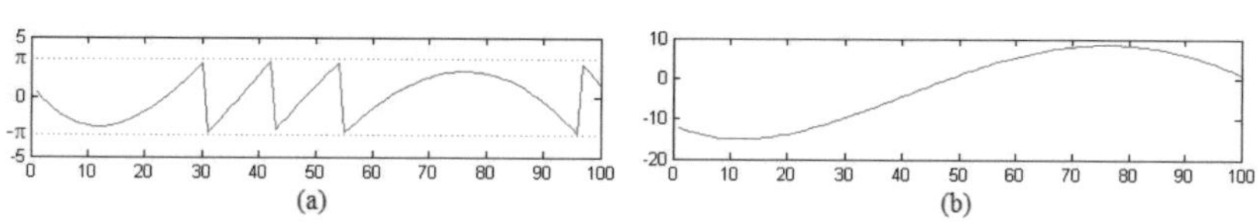

Figure 12. Wrapped (a) and unwrapped phase (b).

An iterative technique that solves the above system of linear equations is the overrelaxation method in which the following equation is iterated until convergence:

$$\phi_{i,j}^{k+1} = \phi_{i,j}^{k} - \frac{(d\phi_{i,j}^{k} - \phi_{i-1,j}^{k+1} - \phi_{i+1,j}^{k} - \phi_{i,j-1}^{k+1} - \phi_{i,j+1}^{k} + L_{i,j})w}{d}, \tag{40}$$

where

$$d = p_{i-1,j} + p_{i+1,j} + p_{i,j-1} + p_{i,j+1}. \tag{41}$$

In the last equations, k is the iteration number and w is a parameter of the overrelaxation method that must be set between 1 and 2. Results of the least square technique are showed in **Figure 13**. A two-dimensional wrapped phase is seen **Figure 13a**, the phases differences are shown in **Figure 13b** and **c**, respectively. The laplacian of the phase is showed in **Figure 13d**,

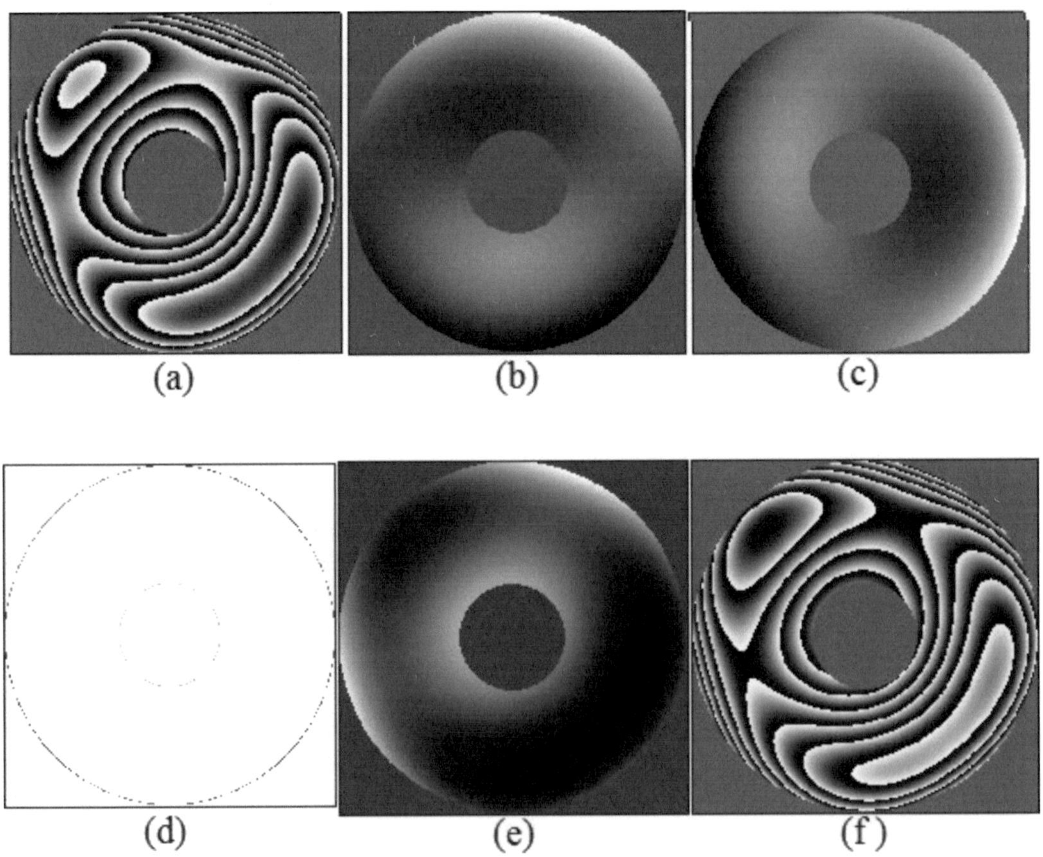

Figure 13. Least square results to unwrap the phase. Wrapped phase (a), phase differences in the x (b) and y (c) directions, laplacian of the phase (d), unwrapped phase (e) and rewrapped phase (f).

the reconstructed phase in a two-dimensional view is observed in **Figure 13e** and finally, for comparison purposes the reconstructed rewrapped phase is also showed in **Figure 13f**. One may observe that the original wrapped phase and the reconstructed wrapped phase are slightly different, this is because the least square method recovers the phase with an arbitrary constant term, however this term is usually neglected since doesn't carry any information of the object being measured.

If desired, the constant term in the retrieved phase may be corrected easily in the following form:

$$\phi_c = \phi_r - atan2[\sin(\phi_r - \phi_w), \cos(\phi_r - \phi_w)]. \tag{42}$$

The wrapped phase seen in **Figure 13a** was constructed as follows:

$$\phi_w = atan2[\sin(\phi), \cos(\phi)], \tag{43}$$

where

$$\phi_{i,j} = 2\pi \left\{ \begin{array}{c} 16\left[(x_i)^2 + (y_j)^2\right]^2 - 18\left[(x_i)^2 + (y_j)^2\right] + \\ 5y_j(x_i)^2 - 4x_i(y_j)^2 + 3y_jx_i + 2x_i - 2y_j + 0.56 \end{array} \right\}. \tag{44}$$

In the above equations x_i and y_j are range variables that vary from −1 to 1, the wrapped phase was multiplied by an annular pupil function with an exterior radius of 198 pixels, while the interior radius was 60 pixels for an image size of 400 × 400. The convergence of the reconstructed phase seen in **Figure 2e** was reached after 700 iterations; the overrelaxation parameter w was set to 1.99, which is usual for large images. The reconstructed phase with the constant term corrected and the phase error, $\varepsilon = \phi = \phi_c$, can be appreciated in **Figure 14a** and **b**. The maximum error was about 0.00005 radians.

Finally, results on a noisy wrapped phase are presented. Random noise with uniform distribution in the range of $-\pi/4$ to $\pi/4$ radians were added to the phase to obtain the wrapped phase

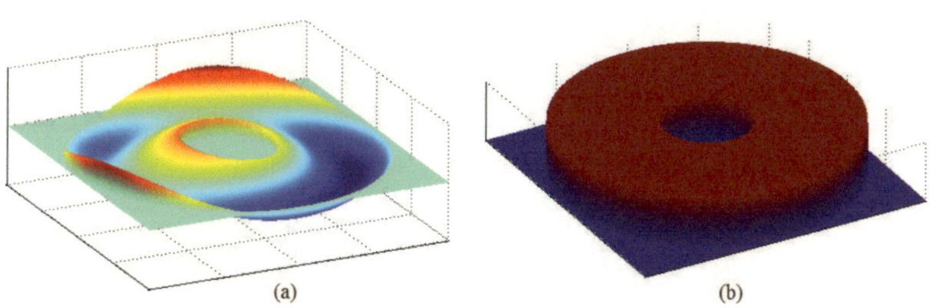

Figure 14. Unwrapped phase and phase error. Three-dimensional view of the reconstructed and corrected phase (a) and phase error (b).

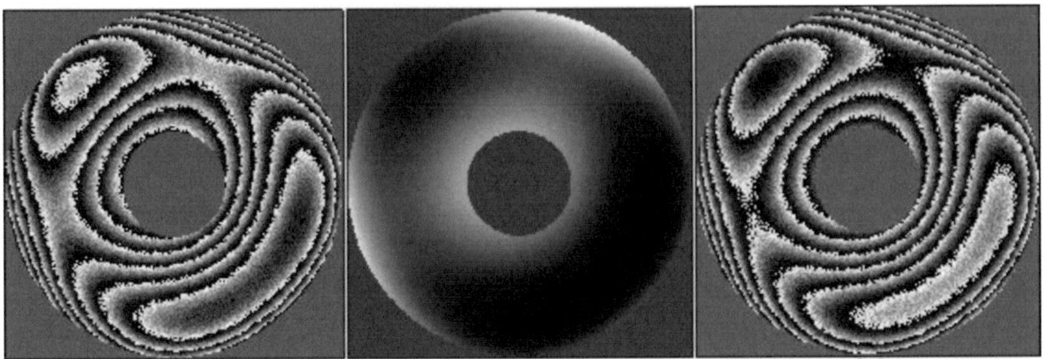

Figure 15. Results on noisy measurements. Wrapped phase with noise (a), retrieved phase (b) and wrapped retrieved phase (c).

seen in **Figure 15a**, the unwrapped phase is shown in **Figure 15b**, and the rewrapped reconstructed phase is observed in **Figure 15c**. As can be noticed the least square method is a very reliable technique that works with any pupil configuration and stands noisy measurements.

7. Phase recovery from lateral shearing interferograms

Lateral shearing interferometry is a very important field in experimental optical measurements, in which, the test beam interferes with a laterally displaced version of itself instead of a reference beam. The resulting fringe patterns are thus related with the object wavefront derivative in a given direction. This is very useful when the object information of interest is related with the derivative as in strain analysis or when the dynamical range of the object wave front is too high that cannot be measured with direct interferometry. Let us consider a laterally shear interferogram with a beam displacement in the x direction a quantity Δx, we obtain:

$$I^x(x,y) = B(x,y) + C(x,y)cos[\psi^x(x,y)], \tag{45}$$

where

$$\psi^x(x,y) = \phi(x,y) - \phi(x-\Delta x, y). \tag{46}$$

In Eq. (45) $B(x, y)$ is the background intensity and $C(x, y)$ is the modulation term. The objective is to retrieve the undisplaced phase $\varphi(x, y)$. To this end, we need, at least, another laterally shear interferogram with a beam displacement in the y direction another quantity Δy, obtaining:

$$I^y(x,y) = B(x,y) + C(x,y)cos[\psi^y(x,y)], \tag{47}$$

where

$$\psi^{y}(x,y) = \phi(x,y) - \phi(x,y-\Delta y). \tag{48}$$

Let us considered that we have retrieved the phase differences $\psi^x(x, y)$ and $\psi^y(x, y)$ by means of a phase-shifting technique as is depicted in **Figure 16** and **Figure 17** for the x and y directions, respectively. A set of four phase-shifted interferograms can be seen in **Figure 16a**. The wrapped phase differences $\psi^{wx}(x, y)$ and the unwrapped phase $\psi^x(x, y)$ are observed in **Figure 16e** and **f**, respectively. The modulation term and a sheared pupil function $p^x(x, y)$ are shown in **Figure 16g** and **h**, respectively. The sheared pupil $p^x(x, y)$ is found after normalization from zero to one and thresholding the normalized modulation term. A second set of phase-shifted interferograms are seen **Figure 17a** and **d**. The wrapped phase differences $\psi^{wy}(x, y)$ and the unwrapped phase $\psi^y(x, y)$ are observed in **Figure 17e** and **f**, respectively. The modulation term and a sheared pupil function $p^y(x, y)$ are shown in **Figure 4g** and **h**, respectively. In a similar way to the first set of phase-shifted interferograms, the pupil $p^y(x, y)$ is found after normalization from zero to one and thresholding the normalized modulation term. The sheared pupils $p^x(x, y)$ and $p^y(x, y)$ are useful to find the undisplaced pupil $p(x, y)$, where the original wave front $\varphi(x, y)$ have valid data, since $p^x(x, y) = p(x, y)p(x - \Delta x, y)$ and $p^y(x, y) = p(x, y)p((x, y) - \Delta y)$

Once the phase differences $\theta^x(x, y)$ and $\theta^y(x, y)$ are known, we can use the next procedure to find the searched phase $\varphi(x, y)$ observing that:

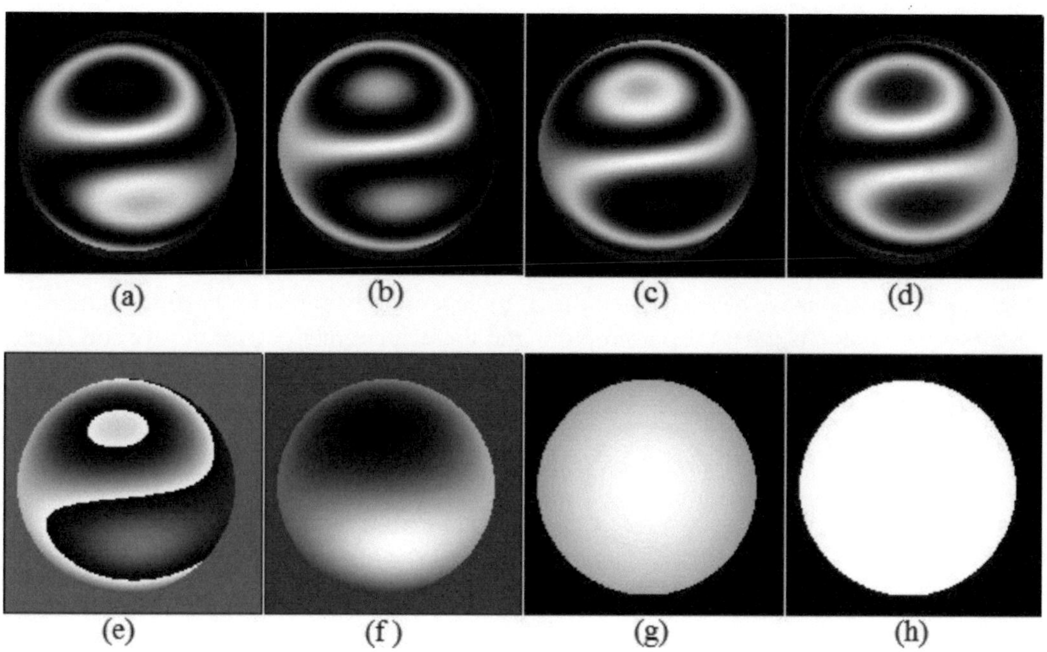

Figure 16. Recovery of the phase differences in the x direction. Set of four sheared interferograms acquired under a phase-shifting technique (a) to (d), wrapped phase differences (e), unwrapped phase differences (f), modulation (g) and recovered sheared pupil (h).

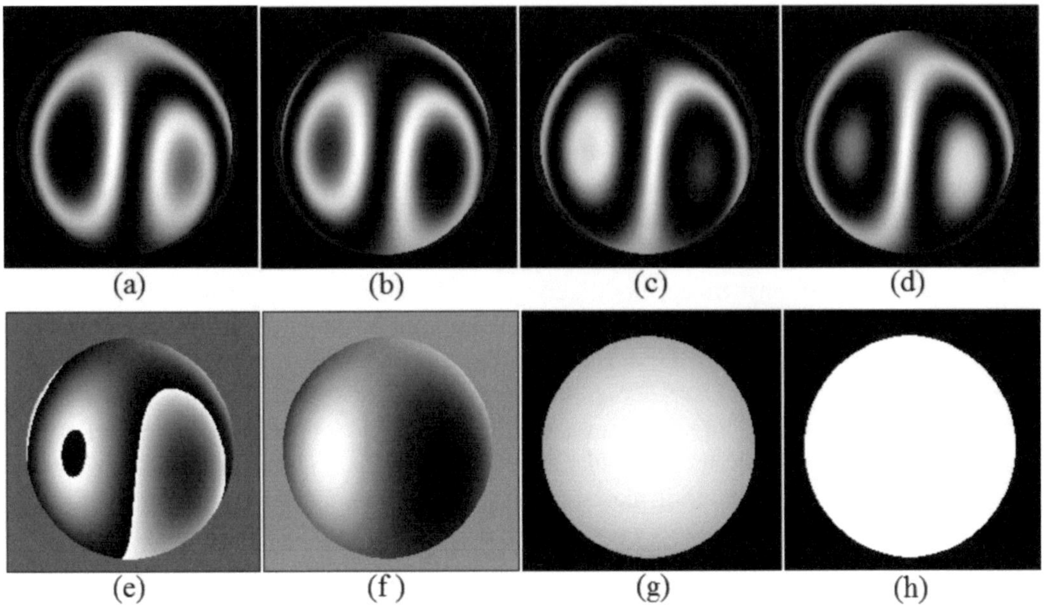

Figure 17. Recovery of the phase differences in the y direction. Set of four sheared interferograms acquired under a phase-shifting technique (a) to (d), wrapped phase differences (e), unwrapped phase differences (f), modulation (g) and recovered sheared pupil (h).

$$\psi^x_{i,j} p_{i-a,j} - \psi^x_{i+a,j} p_{i+a,j} = \phi_{i,j}(p_{i-a,j} - p_{i+a,j}) - \phi_{i-a,j} p_{i-a,j} - \phi_{i+a,j} p_{i+a,j}, \tag{49}$$

and

$$\psi^y_{i,j} p_{i,j-b} - \psi^y_{i,j+b} p_{i,j+b} = \phi_{i,j}(p_{i,j-b} - p_{i,j+b}) - \phi_{i,j-b} p_{i,j-b} - \phi_{i,j+b} p_{i,j+b}. \tag{50}$$

In the above equations we have changed the (x, y) dependence by pixel subscript notation. As described before, $p_{i,\,j}$ is the undisplaced pupil and the displacement quantities Δx and Δy are rounded to the nearest integer in pixel dimensions obtaining a and b, respectively. Adding Eq. (49) and Eq. (50) and solving for the phase, we obtain:

$$\phi_{i,j} = \frac{F + G}{H}, \tag{51}$$

where

$$H = p_{i+a,j} + p_{i-a,j} + p_{i,j+b} + p_{i,j-b}, \tag{52}$$

$$F = \psi^x_{i,j} p_{i-a,j} - \psi^x_{i+a,j} p_{i+a,j} + \psi^y_{i,j} p_{i,j-b} - \psi^y_{i,j+b} p_{i,j+b}, \tag{53}$$

and

$$G = \phi_{i-a,j}p_{i-a,j} + \phi_{i+a,j}p_{i+a,j} + \phi_{i,j-b}p_{i,j-b} + \phi_{i,j+b}p_{i,j+b}. \tag{54}$$

Eq. (51) represents a linear system of equations; however, it is ill posed because there are more unknowns than equations due to the effects of the sheared pupils. Nevertheless, a regularization term may be aggregated to overcome this problem [13–15]. The regularization term is in the form of discrete Laplacians of the phase among adjacent pixels. The following equation that incorporates the regularization term is iterated until convergence:

$$\phi_{i,j}^{k+1} = \phi_{i,j}^{k} - \frac{H\phi_{i,j}^{k} - (F + G) + \alpha(L_{i+1,j}^{x} - 2L_{i,j}^{x} + L_{i-1,j}^{x} + L_{i,j+1}^{y} - 2L_{i,j}^{y} + L_{i,j-1}^{y})}{H}, \tag{55}$$

where

$$L_{i,j}^{x} = (\phi_{i+1,j} - 2\phi_{i,j} + \phi_{i-1,j})p_{i+1,j}p_{i-1,j}, \tag{56}$$

and

$$L_{i,j}^{y} = (\phi_{i,j+1} - 2\phi_{i,j} + \phi_{i,j-1})p_{i,j+1}p_{i,j-1}. \tag{57}$$

Here, α is a parameter that controls the effects of the regularization term. The phase reconstruction is seen in **Figure 18**. A two-dimensional view of the retrieved phase is observed in **Figure 18a**, the same phase but in a three-dimensional perspective is shown in **Figure 18b** and the phase error (the actual phase minus the reconstructed one) can be appreciated in **Figure 18c**. This reconstruction was achieved after 800 iterations with a parameter α = 0.1 obtaining a maximum error of about 0.0004 radians.

The actual phase was constructed as follows:

$$\phi_{i,j} = 2\pi \left\{ \begin{array}{c} 20\left[(x_i)^2 + (y_i)^2\right]^2 - 24\left[(x_i)^2 + (y_i)^2\right] - 6(x_i)^2 y_j \\ +8x_i(y_j)^2 + 5x_i y_j + 4x_i - 5y_j + 0.45 \end{array} \right\}, \tag{58}$$

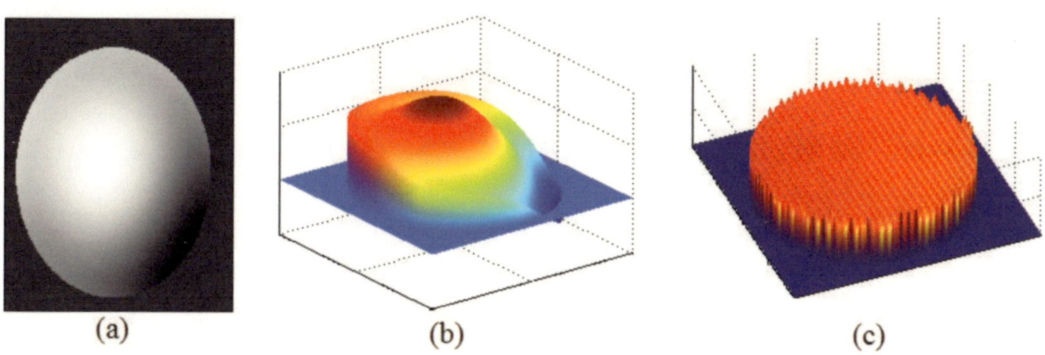

Figure 18. Phase reconstruction from lateral sheared interferograms. Reconstructed phase (a), three-dimensional view of the reconstructed phase (b) and the phase error (c).

where x_i and y_j are range variables that vary from −1 to 1 in both vertical and horizontal directions. The pupil function is a circular one with a radius of 170 pixels. The shear distances were set to $\Delta x = a = 12$ pixels and $\Delta y = b = 12$ pixels.

8. Conclusions

Digital processing techniques applied to interferometric measurements allow to obtain the phase from fringe patterns. The fringe analysis methods described here can be used to recover the phase that is associated with the physical variable under study. Under controlled conditions, phase-shifting techniques are the most used methods to retrieve the wrapped phase. If experimental conditions suffer from vibrations, air turbulences or the object changes dynamically, among other factors, then a Fourier method may be preferable to analyze an open-fringe interferogram. Those procedures deliver a wrapped phase. Then, an unwrapping algorithm is needed to reconstruct a continuous phase related with the object being studied. The aim of this chapter is to present, to the reader, the fundamentals of principal fringe analysis techniques. Numerical simulations are provided, in such way that the reader can reproduce them by its own. The extension of the chapter is insufficient to introduce many important techniques. However, the methods presented here were described as clearly and briefly as we could. We hope that the reader finds this information useful in the interpretation of interferograms obtained in the study of some object or phenomena by using an interferometric setup.

Author details

Jesús Muñoz Maciel*, Francisco Javier Casillas Rodríguez, Miguel Mora González, Francisco Gerardo Peña Lecona and Víctor Manuel Duran Ramírez

*Address all correspondence to: jesnek@hotmail.com

University of Guadalajara, University Center of Lagos, Lagos de Moreno, Mexico

References

[1] Born M., Wolf E. Principles of optics. 7th ed. New York, USA: Cambridge University Press; 2002. 952 p.

[2] Mora-González M., Casillas-Rodríguez F.J., Muñoz-Maciel J., Martínez-Romo J.C., Luna-Rosas F.J., de Luna-Ortega C.A., Gómez-Rosas G., Peña-Lecona F.G. Reducción de ruido digital en señales ECG utilizando filtraje por convolución. Investigación y Ciencia. 2008;**16**(040):26–32.

[3] Mora-Gonzalez M., Muñoz-Maciel J., Casillas F.J., Peña-Lecona, Chiu-Zarate R., Perez-Ladron de Guevara H. Image procesing for optical metrology. In: Ionescu C.M., editor.

MATLAB–a ubiquitous tool for the practical engineer. 1st ed. Rijeka, Croatia: InTech; 2011. pp. 523–546. doi:10.5772/820

[4] Gåsvik K.J. Optical metrology. 3rd ed. West Sussex, UK: John Wiley and Sons, Ltd.; 2002. 372 p.

[5] Malacara D., editor. Optical shop testing. 3rd ed. NJ, USA: John Wiley and Sons, Inc.; 2007. 888 p.

[6] Takeda M., Ina H., Kobayashi S. Fourier-transform method of fringe-pattern analysis for computer-based topography and interferometry. Journal of the Optical Society of America. 1982;**72**:156–160.

[7] Bruning J.H., Herriott D.R., Gallagher J.E., Rosenfeld D.P., White A.D., Brangaccio D.J. Digital wavefront measuring interferometer for testing optical surfaces and lenses. Applied Optics. 1974;**13**:2693–2703.

[8] Servín M., Rodríguez-Vera R. Two-dimensional phase locked loop demodulation of interferogram. Journal of Modern Optics. 1993;**40**:2087–2094.

[9] Shough D.M., Kwon O.Y., Leary D.F. High speed interferometric measurement of aerodynamic phenomena. In: Ulrich P.B., Wilson L.E., editors. Propagation of high-energy laser beams through the earth's atmosphere. Proc. SPIE, LA, USA; 1990. pp. 394–403. doi:10.1117/12.18359

[10] Kujawinska M., Wójciak J. Spatial-carrier phase shifting technique of fringe pattern analysis. In: Jueptner W.P., editor. Industrial applications of holographic and speckle measuring techniques. Proc. SPIE, The Hague, Netherlands; 1991. doi:10.1117/12.47089

[11] Robinson D.W. Phase unwrapping methods. 1st ed. Bristo, UK: CRC Press; 1993. 302 p.

[12] Ghiglia D.C., Pritt M.D. Two-dimensional phase unwrapping: theory algorithms and software. 1st ed. New York, USA: John Wiley and Sons, Inc.; 1998. 512 p.

[13] Legarda R., Rivera M., Rodríguez R., Trujillo G. Robust wave-front estimation from multiple directional derivatives. Optics Letters. 2000;**25**:1089–1091. doi:10.1364/OL.25.001089

[14] Servín M., Malacara D., Marroquín J.L. Wave-front recovery from two orthogonal sheared interferograms. Applied Optics. 1996;**35**:4343–4348. doi:10.1364/AO.35.004343

[15] Villa J., García G., Gómez G. Wavefront recovery in shearing interferometry with variable magnitude and direction shear. Optics Communications. 2001;**195**:85–91. doi:10.1016/S0030-4018(01)01304-9

Chapter 2

Applications of Mach-Zehnder Interferometry to Studies on Local Deformation of Polymers Under Photocuring

Dan-Thuy Van-Pham, Minh-Tri Nguyen,
Hideyuki Nakanishi, Tomohisa Norisuye and
Qui Tran-Cong-Miyata

Additional information is available at the end of the chapter

http://dx.doi.org/10.5772/64611

Abstract

A Mach-Zehnder interferometer (MZI) was built and modified to in situ monitor the deformation of polymers during the photocuring process. In this review, the working principle and method of operation of this MZI were explained together with the method of data analysis. As the examples for the utilization of this modified MZI, measurements of the deformation induced by photopolymerization was demonstrated for three different types of samples: homopolymer in the bulk state, miscible polymer blends and phase-separated polymer blends. Finally, a concluding remark is provided for the usage of MZI in polymer research.

Keywords: Mach-Zehnder interferometry, light scattering, poly(ethyl acrylate), poly(vinyl methyl ether), polystyrene, polymer blends, concentration fluctuations, phase separation, glass transition temperature, photo-cross-linking (photocuring), photodimerization, shrinkage, swelling

1. Introduction

Mach-Zehnder interferometry (MZI) was invented by Ludwig Zehnder in 1891 [1] and was subsequently refined by Ludwig Mach in the year after [2]. The techniques were invented taking advantages of the interference phenomena of light to measure the phase difference between the two light beams in which one is varied by the presence of a sample. As illustrated in **Figure 1**, the basic structure of the interferometry is composed of one beam splitter (half mirror) and two reflecting mirrors. The coherent light beam from a laser after collimation was divided into two beams: the reference beam and the test beam on which the sample is interposed. These two beams serve as two arms, the reference and the test arms of the interferometry. The presence of an object on the test arm will result in the difference in optical path length, thereby changing the interference pattern of the laser at the half mirror [HM2]. The fringe patterns can be monitored and recorded either along the direction of the reference beam or the test beam. Compared to other interferometers like Michelson, the separation of the two arms of MZI can provide a wide application due to large and freely accessible working space though the optical alignment is relatively difficult. Taking advantage of this spacious working place, MZI has been utilized for various experiments: electron interferometer functioning in high magnetic field [3], flow visualization and flow measurements [4], for sensing applications [5]. Furthermore, optofluidic Mach-Zehnder interferometer for sensitive, label-free measurements of refractive index of fluids was also developed [6]. The unique structure of Mach-Zehnder has also been utilized for optical communication as a modulator [7]. On the other hand, a lot of efforts have been made to fabricate microscale optical systems including Mach-Zehnder interferometer modulators using polymeric materials [8, 9]. In this chapter, we focus on studies on the local deformation in polymeric systems undergoing photocuring by ultraviolet (UV) light. Since the polymer mixture undergoes transition from liquid to solid by the reaction and at the same time phase separation takes place, the deformation (shrinkage and/or swelling) would affect the phase separation process and the resulting morphology. Mach-Zehnder interferometry would be useful to monitor the extent of deformation in the nanometer scales during the reaction.

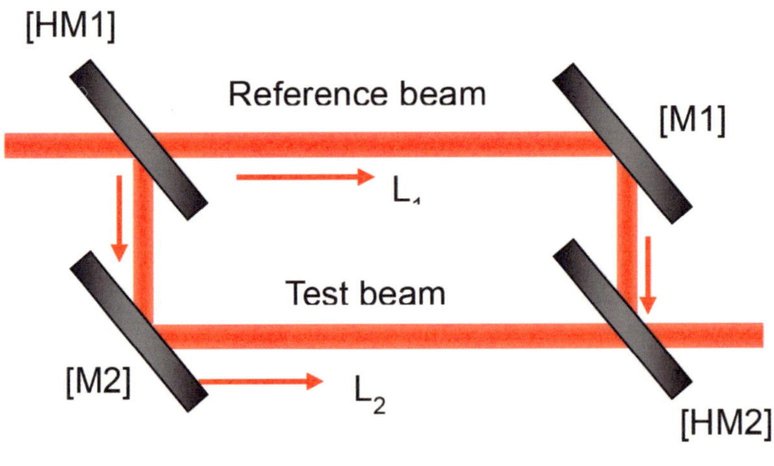

Figure 1. Basic unit of Mach-Zehnder interferometry (MZI).

2. Instrumentation and data analysis

The block diagram of the MZI used in our experiments is illustrated in **Figure 2**.

The details are described in a previous publication [10]. Briefly, a coherent 632.8 nm light beam from a He-Ne laser is passed through a Glan-Thompson prism and is converted into vertically polarized light. The polarized light beam was collimated by using collimating lens before entering the basic unit of the Mach-Zehnder interferometer. After passing the half mirror [HM1], incident light was divided into two beams, forming the reference arm and the test arm of the MZI. A sample was interposed on the test arm and was half hidden by a movable mask to produce the reference part on the sample. The deformation of the sample under curing is obtained from the variation of the part irradiated with UV light compared to the part hidden by a mask on the same sample. In order to observe the deformation caused by the curing reaction, the sample was submitted to the sequence of *masking-irradiating-mask removing-recording*. Movement of this mask is controlled with the precision of a micrometer. Details of the operation are provided in [10]. These two laser beams encounter with each other at the half mirror [HM2] to produce interference patterns which were recorded on a charge-coupled device (CCD) camera. The data were subsequently analyzed using a computer.

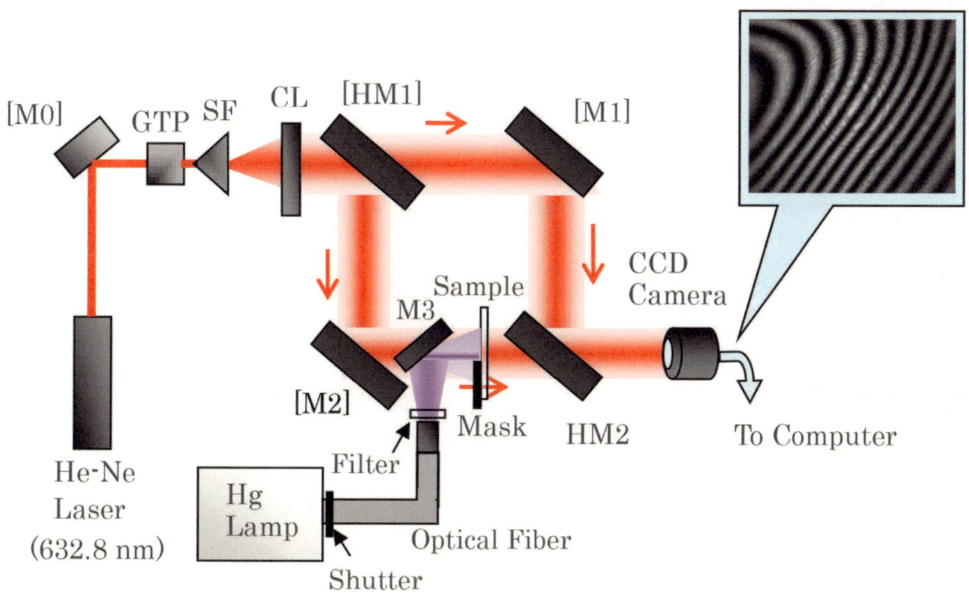

Figure 2. The block diagram of the Mach-Zehnder interferometer (MZI) used in this study for in situ monitoring the deformation induced by photocuring polymers. M, reflecting mirror; HM, half-mirror; GTP, Glan-Thompson prism; SF, spatial filter; CL, collimating lens.

2.1. Basic theory of light interference phenomena

Two monochromatic planar waves E_1 and E_2 traveling along the two arms of the Mach-Zehnder interferometer can be expressed by the following wave equation in complex form:

$$E_1 = E_{01} \exp(i \phi_1) \tag{1}$$

$$E_2 = E_{02} \exp(i \phi_2) \tag{2}$$

Here E and ϕ represent respectively the amplitudes and the phases of the two waves and i is the imaginary number.

The phase of these two waves can be written as

$$\phi_1 = k n_0 L_1 \quad (3)$$

$$\phi_2 = k n_0 L_2 \quad (4)$$

where k and n_0 are respectively the wavenumber of the incident light and the refractive index of the sample before curing. L_1 and L_2 are respectively the path length of light traveling along the reference and the test arms. The phase difference between the two beams can be expressed by

$$\Delta\phi = (\phi_2 - \phi_1) = k(L_2 - L_1) \quad (5)$$

where n_0, the refractive index of air was set equal unity. If the length of two arms is set equal $L_1 = L_2$, $\Delta\phi = 0$. There is no interference for this particular case. In the presence of a sample with refractive index n and the thickness d interposed on the test arm, the phase of the wave traveling along this direction becomes

$$\phi_2 = k \,.\, n_0 \,.\, (L_2 - d) + k \,.\, n \,.\, d \quad (6)$$

The phase difference in the presence of the sample becomes:

$$\Delta\phi = (\phi_2 - \phi_1) = k(n-1)d + k(L_2 - L_1) \quad (7)$$

For the case $L_1 = L_2$, the phase difference becomes

$$\Delta\phi = k(n-1)\,d \quad (8)$$

Therefore, the optical path length (OPL) of the sample *before* irradiation is

$$\mathrm{OPL}_{\mathrm{before}} = \frac{\Delta\,\phi_{\mathrm{before}}}{k} = (n-1)\,d \quad (9)$$

In general, both the refractive index and the thickness of the sample are varied by the reaction:

$$\mathrm{OPL}_{\mathrm{after}} = \left(\frac{\Delta\,\phi_{\mathrm{after}}}{k}\right) = (n + \Delta n)(d + \Delta d) - 1.(d + \Delta d) \quad (10)$$

Since both Δn and Δd are small, $(\Delta n \,.\, \Delta d)$ can be neglected, leading to the final result

$$\text{OPLD} = \left(\text{OPL}_{\text{after}} - \text{OPL}_{\text{before}}\right) = \Delta d\,.\,(n-1) + \Delta n\,.\,d \tag{11}$$

For the case, the change in refractive index is negligible, $\Delta n \cong 0$, the optical path length can be approximately expressed as

$$\text{OPLD} \cong \Delta d\,.\,(n-1) \tag{12}$$

If the initial thickness (before curing) of the sample is d_0, from definition, the deformation ε (either shrinkage or swelling) is given by the below equation:

$$\varepsilon = \frac{\Delta d}{d_0} = \frac{\text{OPLD}}{(n-1)\ d_0} = \frac{\text{OPLD}}{(n-1)\ d_0} \tag{13}$$

The OPLD on the left-hand side can be obtained from MZI experiments. Therefore, if the change in refractive index Δn before and after the curing reaction can be directly measured using some instrument like prism coupler [11], the change in the sample thickness Δd can be obtained.

2.2. Data analysis

The interference patterns obtained for a polymer film under in situ photocuring on the test arm of the MZI are recorded by using a CCD camera. Though the laser beam was passed through the spatial filter to select the best part of the beam and was subsequently collimated before entering the MZI unit, the interference patterns are slightly affected by the spatial distribution of the laser intensity. This effect can be removed by performing some correction assuming that the shape of the laser beam is Gaussian [10].

The interferograms obtained before and after this correction for the intensity distribution of a He–Ne laser (NEC, 1 mW) in the case a polystyrene/poly(vinyl methyl ether) PS/PVME (30/70) blend was used as sample are, respectively, illustrated in **Figures 3** and **4**. To reduce noise, the 2D data (480 pixel × 640 pixel) were divided into 48 horizontal strips with the dimension (10 pixel × 640 pixel) for each strip. Data along the y-axis for each strip were then averaged to provide 1D data as shown in **Figure 3(a)**.

In general, the real part of the intensity of an interferogram is a periodic function of distance and can be expressed in 1D as follows:

$$I(x) = a(x)\cos\left[\phi(x)\right] \tag{14}$$

where $a(x)$ and $\phi(x)$ are the amplitude and phase of the signal, respectively.

The imaginary part $J(x)$ of the intensity can be calculated by using Hilbert transform [10, 12]. From these calculations, the amplitude $a(x)$ and the phase $[\phi(x)]$ are obtained:

$$a(x) = \sqrt{I^2(x) + J^2(x)} \quad (15)$$

$$\text{and } \varphi(x) = \tan^{-1}\left(\frac{J(x)}{I(x)}\right) \quad (16)$$

Finally, the OPLD can be obtained for the left-hand side of Eq. (12).

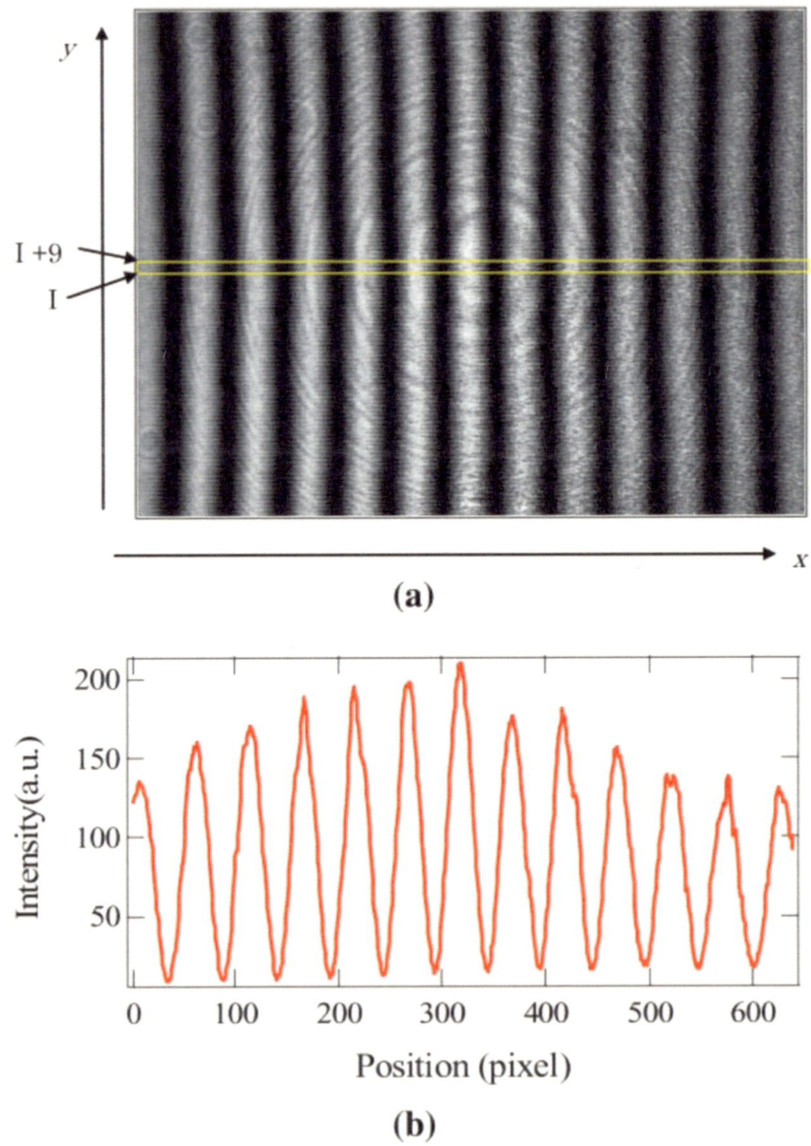

Figure 3. (a) Interferogram obtained at 20°C for a PSA/PVME (30/70) blend; (b) 1D data obtained by averaging along the y direction for each 10 pixels indicated by the two arrows in (a). The peripheral of the intensity distribution in (b) is affected by the Gaussian beam.

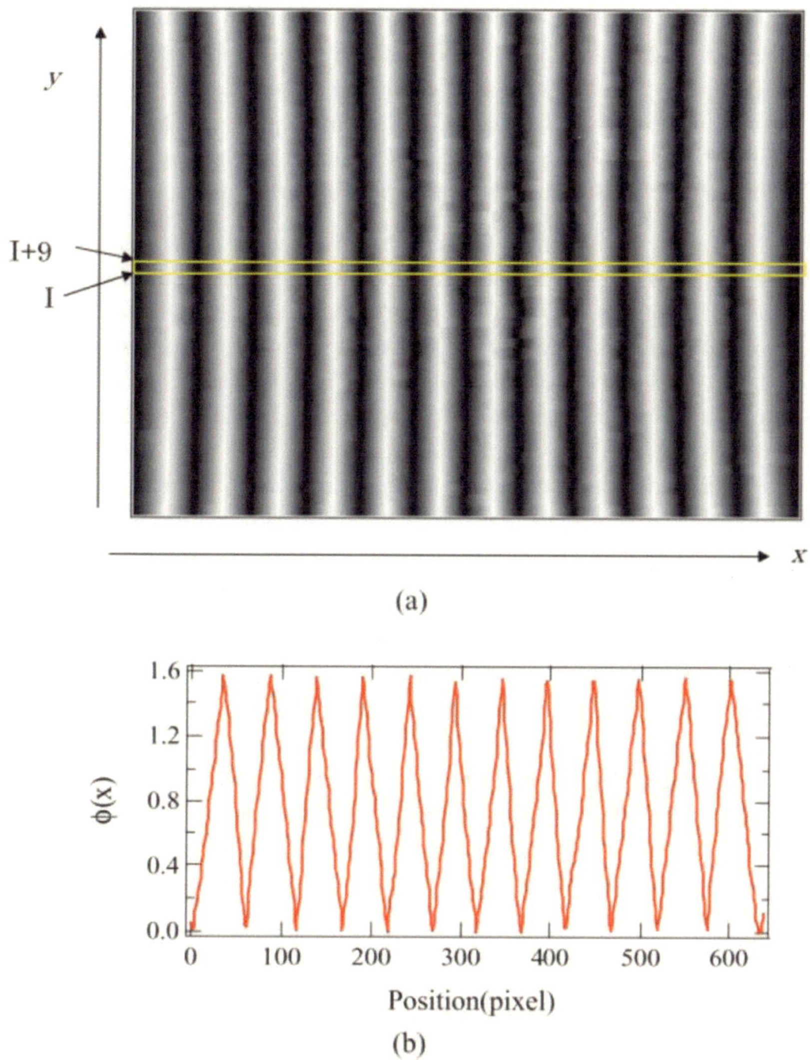

Figure 4. (a) The interferogram shown in **Figure 4** after the Hilbert transformation; (b) the 1D intensity distribution after averaging along the *y*-axis as described for **Figure 3**.

3. Samples

Polymers used in this study are the derivatives of poly(ethyl acrylate) (PEA), polystyrene (PS) and poly(vinyl methyl ether) (PVME). The mixture of PS and PVME exhibits miscibility at room temperature, providing a system for studying the effects of shrinkage on phase separation of polymer blends. All polymers used here have molecular weight larger than 100.000 and the molecular weight distribution around 2.0. The details of chemical synthesis and sample characterization are described in previous publications [13, 14].

Samples for MZI studies were obtained by solvent casting method and were dried under vacuum at least one night. All the samples PS/PVME mixtures with the dimension (20 mm × 20 mm × 10 μm) were annealed under vacuum over 2 h at temperature above the glass transition temperature (T_g) of the blend to erase the thermal history of the preparation process.

3.1. Photodimerization of anthracene as a photocuring reaction

To photo-cross-link a polymer with UV irradiation, photosensitive anthracene was chemically labeled on a given polymer by copolymerizing its monomer with a photoreactive monomer by copolymerization. By doing so, photoreactive anthracene moieties were introduced into the polymer component under examination. The labeling content of anthracene can be adjusted by varying the ingredients of the coupling reactions. Upon irradiation with 365 nm UV light, anthracene undergoes photodimerization as illustrated in **Figure 5** for the case of PEA chains.

3.2. In situ observation of the deformation kinetics in homopolymers undergoing photocuring and relation to physical aging of the photocured polymer

Poly(ethyl acrylate) (PEA, $M_w = 1.6 \times 10^5$, $M_w/M_n = 2.2$) was prepared by conventional free radical polymerization. To be able to cure PEA with UV irradiation, the PEA was chemically labeled with anthracene which served as a cross-linker of the PEA chains as illustrated in **Figure 5**. Upon irradiation with 365 nm UV light, the anthracene moieties labeled on PEA undergoing photodimerization, generating PEA networks in the sample. As a consequence, the sample gradually approaches the glassy state and exhibits shrinkage due to the liquid→solid transition. However, this shrinkage in this particular case is fairly small and could not be observed via monitoring the change in the sample thickness by laser-scanning confocal microscopy as in the case of photopolymerization [15]. As the cross-link density in the sample reaches a critical value, PEA enters the glassy state. Depending on the rate at which PEA enters the glassy state, physical aging [16] could occur. This feature can be observed via the irradiation intensity dependence of the shrinkage associated with irradiation time as shown in **Figure 6**. Here, the deformation ε which was calculated from the OPLD data given in Eq. (13) for the case of negligible change in refractive index is plotted versus irradiation time and elapse time. It is worth noting that the physical aging phenomena are evidenced by the continuation of shrinkage after stopping irradiation. These results suggest that the sample with the cross-link density $\gamma \sim 2$ already enters the glassy state during irradiation, exhibiting the physical aging phenomena. Compared to the result obtained at low light intensity, it was found that the physical aging of photo-cross-linked PEA sample emerges at the cross-link density $\gamma \geq 2$ junctions /chain. This aging process becomes stronger under irradiation with higher light intensity. From the plot of normalized shrinkage ($\varepsilon/\varepsilon_{max}$) vs. non-dimensionalized elapse time defined as ($t_e \cdot k_a$) where t_e is the elapse time and k_a is the characteristic time of the aging process, it was found that all the aging data obtained with a constant irradiation intensity can be expressed by a master curve [17].

3.3. Local deformation of miscible polymer blends under photocuring and relation to physical aging

Mach-Zehnder interferometry was also utilized to detect the local deformation in miscible polymer blends polystyrene derivative (PS) and poly(vinyl methyl ether) (PVME). The curing

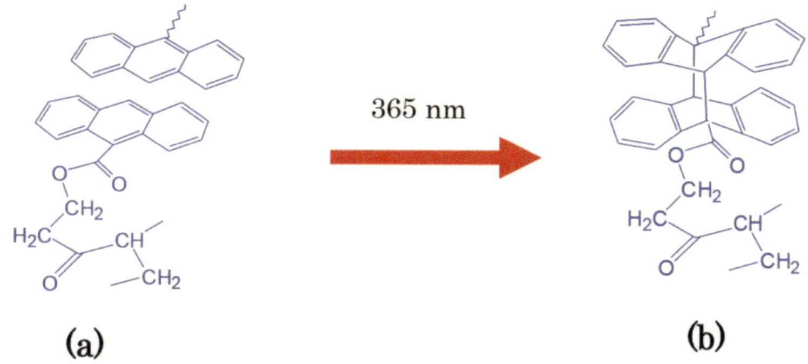

Figure 5. Photodimerization of anthracene chemically labeled on poly(ethyl acrylate): (a) before photodimerization, (b) after photodimerization with the formation of photodimer between two segments of PEA.

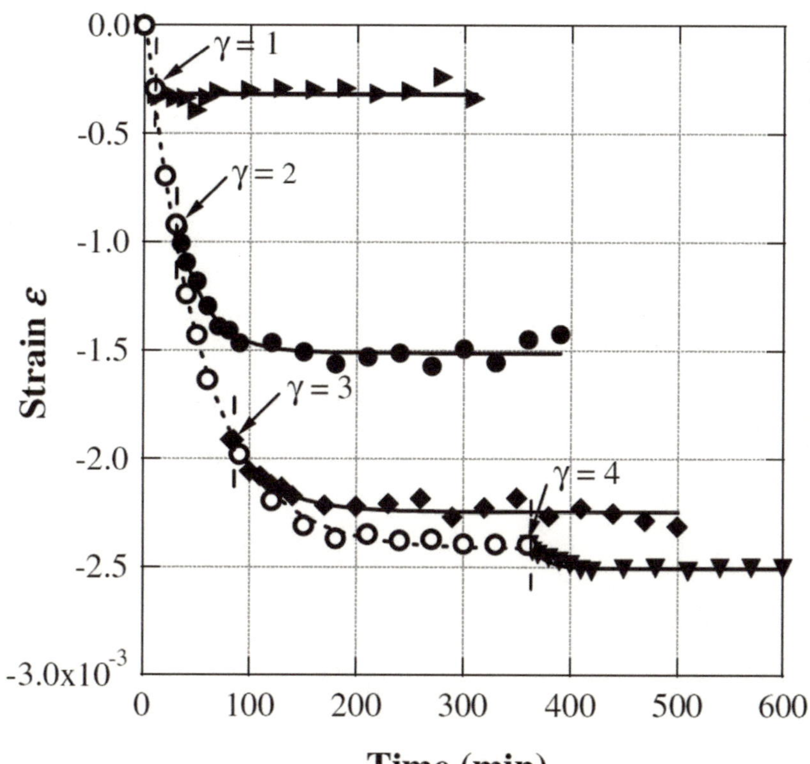

Figure 6. Strain relaxation observed for a PEA sample under curing at different irradiation conditions and the evidence of physical aging phenomena.

reaction was performed taking advantages of the photodimerization of anthracene chemically labeled on the PS chains. The curing reaction was followed by monitoring the change in the absorbance of the photo-cross-linker chemically labeled on PS. It was found that both the curing kinetics and the deformation induced by the curing reaction can be described by the Kohlrausch-Williams-Watts (KWW) kinetics for kinetically inhomogeneous systems [18]:

$$\gamma(t) = A\left[1 - \exp\left\{-k_c t\right\}^{\alpha}\right] \quad (17)$$

$$\varepsilon(t) = A\left[1 - \exp\left\{-k_d t\right\}^{\beta}\right] \quad (18)$$

where A and B are constant, k_c and k_0 are, respectively, rate constant of the reaction and shrinkage. The KWW exponent α and β are less than 1 and in between 0.7 and 0.8.

From the kinetics data expressed by Eqs. (17) and (18), k_d c can be obtained from MZI data and k_c can be deduced from the cross-linking data. It was found that there exists a strong correlation between the curing and the shrinkage processes as illustrated in **Figure 7** [19].

However, the correlation between the cross-link process expressed by the reduced cross-link density $\gamma_r \equiv (\gamma/\gamma_{max})$ and the deformation process indicated by the reduced strain $\varepsilon_r \equiv (\varepsilon/\varepsilon_{max})$ cannot be well expressible by a master curve, particularly at high cross-link density under high light intensity as shown in **Figure 8**. It is worth noting that γ_{max} and ε_{max} are the maximum value at which the cross-link density and the deformation can be achieved at a given light intensity. The increase in the concentration fluctuations in the blends under curing, particularly under high light intensity, would be responsible for the deviation from the master curve.

From the data obtained by MZI, it was also found that the glass transition temperature also plays an important role in the deformation of the cured PS/PVME blends. From the in situ measurements of deformation by MZI under curing, it was found that the photocured sample

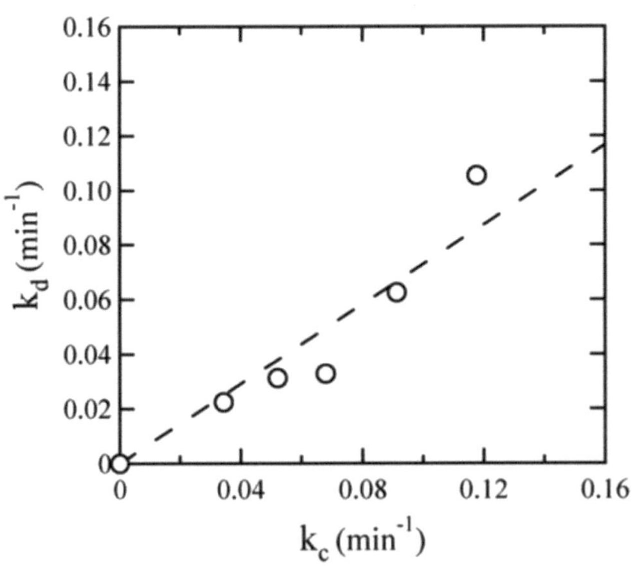

Figure 7. The correlation between the curing reaction kinetics expressed by *k* c and the deformation process revealed by *k* d observed for a PS/PVME (30/70) blend.

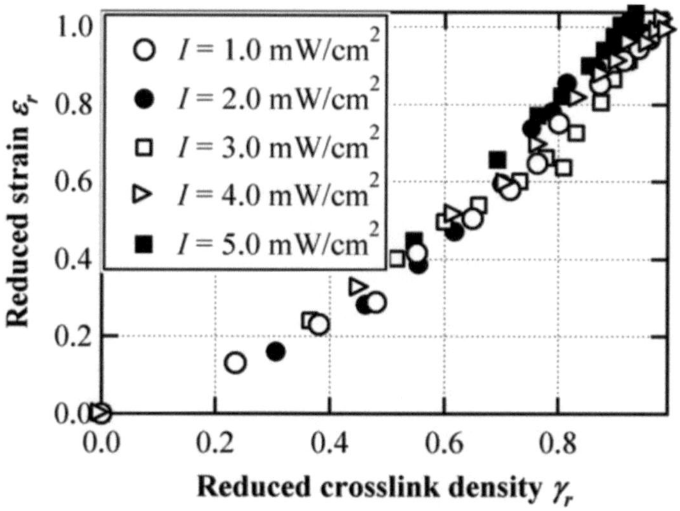

Figure 8. Correlation between the reduced strain and the reduced cross-link density obtained for a PS/PVME (30/70) irradiated with different light intensity ranging from 0.1 to 5.0 mW/cm².

undergoes shrinking during the irradiation process, but the sample also partially recovered by swelling back after stopping irradiation. This process is determined by the difference between the experimental temperature and the resulting glass transition temperature T_g of the cured sample at the time of stopping irradiation. This particular swelling behavior was observed upon raising the experimental temperature [19].

3.4. Local deformation observed by MZI in polymer blends undergoing phase separation in the bulk state

So far, Mach-Zehnder interferometry has been utilized to monitor the local deformation in homopolymers and miscible polymer blends under photocuring. For cured polymer mixtures, phase separation took place as the reaction yield exceeds a critical value. The shrinkage induced by curing reaction not only affects the shape of the blend but also influences the phase separation process. For the polymer mixtures with a lower critical solution temperature (LCST) like PS/PVME, cross-linking the PS component will enlarge the unstable region of the mixture and eventually lead to phase separation. The shrinkage of the mixture reveals some unexpected behavior shown in **Figure 9**, as an example, for a PS/PVME (20/80) blend photo-cross-linked by irradiation with 365 nm UV light. Here, the peak of the scattering intensity in situ monitored during the curing process appears and gradually moves toward the side of *wide angle* (*large wavenumber q*), suggesting that the length scale of the bi-continuous structures resulting from the phase separation gradually decreases, instead of increase as in many cases observed for the conventional phase separation process [20]. Taking into account that polymers often undergo shrinkage upon curing, the deformation of a PS/PVME (20/80) blend was in situ monitored by using a Mach-Zehnder interferometry under irradiation with the same conditions. The period ξ of these spinodal structures induced by photocuring was calculated using the Bragg condition $\xi = 2\pi/q_{max}$, where q_{max} is the wavenumber corresponding to the scattering peak of the intensity

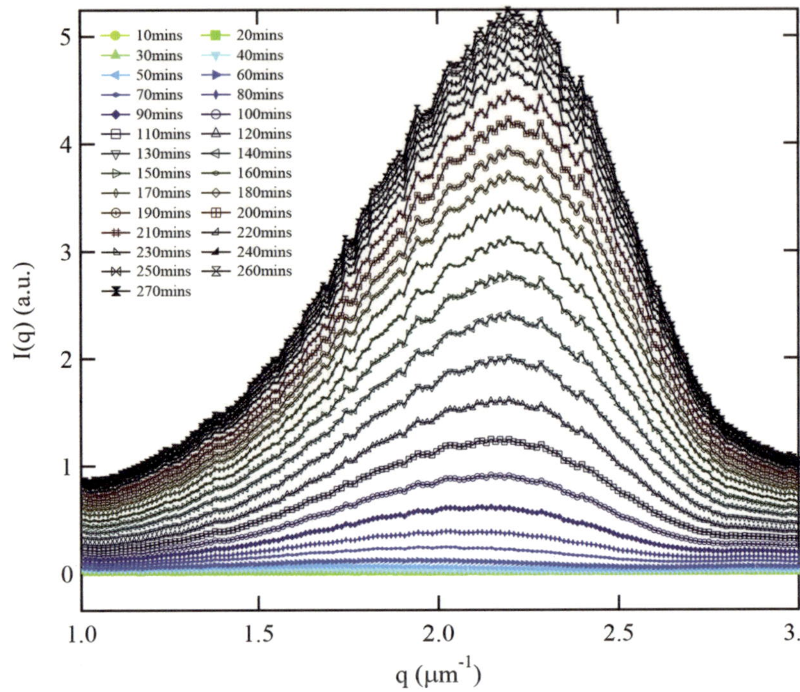

Figure 9. Laser scattering profile obtained for a PS/PVME (20/80) blend photo-cross-link by irradiation with 365 nm UV light. The data were recorded in situ under irradiation. The number in the figure indicates the curing time [21].

distribution shown in **Figure 9**. The plot of the characteristic length scale ξ of the morphology versus the net time of curing clearly shows that ξ decreases with increasing curing (irradiation) time. This result is opposite to the conventional phase separation in the absence of shrinkage, that is, ξ increases as phase separation proceeds. Obviously, as curing time increases, the characteristic length scale gradually decreases and finally reaches a stationary value, indicating the termination of the phase separation process. From the results obtained by Mach-Zehnder interferometry, the elastic strain defined as $(\Delta d/d)$ was calculated and its dependence on curing time was examined. It was found that the elastic strain $(\Delta d/d)$ increases with increasing the irradiation intensity and significantly deviates from these decay curves when phase separation starts. The initial slope of the plot $(\Delta d/d)$ vs. irradiation time, that is $(\tau_i = (d(\Delta d/d))/dt)$ can be used as a measure of the characteristic time of shrinkage process induced by the curing reaction. The rate of shrinkage k_i can be defined as $k_i = (1/\tau_i)$. **Figure 10** depicts the clear correlation between the shrinkage obtained from Mach-Zehnder interferometry and the evolution of morphology detected by light scattering, demonstrating again the effectiveness of the Mach-Zehnder interferometry.

A clear correlation was observed between the apparent rate of the phase separation k_ξ obtained from light scattering experiment and the apparent rate of shrinkage k_i obtained from MZI experiments [20]. The results also reveal the significant roles of Mach-Zehnder interferometry in the kinetic studies on polymer phase separation.

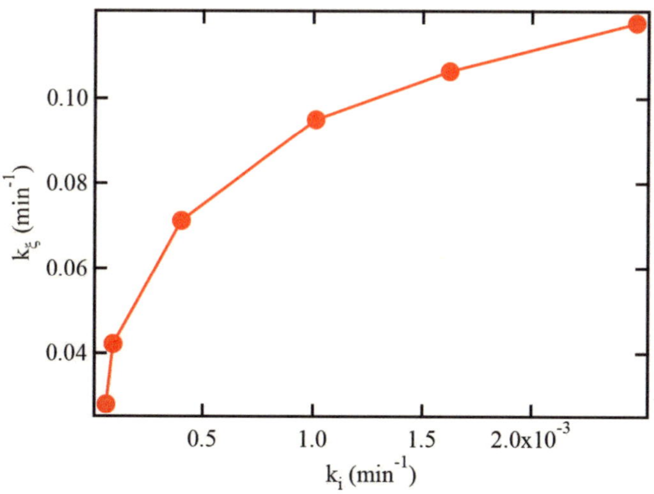

Figure 10. Correlation between the rate of shrinkage caused by photocuring reaction and the apparent rate of phase separation observed for a PS/PVME (20/80) blend photocured with 365 nm UV at 110°C.

Besides the applications of MZI to the polymer researches described above, the effect of polymer molecular weight on the deformation of poly(ethyl acrylate) (PEA) was recently investigated during the photocuring process. The effects of the entanglement molecular weight of PEA on the shrinkage and swelling process were observed by MZI and the results are discussed in terms of polymer diffusion in entangled polymer networks [21].

4. Concluding remarks

For the applications in polymer research, Mach-Zehnder interferometer (MZI) would be a simple instrument to in situ monitor the local deformation in the nanometer scales. The current MZI instrument can be improved in two aspects: accessibility to temperature dependence measurements and improvement of signal-to-noise ratio to increase the data precision. The former requires some careful temperature controls of the experimental environments around the sample and the MZI chamber. On the other hand, the later could be solved by introducing lock-in detection of interference signals. These experiments are underway to prepare for the second generation of Mach-Zehnder interferometry for research on photocuring of polymers.

Acknowledgements

Research described here was financially supported by the Ministry of Education, Japan through Grant-in-Aid for Scientific Research on the Priority-Research-Areas "Molecular NanoDynamics" and "Soft Matter Physics". We greatly appreciate the active participation of former graduate students at the beginning of the project: Kyosuke Inoue, Satonori Komatsu, Ken Ohdomari and Kazuhiro Sorioka. This experiment would not be successful without their great effort and elaboration.

Author details

Dan-Thuy Van-Pham[1], Minh-Tri Nguyen[1, 2], Hideyuki Nakanishi[1], Tomohisa Norisuye[1] and Qui Tran-Cong-Miyata[1]*

*Address all correspondence to: qui@kit.ac.jp

1 Department of Macromolecular Science and Engineering, Graduate School of Science and Technology, Kyoto Institute of Technology, Kyoto, Japan

2 Department of Chemical Engineering, Can-Tho University, Can-Tho, VietNam

References

[1] Zehnder L: "A new interferometer", (*in German)*. Ein neuer interferenzrefraktor. Zeitschrift fur Instrumentenkunde. 1891, **11**: 275–285.

[2] Mach L: "On an interferometer", (*in German)*. Uber einen interferenzrefraktor. Zeitschrift fur Instrumentenkunde. 1892, **12**: 89–93.

[3] Ji Y, Chung Y, Sprinzak D, Heiblum M, Mahalu D, Shtrikman H: An electronic Mach-Zehnder interferometer. Nature 2003, **422**:415. DOI:10.1038/nature01503.

[4] Goldstein RJ, Kuehn TH. Optical systems for flow measurement: Shadowgraph, Schlieren and interferometric techniques. In: 2nd Goldstein RJ. Fluid Mechanics Measurements. Philadelphia, PA: Taylor & Francis; 1996. 451 p.

[5] Li L, Xia L, Xie Z, Liu D. All-fiber Mach-Zehnder interferometers for sensing applications. Opt. Express 2012, **20**, 11109–11120. DOI: 10.1364/OE.20.011109.

[6] Lapsley M, L, Chiang IK, Zheng YB, Ding XY, Mao X, Huang TJ. A single-layer, planar, optofluidic Mach-Zehnder interferometer for label-free detection. Lab Chip 2011, **11**, 1795–1800. DOI:10 1039/c01c00707b.

[7] Huang Y, Paloczi GT, Yariv A, Zhang C, Dalton LR. Fabrication and replication of polymer integrated optical devices using electron-beam lithography and soft lithography. J. Phys. Chem. B 2004, **108**, 8606–8613. DOI: 10.1021/jp049724d.

[8] Zhao X-M, Smith SP, Waldman SJ, Whitesides GM, Prentiss M: Demonstration of waveguide couplers fabricated using microtransfer molding. Appl. Phys. Lett. 1997, **71**, 1017. DOI: 10.1063/1.119713

[9] Xia Y, Whitesides GM. Soft lithography. Angew. Chem. Intl. Ed. 1998, **37**, 550–575, DOI: 10.1002(SICI)1521-3773(19980316)37:5.

[10] Inoue K, Komatsu S, Trinh XA, Norisuye T, Tran-Cong-Miyata Q. Local deformation in photo-crosslinked polymer blends monitored by Mach-Zehnder interferometry. J. Polym. Sci. Polym. Phys. 2005; **43**, 2898–2913. DOI: 10.1002/polb.20593

[11] Tien PK: Light waves in thin films and integrated optics. Appl. Optics. 1971; **10**, 2395–2413.

[12] Taylor ED, Cates C, Mauel ME, Maurer DA, Nadle D, Navratil GA, Shilov M: Nonstationary signal analysis of magnetic islands in plasmas. Rev. Sci. Instrum. 1999; **70**, 4545–4551. DOI: 10.1063/1.1150110.

[13] Harada A, Tran-Cong Q. Modulated phases observed in reacting polymer mixtures with competing interactions. Macromolecules. 1997, **30**, 1643–1650. DOI: 10.1021/ma961542x

[14] Nakanishi H, Satoh M, Norisuye T, Tran-Cong-Miyata Q. Generation and manipulation of hierarchical morphology in interpenetrating polymer networks by using photochemical reactions. Macromolecules. 2004, **37**, 8495–8498. DOI: 10.1021/ma048657i

[15] Tran-Cong-Miyata Q, Kinohira T, Van-Pham D-T, Hirose A, Norisuye T, Nakanishi H. Phase separation of polymer mixtures driven by photochemical reactions: Complexity and fascination. Curr. Opin. Solid State Mater. Sci. 2011, **15**, 254–261. DOI: 10.1016/j.cossms.2011.06.004

[16] Hutchinson JM. Relaxation processes and physical aging. In: Haward RN, Young RJ, editors. The Physics of Glassy Polymers. London: Chapman & Hall; 1997. pp. 85–153.

[17] Van-Pham D.-T, Sorioka, K, Norisuye T, Tran-Cong-Miyata, Q. Physical aging of photo-crosslinked poly(ethyl acrylate) observed in the nanometer scales by Mach-Zehnder interferometry. Polymer J. 2009, **41**, 260–265. DOI:10.1295/polymj.PJ2008202

[18] Williams G, Watts DC. Non-symmetrical dielectric relaxation behaviour arising from a simple empirical decay function. Trans. Faraday Soc. 1970, **66**, 80–85. DOI: 10.1039/TF9706600080

[19] Van-Pham DT, Sorioka K, Norisuye T, Tran-Cong-Miyata Q. Formation and relaxation of the elastic strain generated by photocuring in polymer blends monitored by Mach-Zehnder interferometry. Polymer 2011, **52**, 739–745. DOI: 10.1016/j.polymer.2010.12.028

[20] Trinh X-A, Fukuda J, Adachi Y, Nakanishi H, Norisuye T, Tran-Cong-Miyata Q. Effects of elastic deformation on phase separation of a polymer blend by a reversible photo-cross-linking reaction. Macromolecules. 2007, **40**, 5566–5574. DOI: 10.1021/ma0705220

[21] Kawamoto T, Van-Pham D-T, Nakanishi H, Norisuye T, Tran-Cong-Miyata Q. Effects of molecular weight on the local deformation of photo-cross-linked polymer blends studied by Mach-Zehnder interferometry. Polymer J. (Tokyo). 2014, **46**, 819–822. DOI:10.1038/pj.2014.63.

Chapter 3

Interferometry for Online/In-Process Surface Inspection

Feng Gao

Additional information is available at the end of the chapter

http://dx.doi.org/10.5772/66530

Abstract

Interferometers are normally operated in environment-controlled optical laboratories because vibrations will induce errors in measurement results. In order to extend the application of interferometry to shop floor inspection, two methods are adapted: one method is to introduce a reference interferometer and vibration compensation system to the main interferometer, to compensate for the environmental disturbance; and the other method is to realize the data sample in just one image shot. Each method has its own applications. With the advances of these technologies, the use of interferometry as a highly accurate and fast measurement method will become more common in shop floor measurements and inspections.

Keywords: online/in-process measurement, surface measurement, environmental disturbance, vibration compensation, reference interferometer, common path

1. Introduction

The applications of optical interferometry are very extensive. It can be used to measure displacement, vibration, angle, distance, pressure, refractive index and temperatures, to just name a few. As a high accuracy and high sensitivity measurement method, interferometers are normally operated in optical laboratories in which operational environments are well controlled. Any variations such as temperature, air flow, air pressure and vibrations in the environment will induce errors to the measurement results and even make a measurement impossible to conduct. In this chapter, we will focus on discussions of the research conducted in our research centre on the applications of optical interferometry at a shop floor environment, which is normally subjected to environmental disturbances and vibrations. These researches mainly focus on surface measurement and inspection, and this is the main focus of the research centre. The methods discussed here may also be applied to other application areas depending on the optical set-up of the interferometer system and the measurement objects.

The manufacture of highly added value components in developed economies is rapidly shifted to the design and production of micro-/nano-structured and free-form surfaces. The application of the use of micro-/nano-scale and ultra-precision structured surfaces is broad and covers the optics, silicon wafer, hard disks, MEMS/NEMS, micro-fluidics and micro-moulding industries. All these industries critically rely on ultra-precision surfaces. However, there is an essential factor to limit of the manufacture of these kinds of surfaces, that is to say the ability to measure the product quickly and easily within the manufacturing environment. According to the report, currently the quality of fabrication of these kinds of products depends mainly on the experience of process engineers backed up by an expensive trial-and-error approach. Subsequently, the scrap rates of these manufactured items are 50–70% high [1].

Optical interferometry has been extensively studied for surface measurement due to the advantages of non-contact measurement and high measurement accuracy. Nevertheless, conventional optical interferometry techniques are exceptionally sensitive to environmental noises such as air turbulence, temperature drift and mechanical vibration. Such noises can cause errors during surface measurement and produce void measurement results. There are a number of methods to reduce the influence of these noises. While controlling the measurement environment by using a vibration isolation stage and retaining a stable temperature is an effective way to reduce noise for laboratory and offline applications, it may not be practical in manufacturing conditions, for instance, when a measurement part is too large to be mechanically insulated.

In order to extend the application of interferometry to shop floor inspection, two methods can be adapted. One method is by introducing a reference interferometer and vibration compensation system to the main interferometer to compensate the environmental disturbance [2–4]. Complete common-path interferometers such as the scatterplate interferometer are insensitive to noises as well [5–7]. These noise reduction approaches are generally used for laser-based phase-shifting interferometry, for which the applications are limited to measurement of relatively smooth surfaces due to the well-known 2π phase ambiguity problem of monochromatic interferometry. The other method is to realize the data sampling in just one image shot [8–10]. This kind of so-called one shot interferometer is immune to the environmental noise and vibrations.

We will discuss the above two kinds of interferometry through two case studies: the common-path wavelength scanning interferometer and the single-shot line-scan dispersive interferometer.

http://dx.doi.org/10.5772/66530

2. Common-path wavelength scanning interferometer

2.1. Measurement principle

White light vertical scanning interferometry (WLSI) is able to overcome the 2π phase ambiguity problem and extend its application to rough surfaces and structured surfaces with large step heights [11–13]. It measures the optical path difference (OPD) by determining the peak position from the interferograms. It can be used to measure optically smooth surfaces as well as the optically rough surfaces. (The definition of an optically smooth surface is the surface height variation within the resolution cell of the imaging system is not exceed one-eighth of the wavelength of the light used. For optically rough surface which is defined as the height variations within the resolution cell of the imaging system is exceed one-fourth of the wavelength of the light used [13]). However, the need to perform a mechanical scanning of the heavy probe head or the specimen stage limits the measurement speed. In addition, the data acquisition procedure and processing are more complicated than monochromatic interferometry.

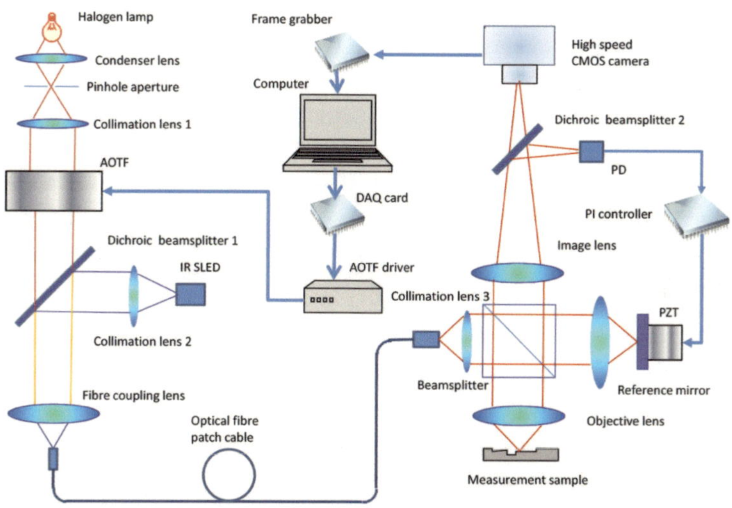

Figure 1. Schematic diagram of the proposed surface measurement system. AOTF–acousto-optic tuneable filter; PD–photo diode; IR SLED–inferred superluminescent light-emitting diode; DAQ–data acquisition card.

Wavelength scanning interferometry (WSI) has been reported by many researchers worldwide in areal surface measurement by using a two-dimensional CCD detector [4, 14–16]. Compared with WLSI, surface topography measurements are based on the phase shifts due to wavelength variations, thus avoiding any mechanical scanning process. Absolute optical path difference can be measured without any 2π phase ambiguity. WSI has advanced itself to dispersive white light interferometry [17, 18], measuring full field of surface instead of a single point of the surface by means of spectrometry.

We have proposed an environmentally robust fast surface measurement system by means of wavelength scanning interferometry and active servo control techniques. The basic configuration of the proposed surface measurement system is illustrated in **Figure 1**. The measurement system is composed of two Linnik interferometers that share a common optical path. The measurement interferometer is illuminated by a white light source through an acousto-optic tuneable filter (AOTF) to filter the light from the white light source to the main interferometer. This is to select a specific wavelength for the interferometer, thus producing an interferogram at the CCD sourced only by that specific wavelength. The selected light wavelength is determined by:

$$\lambda = \Delta n\alpha \frac{v_a}{f_a} \tag{1}$$

where Δn is the birefringence of the crystal used as the diffraction material, α is a complex parameter subject to the design of the AOTF, and v_a and f_a are the velocity and frequency of the driving acoustic wave, respectively. The wavelength of the light which is selected by the AOFT can therefore be varied just by changing the driving frequency f_a. Consequently, different wavelengths of light will pass through the AOTF in sequence so a series of interferograms of different wavelengths will be detected by the CCD camera. The absolute optical path difference can be calculated in real time through analyzing interferograms captured by the CCD camera. The reference interferometer, which is illuminated by an inferred superluminescent light-emitting diode (SLED), is used to observe and compensate for the environmental noise, for example, mechanical vibration, temperature drift and air turbulence. Because the two interferometers undergo similar environmental noise, the measurement interferometer will be capable of measuring surface information once the reference interferometer is 'locked' into the compensation mode.

The light beams from the AOTF and the IR SLED are combined by a dichroic mirror that is highly reflective in the inferred wavelength and transmissive in the visible light wavelength range. After passing through the dichroic mirror, the light beam is coupled into an optical fibre patch cable. By separating the light source and AOTF from the interferometers, not only the size and weight of the interferometers have been greatly reduced, but also the thermal influence from the light source has also been eliminated.

It is well known that surface measurement in the workshop/manufacturing environment has been challenging to achieve using interferometric techniques since they are very sensitive to environmental vibrations, in particular, axial (vertical) vibration [5]. In addition, measurement noise can be induced by air turbulence and temperature drift as well. In this experimental study, the reference interferometer is illuminated by a SLED, which made by EXOLES (EXS2100068–01, 850 nm centre wavelength with 50 nm bandwidth) together with a servo feedback electronic unit to compensate the environmental noise effectively. Output light from the SLED is combined with the measurement light and travels virtually the same optical path as the measurement interferometer. The interference signal of the reference interferometer is picked up by a photodiode after being filtered off by dichroic beamsplitter 2 (Thorlabs, DMSP805 short pass dichroic mirror with 805 nm cut-off wavelength). As a result of a shared optical path, it is intended that if the noise happening in the reference interferometer is monitored and compensated for; the measurement interferometer will not suffer any noise disturbance during measurement. The reference interferometer will be locked at about quadrature to maximize sensitivity to environmental instabilities through a piezoelectric transducer (PZT). The resolution of modern PZT can be up to 0.05 nm with a frequency response of 35 kHz (e.g. P-249.10, PI Company); the noise compensation can be quick and accurate provided the load is light. This technique has been tested effectively and proved in our previous research [6, 7].

2.2. Phase calculation principle

Intensities detected by pixel (x, y) of the CCD camera that correspond to one point on the test surface can be expressed by

$$I(x,y;k) = a(x,y;k) + b(x,y;k)\cos(2\pi kh(x,y)) \tag{2}$$

where *a(x, y; k)* and *b(x, y; k)* are the background intensity and fringe visibility, respectively, *k* is the wavenumber which is the reciprocal of wavelength, *h(x, y)* and is the absolute optical path difference of the interferometer.

The phase of the interference signal *φ(x, y; k)* is given by

$$\phi(x,y;k) = 2\pi kh(x,y). \tag{3}$$

The phase shift of the interference signal owing to the wavenumber shift is given by

$$\Delta\phi(x,y;\Delta k) = 2\pi\Delta kh(x,y). \tag{4}$$

The phase change of the interference signal is proportional to the wavenumber k change. Then, the optical path *h(x, y)* difference is given by

$$h(x,y) = \frac{\Delta\phi(x,y,\Delta k)}{2\pi\Delta k}. \tag{5}$$

The change of *k* can be calibrated by using an optical spectral analyzer, and the key issue here is how to determine the phase change. There are many phase calculation algorithms that may be used in spectral scanning interferometry. These algorithms include phase demodulation by a lock in amplifier [19], and phase calculation by a seven-point method used in classical phase-shifting interferometry [20], extremum position counting [17] and Fourier transform-based techniques [21]. In this paper, we use phase calculation by Fourier transform because it is fast, accurate and insensitive to intensity noise.

In Eq. (2), as mentioned, $a(x, y; k)$ and $b(x, y; k)$ are slowly variable with respect to due to the response of k the CCD camera and the spectrum intensity of the light source. Practically, the optical path difference of the interferometer is adjusted to be large enough with the intention that the frequency of the cosine term is higher than the variation frequency of terms $a(x, y; k)$ and $b(x, y; k)$ so that they can be easily separated from each other. Eq. (2) can be rewritten as

$$I(x,y;k) = a(x,y;k) + \frac{1}{2}b(x,y;k)\exp[2\pi ikh(x,y)] + \frac{1}{2}b(x,y;k)\exp[-2\pi ikh(x,y)]. \tag{6}$$

The Fourier transform of Eq. (6) with respect to k can then be written as

$$\hat{I}(x,y;\xi) = A(x,y;\xi) + B(x,y;\xi - h(x,y)) + B(x,y;\xi + h(xy)) \tag{7}$$

where the uppercase letters denote the Fourier spectra of the signal expressed by the corresponding lower-case letters. If $h(x, y)$ is selected to be higher than the variation of $a(x, y, k)$ and $b(x, y, k)$, the three spectra can be separated from one another. To retrieve the phase distribution of the fringe, the term $B(x, y; \xi-h(x, y))$ is selected, and therefore, the background intensities $a(x, y, k)$ are filtered out. The inverse Fourier transform of $B(x, y; \xi- h(x, y))$ is

$$IFFT(B(x,y;\xi - h(x,y))) = \frac{1}{2}b(x,y;k)\exp[ikh(x,y)]. \tag{8}$$

Taking the natural logarithm of this signal yields,

$$\log\left\{\frac{1}{2}b(x,y;k)\exp[2\pi ikh(x,y)]\right\} = \log\left[\frac{1}{2}b(x,y;k)\right] + i\phi(x,y;k) \tag{9}$$

from which the imaginary part of Eq. (9) is exactly the phase distribution to be measured. Following the above processes, the phase distribution of each CCD pixel, as well as the height map of the surface to be measured, can be acquired. Because the main processes here are fast Fourier transform (FFT) and inverse fast Fourier transform (IFFT), the data processing is very fast.

2.3. Experimental results

The method described in the previous section was used to measure standard height specimens and different types of structured surfaces, for which two examples are presented here. In these experimental studies, the radio frequency applied to the AOTF (Model LSGDN-1, SIPAT Co.) was scanned from 80 to 110 MHz in steps of 10 kHz, equivalent to a wavelength interval of 0.48 nm. This range of radio frequencies provides a range of scanning wavelengths from 680.8 to 529.4 nm. Throughout the wavelength scanning process, 300 interferograms were captured by a high-speed CCD camera (Model OK-AM1131, JoinHope Image Tech. Ltd.) at a frame rate of 100 frames/s. **Figure 2** shows the intensity distribution recorded by one of the CCD pixels (100, 100), and **Figure 2** shows the corresponding retrieved phase of this intensity distribution. It can be seen from **Figure 2** that this result suffers from discontinuities, where the values are in the range from $-\pi$ to π. These discontinuities have been amended by adding 2π at the discontinuous points to achieve a continuous phase distribution as shown in **Figure 2**. By using the continuous phase distribution, the sample step height of this pixel can be computed.

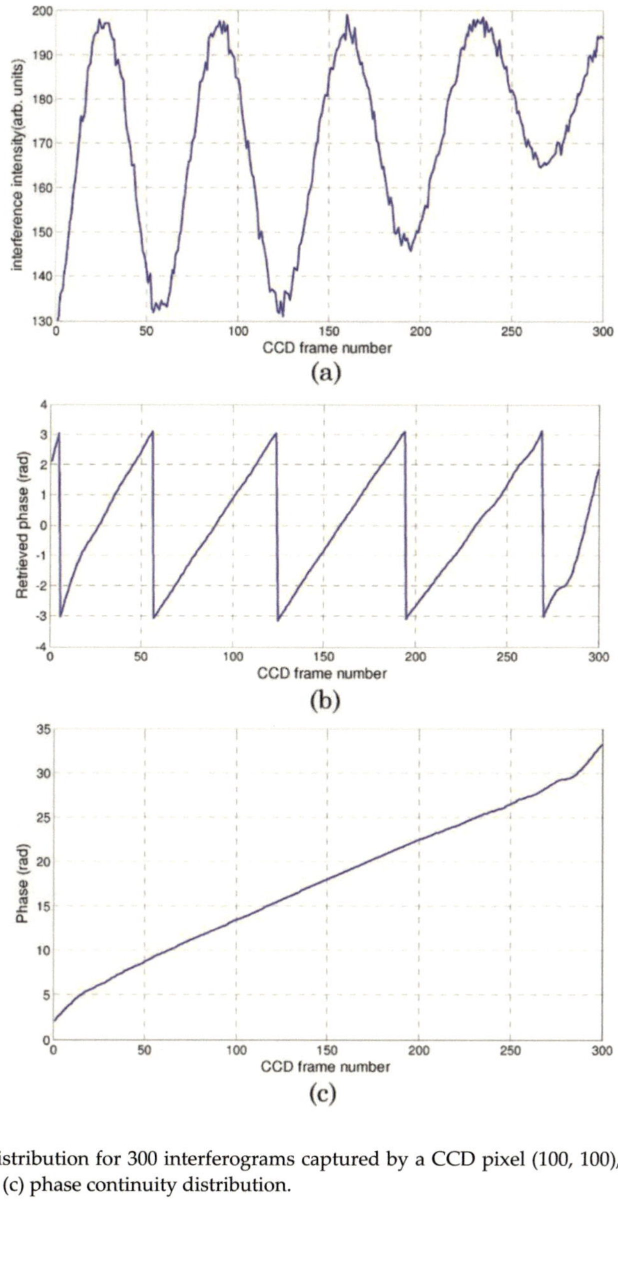

Figure 2. (a) Intensity distribution for 300 interferograms captured by a CCD pixel (100, 100), (b) retrieved phase discontinuous distribution, (c) phase continuity distribution.

A standard step height specimen with a calibrated step of 2.970 μm, provided by the National Physical Laboratory (NPL), UK, was used as a measurement sample. This specimen has been processed according to the above proposed measurement procedure and an areal surface view obtained as shown in **Figure 3**. The measured average step height is 2.971 μm with a measurement error of 1 nm.

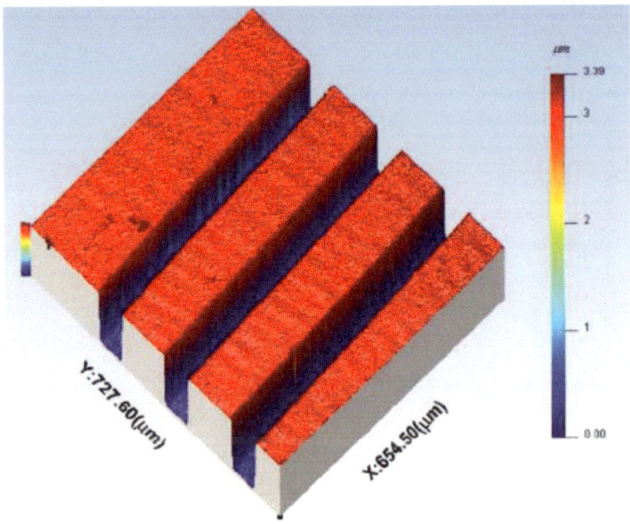

Figure 3. Measurement results of a 2.970 μm standard step.

The second measurement example is to investigate the effectiveness of the instrument's vibration compensation as follows:

1. A semiconductor daughterboard sample was measured without inducing mechanical disturbance. The measured surface step height is 4.756 μm (**Figure 4**);

2. A 40 Hz and 400 nm peak-to-peak sinusoidal mechanical disturbance using a PZT was applied to the reference mirror. During the disturbance, the measured surface step height is 11.711 μm (**Figure 5**). The surface roughness signal is completely distorted;

3. When the vibration compensation system is switched on, a decrease in the disturbance of the fringe pattern is clearly observed. A measurement of the sample is carried out with the vibration compensation system on. The data were retrieved as the original measurement and illustrate that the compensation vibration can be used to overcome environmental disturbance. The measured step height is 4.743 μm (**Figure 6**).

The difference between the two measured step height values is 13.5 nm. The measurement results are shown in **Figures 4–6**, respectively. The observed disturbance attenuation between parts (2) and (3) of the experiment was 12.2 dB, according to the reference interferometer signal output, which is in good agreement with the measured standard sample error.

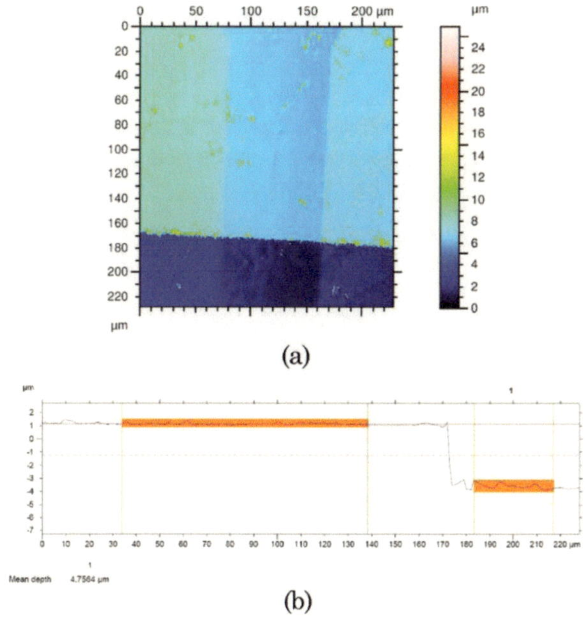

Figure 4. Measurement results of a semiconductor daughterboard sample without an induced mechanical disturbance: (a) measured surface and (b) cross-sectional profile.

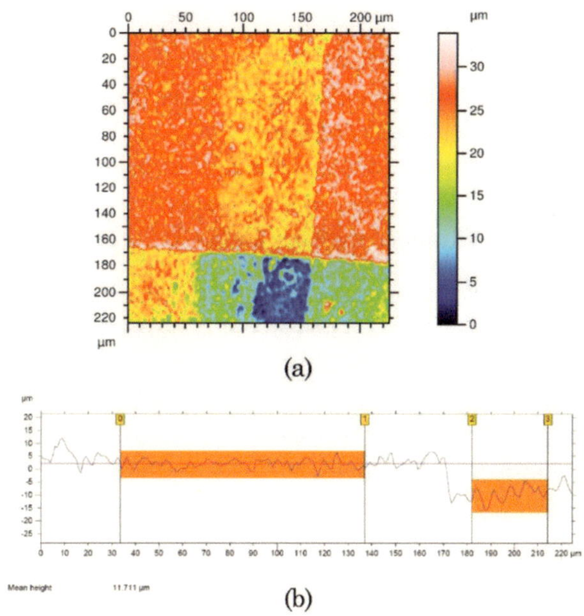

Figure 5. Measurement results of a semiconductor daughterboard sample with a sinusoidal mechanical disturbance of 40 Hz: (a) measured surface and (b) cross-sectional profile.

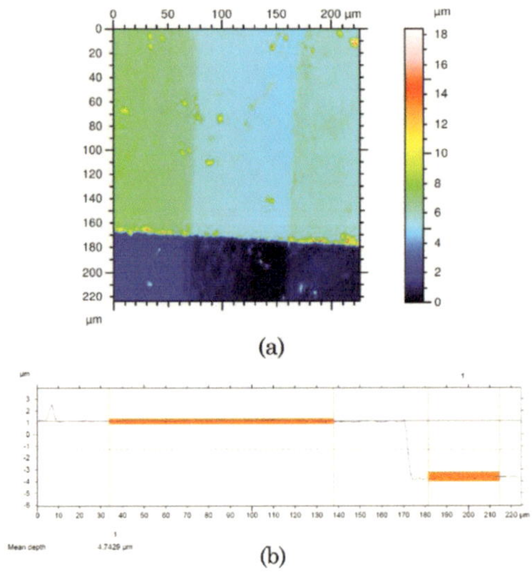

Figure 6. Measurement results of a semiconductor daughterboard sample with vibration compensation: (a) measured surface and (b) cross-sectional profile.

From the above case study, we can conclude that by using a common-path reference interferometer together with an active environmental noise compensation system, a wavelength scanning interferometer can be used for shop floor surface measurement.

Roll-to-roll (R2R) processing is a fast and economic processing method for manufacturers that produce high-volume products using large area foils such as packaging products, photovoltaic films and emerging market sectors such as flexible electronics. However, there is an increased risk of defects forming as the number of interfaces increases in the multi-layer films, and the size and nature of those defects change as the layer thicknesses shrink to the nanoscale [22, 23]. Because of the nature of these practices, the inspection methods have to be in non-contact with the film surfaces. Effective surface inspection is the key for further processes such as applying local repair techniques to eliminate the defects from the film surface.

The above system has been implemented into a R2R system for demonstration of online surface inspection [24]. Nevertheless, a measurement can only be achieved if the measured sample is at a standstill. However, the tested sample surface is in constant movement during inspection, due to the nature of R2R processing. In order to achieve dynamic inspection, all the measurement information must be sampled in just one sample, and the sampling rate should be greater than a few hundred Hz to reduce the effect of mechanical vibration. In this case, a dynamic interferometer [8–10] is the solution. However, these approaches have one drawback for the application of online surface measurement: the 2µ phase ambiguity for phase-shift interferometer, which restricts the vertical measurement range to a few hundred nanometres. It is short of the demands of most surface measurements and inspections. We have introduced a single-shot line-scan dispersive interferometer [25] which is able to perform dynamic surface inspection and has a vertical measurement range over a few hundred micrometers.

http://dx.doi.org/10.5772/66530

3. Single-shot line-scan dispersive interferometer

3.1. Experimental set-up

The basic configuration of the proposed line-scan dispersive interferometer (LSDI) is shown in **Figure 7**. A halogen bulb with broadband spectrum is used to generate the white light illumination for the system. This low-coherence white light is coupled into a multi-mode optical fibre patch cable with a numerical aperture of 0.39. The tested surface is observed through a 4 × Michelson interferometric objective. The interfered beam is focused using a tube lens with a focal length of 250 mm and split into two parts by a pellicle beamsplitter coated with a 45:55 (reflectance: transmission) split ratio. The reflectance part is received by camera 1 which provides real-time imaging of the observed surface. The other part is brought to a spectrometer to generate spectral interferograms for surface measurement. More specifically, this transmission beam first passes through a slit to block the light redundant for measurement, after that a narrow line of light which represents an interference signal of a surface profile is selected and diffracted by the grating and finally received on camera 2. The direction of the slit is set to be parallel to the columns of camera pixels, so that the dispersion axis is along the rows.

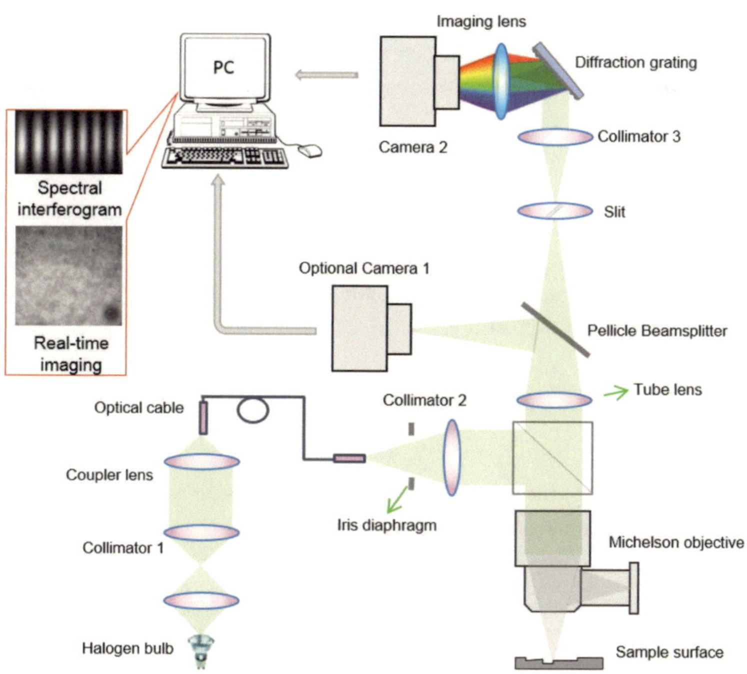

Figure 7. Schematic diagram of the optical set-up.

3.2. Interferograms analysis

Many algorithms have been developed to extract the phase from the spectral interferograms, including the techniques based on fast Fourier transform (FFT) [26], convolution [27] and Hilbert transform [28]. We developed a FFT-based algorithm for the proposed system because it is effective, accurate and insensitive to intensity noise [5]. The developed algorithm for interpreting the captured interferograms contains five sections, as shown in the flowchart which follows (**Figure 8**).

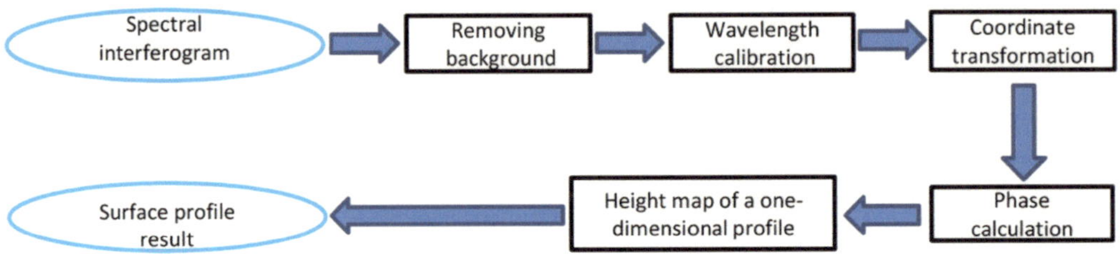

Figure 8. Flowchart of the FFT-based algorithm.

3.2.1. Removing background

The recorded interferogram contains a background intensity variation which is generated by the spectral distribution of the light source and the spectral response of the CCD camera. By blocking the measurement arm of the interferometer, a reference frame without interference effect can be captured and afterwards removed from each spectral interferogram.

3.2.2. Wavelength calibration

The output of the spectrometer is spectrally decomposed along the rows (or columns) of CCD pixels. For example, if a diffraction grating has space d, when a plane wave is incident with an angle θ_i, and the diffraction angle is θ_m to the order m, then for a beam with wavelength λ, the following equation should be satisfied

$$m\lambda = d(\sin\theta_i + \sin\theta_m). \tag{10}$$

Grating Eq. (10) shows that the relationship between the dispersed wavelength and pixel number is not linear. Therefore, the exact relationship between the pixel number and the specific wavelength needs to be calibrated. A white light laser source (WhiteLase^{TM} micro) in conjunction with the acoustic-optical tuneable filter (AOTF) is used for the calibration. The specific wavelength selected by the AOTF is determined by Eq. (11)

$$\lambda = \Delta n \alpha v_a {}^*(f_a)^{-1}. \tag{11}$$

http://dx.doi.org/10.5772/66530

A second-order polynomial is used to represent the relationship between pixel number and wavelength, as shown in Eq. (12)

$$\lambda_p = Ap^2 + Bp + C \tag{12}$$

where λ_p is the wavelength of pixel p, C is the wavelength of pixel 0, B is the first-order coefficient (nm/pixel), and A is the second-order coefficient (nm/pixel^2). More than 30 groups of experimental data (spectral lines) have been recorded to calculate the values for A, B and C through the least squares equation. The calibration result is depicted in **Figure 9**.

3.2.3. Coordinate transformation

The original signal obtained is a curve of irradiance with respect to wavelength λ (or pixel number). However, the phase variation extracted from the channelled spectrum is linearly related to the wavenumber $\sigma(\sigma = 1/\lambda)$, which makes the coordinate transformation necessary. The original sinusoidal signals are resampled in equal intervals through spline interpolation and finally reconstructed to a wavenumber-related curve.

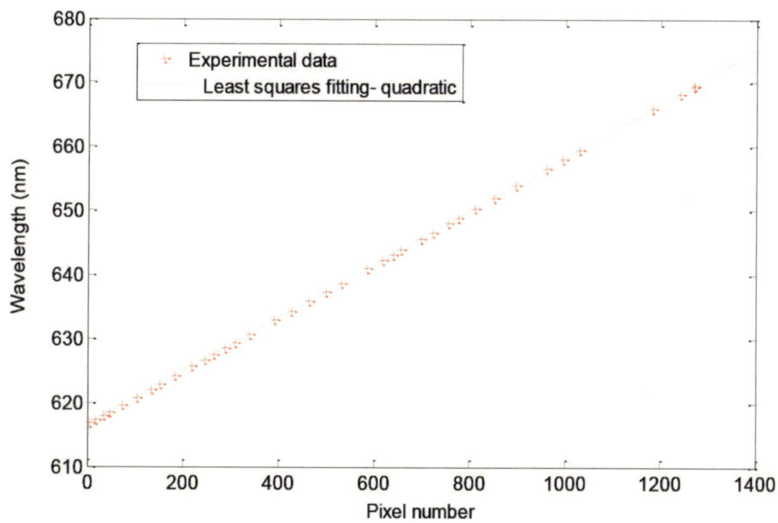

Figure 9. Wavelength calibration of the CCD camera.

3.2.4. Phase calculation

As mentioned above, we use FFT to extract the phase from the spectral interferogram. The spectral intensity $I(\text{h}, \sigma)$ recorded at the output of the spectrometer can be expressed as

$$I(x,y;\sigma) = a(x,y;\sigma) + b(x,y;\sigma)\cos[\varphi(x,y;\sigma)] \tag{13}$$

where $I(x, y; \sigma)$ is the normalized intensity with background intensity variation removed, (x, y) denotes the spatial coordinates of the interferogram, $a(x, y; \sigma)$ and $b(x, y; \sigma)$ represent the background intensity and fringe visibility, respectively, and the phase $\varphi(x, y; \sigma)$ is defined by formula

$$\varphi(x,y;\sigma) = 4\pi\lambda^{-1} \times h(x,y) + \varphi_0 = 4\pi\sigma h(x,y) + \varphi_0 \tag{14}$$

where $h(x, y)$ represents the surface elevation, and ϕ_0 is the initial phase. Eq. (13) can be written in another form as Eq. (15)

$$I(x,y;\sigma) = a(x,y;\sigma) + c(x,y;\sigma) + c^*(x,y;\sigma) \tag{15}$$

with

$$c(x,y;\sigma) = \frac{1}{2}b(x,y;\sigma)\exp[i\ \varphi(x,y;\sigma)] \tag{16}$$

where * denotes a complex conjugate. Fourier transform on the original spatial signals can be expressed in frequency domain as Eq. (17)

$$\tilde{I}(x,y;f) = A(x,y;f) + C(x,y;f) + C^*(x,y;f) \tag{17}$$

where the capital letters denote the Fourier spectra, and f is the spatial frequency. The unwanted background variation is removed by setting a filtration window, and then, the desired term $C(x, y, f)$ is selected to compute the inverse fast Fourier transform (IFFT). Taking the natural logarithm of the IFFT $[C(x, y; f)]$, the phase $\varphi(x, y; \sigma)$ of each point is extracted as the imaginary part of the Eq. (18)

$$log\left\{\frac{1}{2}b(x,y;\sigma)\exp[i\ \varphi(x,y;\sigma)]\right\} = log\left[\frac{1}{2}b(x,y;\sigma)\right] + i\ \varphi(x,y;\sigma). \tag{18}$$

Finally, the phase unwrapping process, illustrated in detail by Takeda et al. in 1982 [26], is performed to the discontinuities phase distribution.

3.2.5. *Height map of a one-dimensional profile*

The vertical axis of the spectral interferogram represents one dimension of lateral resolution, which means each row signal contains the height information of one point. By using the phase slope S obtained from the phase calculation process, the height value can be acquired using Eq. (19). After analysis of a series of row signals, the height map of a surface profile can be obtained.

$$h = \frac{S}{4\pi} = \Delta\varphi^{*}[4\pi(\sigma_m - \sigma_n)]^{-1} \tag{19}$$

where σ_m and σ_n> are the wavenumbers corresponding to the phase difference $\Delta\phi$.

3.3. Results and discussion

Two surface samples were measured to evaluate the performance of the prototype. First, to verify a good accuracy and measurement repeatability of the system, a standard sample from the National Physical Laboratory (NPL) with a step height of 100.0 nm was measured 24 times as shown in **Figure 10**. The measured average heights were recorded. It was found that the mean value of these 24 values is 96.6 nm, and the standard deviation was then calculated as 8.2 nm using Eq. (20).

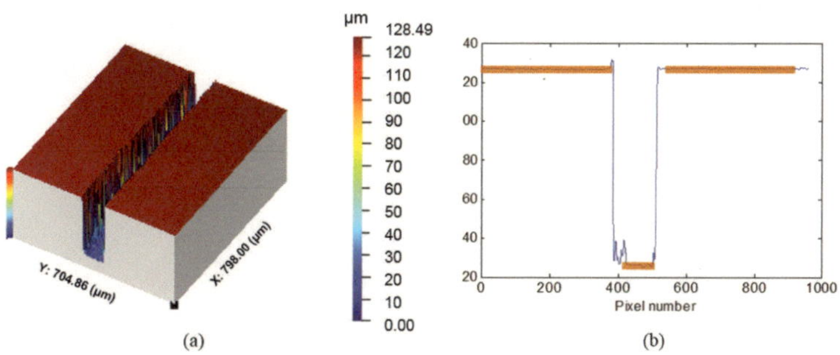

Figure 10. Measurement result of a micro-fluid chip with 100 µm step height: (a) 3D surface map, (b) cross-sectional profile.

$$\sigma = \left(\sum_{i=1}^{N} (H_i - \bar{H})^2 {*} (N-1)^{-1} \right)^{\frac{1}{2}}. \tag{20}$$

The second sample we measured was a multi-layer polyethylene naphthalate (PEN) film manufactured by the Centre for Process Innovation (CPI). This thin film is composed of three layers, namely a PEN substrate layer (125 µm), a planarization layer (3–4 µm) for planarising the pit and spike features on the PEN substrate, and an atomic layer deposition (ALD) barrier (40 nm) for prevention of moisture and oxygen ingress. The results acquired by a Taylor-Hobson Coherence Correlation Interferometry (CCI) instrument (Talysurf CCI 3000) and LSDI are shown in **Figure 11** and **Figure 12**. The 3D surface map shown in **Figure 11** is generated by 1400 profiles, which represents a scanning length of 2.31 mm. The surface topography results shown in **Figure 11** have different forms because different surface tensions are generated when fixing this film sample in these two separate measurements. Both 3D and 2D surface topography results demonstrate that LSDI is capable of detecting most of defects as the CCI does and the relative positions between each defect are well matched.

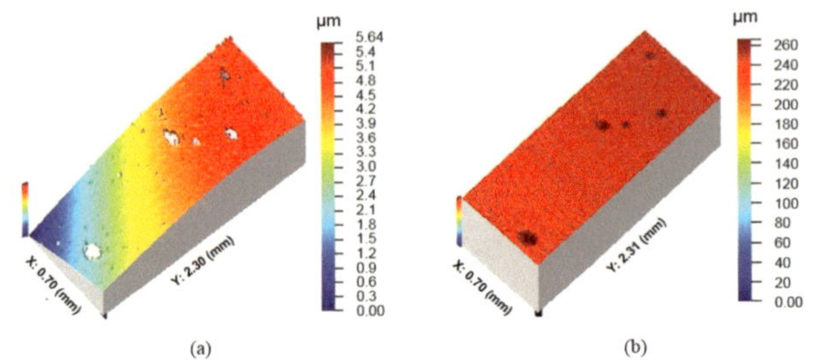

Figure 11. Defects detection on the Al_2O_3 ALD barrier film surface: (a) 3D surface map result using CCI and (b) result using LSDI.

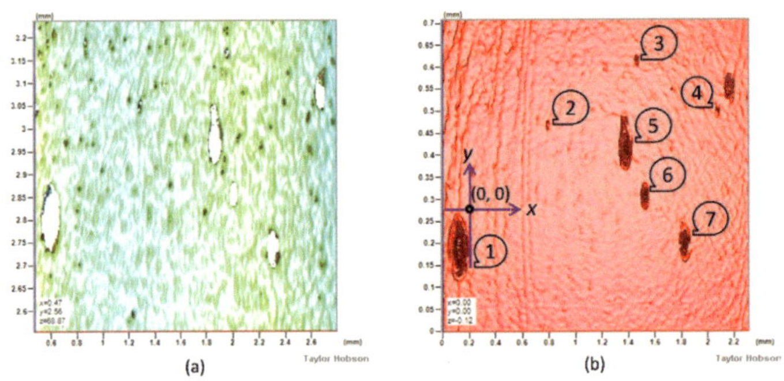

Figure 12. 2D view of defects on the Al_2O_3 ALD barrier film surface: (a) CCI result and (b) LSDI result.

	Size width × length (mm)		Position (x, y) (mm)	
	WLCSI	**CCI**	**WLCSI**	**CCI**
Defect 1	0.17 × 0.15	0.16 × 0.15	Reference point (0, 0)	
Defect 2	0.03 × 0.04	0.03 × 0.04	(0.61, 0.21)	(0.62, 0.19)
Defect 3	0.05 × 0.03	0.05 × 0.03	(1.29, 0.36)	(1.30, 0.30)
Defect 4	0.05 × 0.03	0.04 × 0.03	(1.90, 0.24)	(1.92, 0.16)
Defect 5	0.12 × 0.14	0.12 × 0.14	(1.24, 0.23)	(1.25, 0.18)
Defect 6	0.08 × 0.06	0.08 × 0.05	(1.36, 0.07)	(1.37, 0.00)
Defect 7	0.10 × 0.09	0.10 × 0.09	(1.68, −0.03)	(1.68, −0.10)

Table 1. Defects' specifications (size, location).

http://dx.doi.org/10.5772/66530

Analysis results of defects specifications for both CCI and LSDI are listed in detail in **Table 1**. The sizes and positions (x axis) of the defects correlate well between the CCI measurement of the sample and the measurement performed using LSDI. The slight difference of the relative positions along the y axis is due to different surface tensions as mentioned above, which does not affect the assessment of LSDI's performance.

The results above verified that LSDI performs well in defects detection. Only one shot for a surface profile makes it a fast and environmentally robust measurement system.

4. Conclusion

Two case studies for online surface measurement by means of interferometry have been discussed. The first case study presents a surface measurement technique with active control of environmental noise and disturbance. Nanometre accuracy surface measurement results have been obtained for micrometre step height samples. Disturbance has been reduced by 12.2 dB at 40 Hz vibration frequency. The second case study presents a line-scan dispersive interferometer with a good lateral resolution and nano-scale measurement repeatability. It has an extended axial measurement range by dispersing the output of the spectrometer along the camera pixels, without the 2π phase ambiguity. Free from any mechanical scanning and obtaining a surface profile in a single shot allows this set-up to minimize the effect of external perturbations and environmental noise.

Acknowledgements

The author gratefully acknowledges the UK's Engineering and Physical Sciences Research Council (EPSRC) funding of the EPSRC Centre for Innovative Manufacturing in Advanced Metrology (Grant Ref: EP/I033424/1), the funding with Grant Ref: EP/K018345/1 and the funding of the EPSRC HVMC Fellowship.

Author details

Feng Gao

Address all correspondence to: f.gao@hud.ac.uk

Centre for Precision Technologies, University of Huddersfield, UK

References

[1] L. Singleton, R. Leach, A. Lewis and Z. Cui, "Report on the analysis of the MEMSTAND survey on Standardisation of MicroSystems Technology," MEMSTAND Project IST-2001-37682 (2002), Proceedings of the International Seminar MEMSTAND - "Standardisation for Microsystems: the Way Forward"., 24–26 February 2003, Barcelona, Spain, 11–31.

[2] X. Jiang and D. Whitehouse, "Miniaturized optical measurement method for surface nanometrology," Annals of the CIRP,55, 1 (2006).

[3] X. Jiang, K. Wang and H. Martin, "Near common-path optical fiber interferometer for potential fast on-line microscale-nanoscale surface measurement," Optics Letters, 31, 3603–3605 (2006).

[4] X. Jiang, K. Wang, F. Gao and H. Muhamedsalih, "Fast surface measurement using wavelength scanning interferometry with compensation of environmental noise," Applied Optics, 49(15), 2903–2909 (2010). ISSN 1559–128X.

[5] M. North-Morris, J. Van Delden and J. Wyant, "Phase-shifting birefringent scatterplate interferometer," Applied Optics, 41(4), 668–677 (2002).

[6] J. Huang, T. Honda, N. Ohyama, and J. Tsuiiuchi, "Fringe scanning scatter plate interferometer using a polarized light," Optics Communications, 68, 235–238 (1988).

[7] D. Su and L. Shyu, "Phase-shifting scatter plate interferometer using a polarization technique," Journal of Modern Optics, 38, 951–959 (1991).

[8] C. Koliopoulos, "Simultaneous phase-shift interferometer," Proceedings of SPIE, 1531, 119–127 (1992).

[9] J. E. Millerd and N. J. Brock, "Methods and apparatus for splitting, imaging, and measuring wavefronts in interferometry," U.S. Patent 6,304,330 (2001).

[10] N. I. Toto-Arellano, G. Rodrĺguez-Zurita, C. Meneses-Fabian, and J. F. Vazquez-Castillo, "Phase shifts in the Fourier spectra of phase gratings and phase grids: an application for one shot phase-shifting interferometry," Optics Express, 16, 19330–19341 (2008).

[11] G.KinoandS.Chim,"Miraucorrelationmicroscope,"AppliedOptics,29,3775–3783(1990).

[12] L. Deck and P. de Groot "High-speed noncontact profiler based on scanning white light interferometry," Applied Optics, 33, 7334–7338 (1994).

[13] T. Dresel, G. Hausler, and H. Venzke, "Three-dimensional sensing of rough surfaces by coherence radar," Applied Optics 31, 919–925 (1992).

[14] S. Kuwamura, and I. Yamaguchi, "Wavelength scanning profilometry for real-time surface shape measurement microscope," Applied Optics 36, 4473–4482 (1997).

[15] A. Yamamoto, C. Kuo, K. Sunouchi, S. Wada, I. Yamaguchi and H. Tashiro, "Surface shape measurement by wavelength scanning interferometry using an electronically tuned Ti: sapphire laser", Optical Review, 8, 59–63 (2001).

[16] A. Yamamoto and I. Yamaguchi, "Surface profilometry by wavelength scanning Fizeau interferometer", Optics Laser and Technology, 32, 261–266 (2000).

[17] J. Schwider and L. Zhou, "Dispersive interferometric profilometer, " Optics Letters, 19(13), 995–997 (1994).

[18] U. Schnell, R. Dandliker and S. Gray, "Dispersive white-light interferometry for absolute distance measurement with dielectric multilayer systems on the target", Optics Letters, 21(7), 528–530 (1996).

[19] X. Dai, and S. Katuo, "High-accuracy absolute distance measurement by means of wavelength scanning heterodyne interferometry," Measurement Science and Technology, 9, 1031–1035 (1998).

[20] P. Sandoz, G. Tribillon and H. Perrin, "High-resolution profilometry by using phase calculation algorithms for spectroscopic analysis of white-light interferograms," Journal of Modern Optics, 43(4), 701–708 (1996).

[21] M. Takeda and H. Yamamoto, "Fourier-transform speckle profilometry: three-dimensional shape measurements of diffuse objects with large height steps and/or spatially isolated surfaces," Applied Optics, 33(34), 7829–7837 (1994).

[22] B. Parida, S. Iniyan and R. Goic., "A review of solar photovolataic technologies," Renewable and sustainable Energy Reviews, 15, 1625–1636 (2011).

[23] F. C. Krebs, "Fabrication and processing of polymer solar cells: a review of printing and coating techniques," Solar Energy Materials & Solar Cells, 93, 394–412 (2009).

[24] F. Gao, et al. "A flexible PV barrier films defects detection system for in-situ R2R film processing". In: Special Interest Group Meeting: Structured Freeform Surfaces 2014 Programme, 19–20 Nov 2014, Padova, Italy (2014).

[25] D. Tang, F. Gao and X. Jiang, "On-line surface inspection using cylindrical lens-based spectral domain low-coherence interferometry" Applied Optics, 53(24), 5510–5516 (2014). ISSN 1559–128X.

[26] M. Takeda, H. Ina and S. Kobayashi, "Fourier-transform method of fringe-pattern analysis for computer-based topography and interferometry", Journal of the Optical Society of America, 72, 156–160, (1982).

[27] C. Sainz, J. Calatroni, and G. Tribillon, "Refractometry of liquid samples with spectrally resolvedwhitelightinterferometry,"MeasurementScienceandTechnology,1,356–61(1990).

[28] S. Debnath and M. Kothiyal, "Analysis of spectrally resolved white light interferometry by Hilbert transform method", SPIE Optics + Photonics. International Society for Optics and Photonics (2006), Proceedings of SPIE Volume 6292.

Chapter 4

Application of Optical Interferometry for Characterization of Thin-Film Adhesion

Sanichiro Yoshida , David R. Didie ,
Jong-Sung Kim and Ik-Keun Park

Additional information is available at the end of the chapter

http://dx.doi.org/10.5772/66205

Abstract

In this chapter, application of optical interferometry for the characterization of thin-film adhesion to the substrate is discussed. The thin-film system is configured as one of the end mirrors of a Michelson interferometer and oscillated with an acoustic transducer from the substrate side. The oscillation causes sinusoidal displacement of the film surface around the initial (neutral) position, and the interferometer detects its amplitude as the relative phase difference behind the beam splitter. When the driving frequency of this oscillation is tuned to a range where the film-substrate interface is dominantly oscillated, the elasticity of the interface can be analyzed from the oscillation amplitude. The principle of this method is straightforward but in reality, fluctuation of the initial phase (the relative phase corresponding to the initial film position) compromises the signal. A technique known as the carrier fringe method along with spatial frequency domain analysis is employed to reduce the noise associated with the initial phase fluctuation. The possibility of the present method to analyze the so-called blister effect on thin-film adhesion is discussed.

Keywords: optical interferometry, opto-acoustic technique, thin-film adhesion, blister effects, non-destructive evaluation

1. Introduction

Thin-film systems are used in a variety of applications ranging from micro-electro-mechanical-systems (MEMS) to artificial joints. Poor adhesion of the film material to the substrate leads to delamination or other modes of coating failure, and is an important factor of quality control in the manufacturing stage. However, detection of poor adhesion is not easy. In particular, when the film material is poorly adhered but not causing a structural abnormality, detection is very difficult. Static methods such as acoustic imaging microscopy or X-ray diffractometry cannot be used to identify the problem. Dynamic analysis capable of characterizing the elastic behavior of the interface is essential.

Among dynamic techniques to evaluate the film adhesion strength, ultrasonic techniques [1–5] are the prevailing methods. In these methods, ultrasonic waves are excited in the substrate and film materials, and abnormality is detected from the propagation characteristics of the ultrasonic wave. The recent trend indicates that the film thickness is reduced for better performance of the thin-film system. This forces the ultrasonic wavelength to be shorter, hence the frequency to be higher. Reduction in the wavelength works well for the purpose of detecting defects or other nonuniform issues in the interface, as the spatial resolution is increased. However, for characterization of elasticity of the film-substrate interface, an increase in the frequency makes the analysis difficult. This is because normally, poor adhesion has lower elastic modulus than the healthy adhesion. Consequently, the frequency is too high to oscillate the poor adhesion effectively, and the signal representing the poor adhesion tends to be small. In other words, the detection system tries to probe the oscillatory behavior caused by the poor adhesion at a frequency on the blue side of the spectrum, as schematically illustrated in **Figure 1**. The transmissibility (the transfer function) of a mechanical oscillator decreases with a quadratic dependence on the frequency (f^{-2}) on the high frequency side of the resonance.

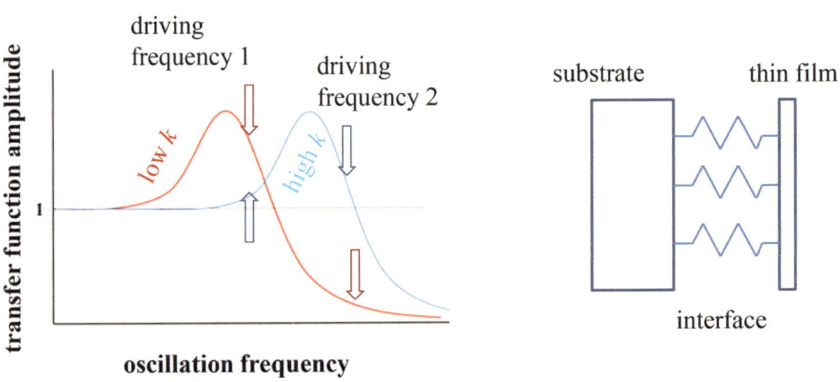

Figure 1. Resonance curves with different resonant frequencies.

Considering the above situation, we have devised an optical interferometric system to characterize the adhesion of thin films to their substrate [6]. A Michelson interferometer is used to analyze harmonic response of thin-film specimens when they are oscillated with an acoustic transducer. The film surface displacement resulting from the acoustic oscillation is detected as relative optical path changes behind the beam splitter. With the assumption that the film-substrate interface has a lower elastic modulus than the film or the substrate material and by choosing the acoustic frequency appropriately, it is possible to characterize the elastic behavior of the interfaces. For several thin-film specimens, resonance-like behaviors of the film-substrate interface have been found [7].

In the course of this research, we have learned much about practical issues of Michelson interferometers. The operation principle of a Michelson interferometer is straightforward. However, in reality, its application to engineering is not as simple as it sounds. Especially, when the interferometer is used in air, environmental disturbance can easily affect the measurement. It is always possible to place the entire interferometric paths in a vacuum, but it causes extra costs and handling procedures. In many engineering applications, it is not favorable.

In this paper, after various findings from the above research being discussed, a method is proposed to reduce environmental disturbance that compromises the optical phase signal representing the oscillation at the acoustic frequency. In this method, a known optical-path variation is introduced in a direction lateral to the interferometer axis so that the Initial Phase Difference is visualized as a pattern of mutually parallel dark and bright stripes (known as a carrier-fringe pattern [8]), and the data is processed in the spatial-frequency domain. With this configuration, a change in the relative optical path shifts the fringe locations in a plane normal to the interferometric axis. This allows us to perform two-dimensional analysis on the image plane of the imaging device, and thereby find a weakly-adhered spot. The method is especially useful when the frame rate of the imaging device is significantly lower than the acoustic frequency, as is normally in this case. Under this condition, the relative phase change at the acoustic frequency is detected as the corresponding reduction in the fringe contrast that can be related to the height of the main peak in the Fourier spectrum. Since the spatial frequency corresponding to the main peak is determined by the spacing of the carrier fringe pattern, the peak value does not depend on slow shift of the entire fringe pattern due to an environmental disturbance.

2. Michelson interferometer for thin film analysis

2.1. Why optical interferometer?

The substrate of typical thin film systems is of the order of 100 μm or less in thickness. In order to excite 10 waves in the substrate, the wavelength of the acoustic signal must be 10 μm or shorter. The acoustic velocity in silicon (a typical substrate) is 8 km/s. This results in the acoustic frequency higher than 800 MHz, leading to the situation where detection of elastic behavior associated with resonant frequency of the order of 100 MHz or less is difficult. Poor adhesion normally has a resonant frequency substantially lower than 100 MHz. In addition, most thin-film systems are subject to environmental disturbance of much lower frequency. Therefore, the use of high acoustic frequency is unrealistic.

2.2. Optical configurations

Figure 2 illustrates the principle of operation of the present method. A Michelson interferometer is configured with one end mirror replaced by the thin-film specimen. Call this interferometric arm the signal arm, and the other the reference arm. In the signal arm, the specimen is placed with the film side facing the beam splitter. The specimen is oscillated with an acoustic transducer from the rear (substrate) side, and the resultant oscillation of the film surface is detected as the corresponding change in the optical path length relative to the reference path. It is postulated that the elastic modulus of the interface is lower than that of the film or the substrate material, as the right drawing of **Figure 1** illustrates. With this postulate, it follows that the resonant frequency of the interface is lower than that of the film or substrate. In other words, when the acoustic frequency is tuned in a frequency range where the interface can possibly have resonant points, both the film and substrate oscillate as rigid bodies. Therefore, it is possible to detect the differential displacement of the film surface that represents the dynamics of the interface.

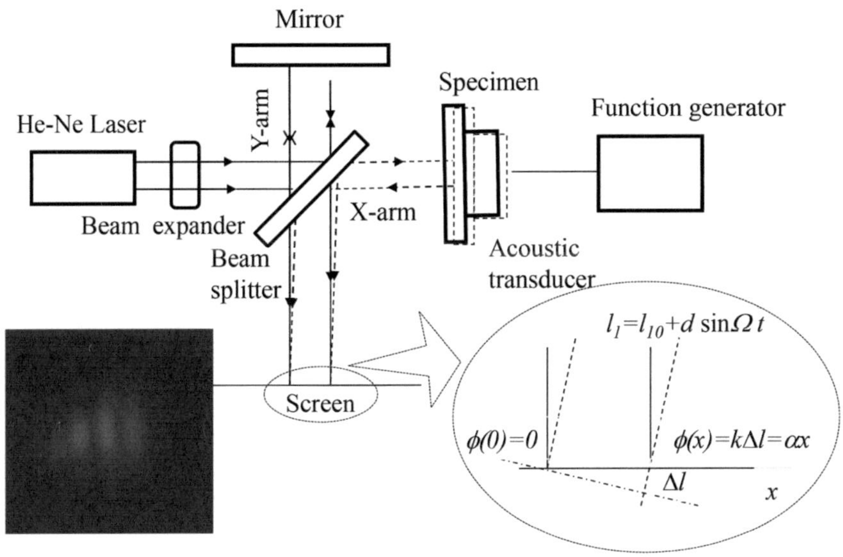

Figure 2. Michelson interferometer and experimental arrangement.

2.3. Optical intensity behind beam splitter

The light intensity on the image plane behind the beam splitter can be expressed as follows.

$$I(t) = 2I_0 + 2I_0 \cos\left[k\left(l_{s0} - l_{r0}\right) + kd\sin\omega t\right] = 2I_0 + 2I_0\cos\left[\delta_0 + \delta\sin\omega t\right] \quad (1)$$

Here I_0 is the intensity of the reference and signal beams, k is the wave number of the laser light in (rad/m), l_{s0} and l_{r0} are respectively, the initial (physical) length of the signal and reference arms, $\delta_0 = l_{s0} - l_{r0}$ is the arm length difference, $\delta = kd$ and d are the oscillation amplitude of the film surface in (rad) and (m), and ω is the oscillation (driving) angular frequency of the film specimen. Here the reference and signal beams are the noninterfering light beams in the reference and signal arms, respectively; their intensities are assumed to be equal to each other. (In reality, they are not equal to each other but the gist of the argument here is not affected by inequality. The error associated with the inequality is discussed later in this paper.) The second term in the right-hand side of Eq. (1) with $\cos[\delta_0 + \delta \sin \omega t]$ is called the interference term. This term is important as it contains the relative phase change information. $I(t)$ is captured by a photodetector or an imaging device placed behind the beam splitter.

Two methods are possible to detect the relative optical path change behind the beam splitter. Call them the total-intensity and two-dimensional methods. Both methods have advantages and disadvantages. In the total-intensity method, the total intensity $I(t)$ is captured by a fast photodetector such as a silicon PIN photodiode. The advantage of this method is that the response time of the detector is fast enough to analyze $I(t)$ at the same frequency as the acoustic transducer (ω in Eq. (1)). A disadvantage of this method is that it detects the interference term representing the entire cross-sectional area of the laser beam reflected off the specimen. It is unable to resolve the intensity over the plane of the specimen. Another disadvantage of this method is that it is vulnerable to unwanted optical path changes due to environmental disturbance.

In the two-dimensional method, an imaging device (an array of photodetector such as a CCD (Charge Coupled Device)) is used for the photodetector behind the beam splitter. (Hereafter, this type of imaging device is referred to as a CCD.) A typical CCD consists of approximately 500 rows and 500 columns of pixels. It is possible to detect the relative phase change two dimensionally on a pixel-by-pixel basis. Another advantage of this method is that by introducing a so-called carrier fringe system, it is possible to reduce the influence of the unwanted optical path length change due to environmental disturbance. The disadvantage of this method is that the frame rate (the sampling rate) of a CCD has normally orders of magnitude lower than the acoustic frequency. Consequently, the detected signal is greatly down-sampled. Below we discuss the two configurations in more detail.

2.3.1. Total intensity configuration

In this configuration, the light beams from the two arms are aligned so that they overlap each other for the entire path they share, i.e., the optical path from the beam splitter to the photodetector. Under these conditions, the film surface displacement due to the acoustic oscillation changes the relative path length difference commonly to all points on the x-y plane (the plane of the specimen). The photodetector signal is proportional to the total intensity expressed by Eq. (1), where the oscillation comes from the interference term. By measuring the intensity of the signal beam and the reference beam separately, we can evaluate $2I_0$ and express the interference as follows.

$$\cos\left[\delta_0 + \delta \sin \omega t\right] = \frac{I(t) - 2I_0}{2I_0} \tag{2}$$

In evaluating the oscillation amplitude δ from Eq. (2), the initial phase difference δ_0 plays an important role. The phase oscillation due to the acoustic transducer occurs around δ_0. Since the function $\cos\theta$ has the greatest slope at $\theta = 0$, the oscillation amplitude of Eq. (2) is maximized when $\delta_0 = 0$. In other words, the amplitude of the oscillation due to δ is maximized when the initial interference is totally destructive. **Figure 3** illustrates this situation for $\delta_0 = 0, \frac{\pi}{4}, \frac{\pi}{2}$, when the oscillation amplitude δ is 0.2 as an example. The issue is that when an environmental factor such as changes in the refractive index of the air due to temperature fluctuations vary the value of δ_0 in a random fashion, it becomes impossible to distinguish whether observed interferometric intensity variation is due to δ or δ_0 in Eq. (2).

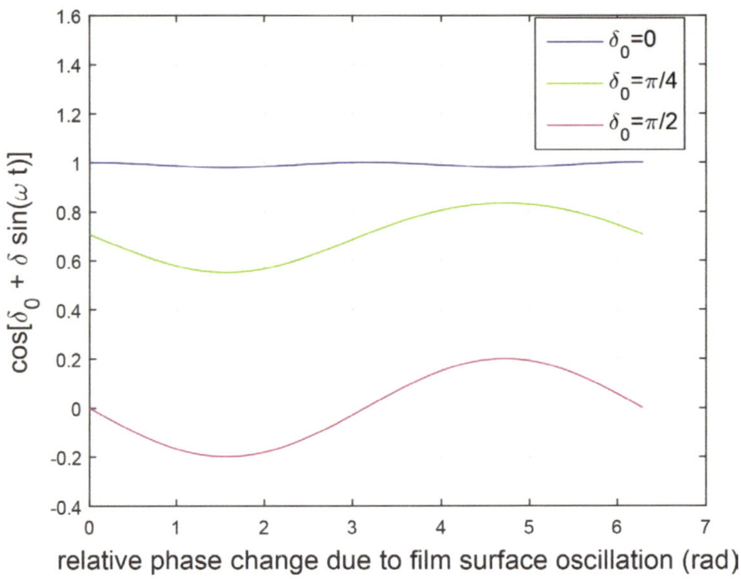

Figure 3. Interference term with three different initial phases.

2.3.2. Two-dimensional configuration

In this configuration, a CCD is used to capture the intensity represented by Eq. (1). Normally, the frame rate of a CCD is significantly lower than the acoustic frequency. A CCD with a frame rate comparable to the acoustic frequency is available but it is expensive and sometimes the number of pixels is limited in exchange for a higher frame rate. So, here we discuss image analysis for a low frame rate case.

The issue of environmental relative phase fluctuation discussed above applies to the two-dimensional configuration as well. However, the introduction of a carrier fringe system in conjunction with frequency domain analysis greatly overcomes this issue. Below we first consider the case when a carrier fringe system is not introduced (called the simple Michelson method) followed by the carrier fringe method.

(a) Simple Michelson method

Eq. (1) represents the instantaneous intensity observed behind the beam splitter. To discuss the case where the CCD's frame rate is much lower than the signal frequency, it is convenient to rewrite Eq. (1) in terms of Bessel functions of first kind. Using the following identities,

$$\cos(\delta \sin \omega t) = J_0(\delta) + 2J_2(\delta)\cos 2\omega t + 2J_4(\delta)\cos 4\omega t + \cdots$$

$$\sin(\delta \sin \omega t) = 2J_1(\delta)\sin \omega t + 2J_3(\delta)\sin 3\omega t + \cdots$$

Eq. (1) can be rewritten as follows.

$$\begin{aligned} I(t) &= 2I_0 + 2I_0\cos(\delta_0 + \delta\sin\omega t) = 2I_0 + 2I_0\cos\delta_0\cos(\delta\sin\omega t) - \\ &2I_0\sin\delta_0\sin(\delta\sin\omega t) = 2I_0 + 2I_0\cos\delta_0\{J_0(\delta) + 2J_2(\delta)\cos 2\omega t\} + \\ &2J_4(\delta)\cos 4\omega t + \cdots\} - 2I_0\sin\delta_0\{2J_1(\delta)\sin\omega t + 2J_3(\delta)\sin 3\omega t + \cdots\} \end{aligned} \tag{3}$$

In the present case, the acoustic frequency is in the range of 1–20 KHz, and the CCD has a frame rate of 30 fps (frames per second), or three orders of magnitude lower than the acoustic frequency. In other words, the data taken by the digital camera is greatly down-sampled. Under these conditions, the output of the CCD $S(\tau) = \int_0^\tau I(t)dt$ can be expressed as follows.

$$\begin{aligned} &2I_0\int_0^\tau\{1 + \cos\delta_0 J_0(\delta)\}dt + 4I_0\cos\delta_0\sum_{N=1}^{\infty}\int_0^\tau J_{2N}(\delta)\cos(2N\omega t)dt - \\ &4I_0\sin\delta_0\sum_{N=1}^{\infty}\int_0^\tau J_{2N-1}(\delta)\sin(2N-1)\omega t \end{aligned} \tag{4}$$

Here τ is the exposure time of the CCD and N is an integer. Of the terms on the right-hand side of Eq. (4), those terms that contain the summation over N oscillate. On the other hand, the first integral is constant with respect to time and therefore increases in proportion to τ. Consequently, under the condition where the exposure time is much greater than the period of oscillation, i.e., $\tau \gg 2\pi/\omega$, the signal $S(\tau)$ can be approximated by the first integral.

$$S(\tau) \cong 2I_0\tau\{1 + \cos\delta_0 J_0(\delta)\} \tag{5}$$

Figure 4 compares the signal evaluated by Eq. (4) and the approximate signal by Eq. (5) for the driving frequency of 11 KHz as a function of the exposure time. It is seen that for τ = 1/30 =33 (ms), the exposure time corresponding to the frame rate of 30 fps (the frame rate used in the present study), the approximation by Eq. (5) is accurate.

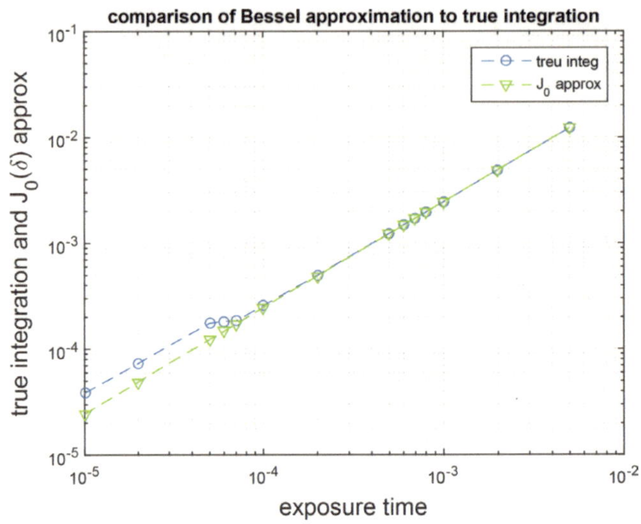

Figure 4. Comparison of $J_0(\delta)$ approximation and complete integral.

In Eq. (5), the first term $2I_0\tau$ is the sum of the optical intensity of the signal and reference arms. Experimentally, these intensities can be easily obtained by blocking one of the arms at a time and using the CCD behind the beam splitter. By subtracting this term from the total signal and dividing the result by $2I_0\tau$, we can derive expression of the interference term (Eq. (2)) for the total intensity configuration as follows.

$$\cos\delta_0 J_0(\delta) = \frac{S(\tau) - 2I_0\tau}{2I_0\tau} \tag{6}$$

In principle, by knowing the initial relative phase δ_0 from an independent experiment (such as changing the reference arm length through fringes with the acoustic transducer turned off), we can even estimate the value of δ from the known curve of $J_0(\delta)$ and $d = \delta/k$. However, in reality, environmental noise causes fluctuations in the optical path length. As will be discussed later, a temperature change of 0.1°C in the air in the beam path can cause a considerable change in the relative phase δ_0. Also, angular misalignment such as the one due to seismic disturbance reduces the accuracy in the subtraction of the I_0 in Eq. (6); since the intensity of the reference and signal beams is measured at different times from the total intensity, any angular misalignment shifts the beam center on the image plane of the CCD, and that reduces the accuracy of Eq. (6).

(b) Carrier fringe method

http://dx.doi.org/10.5772/66205

In this method a linear spatial phase-variation is introduced so that the relative phase changes over several periods of 2π across the cross-section of the laser beam. This technique is known as the introduction of carrier fringes and widely used in ESPI (Electronic Speckle–Pattern Interferometry) [7]. Carrier fringes can be introduced by slightly tilting the specimen or the reference mirror, or by inserting a mechanism to introduce a linear phase variation such as an optical wedge. **Figure 2** shows an example where the specimen is slightly tilted.

We can express the CCD's output in this case by replacing $\cos\delta_0$ with $\cos\alpha x$ as follows.

$$S(\tau) \cong 2I_0\tau\left\{1+\cos\alpha x J_0(\delta)\right\} \tag{7}$$

Here, x is the coordinate axis set up on the specimen's surface and α is the angle of tilt.

The advantage of this technique in the present context is as follows. In the simple Michelson method, if environmental disturbance changes the relative phase, it directly affects the signal expressed by Eq. (5). There is no way of knowing whether the change in the signal is due to the oscillation amplitude or the environmental noise unless the sampling rate of the optical detector is higher than the acoustic frequency. When the sampling rate is lower than the acoustic frequency, the detector's signal passes though a number of maxima and minima corresponding to the constructive and destructive interference. On the other hand, if carrier fringes are introduced, it is always possible to capture the constructive and destructive interferences; the former corresponds to the bright fringes and the latter to the dark fringes. An optical path length change causes a shift of the fringe pattern in a lateral direction at the same frequency as the optical path length change. If the optical path length change is due to the acoustic oscillation of the specimen, the fringes move back and forth transversely to the beam (dither) at the acoustic frequency. The CCD cannot resolve this fast dithering motion. Consequently, the fringe contrast is reduced. If the optical path length change is due to an environmental effect, the fringe shift is most likely slower than the sampling rate and can be resolved as a change in the fringe location by the CCD with the fringe contrast unchanged. A slight angular misalignment changes the fringe spacing, and causes some error in the frequency-domain analysis as will be discussed later. However, the error is much smaller than the simple Michelson method.

The carrier fringe method is especially effective if the fringe data is analyzed in the spatial-frequency domain. The reduction in the fringe contrast due to fast dithering is detected in the Fourier spectrum as a reduction in the height of the main peak at the frequency determined by the fringe spacing associated with $\cos\alpha x$ in Eq. (7). Thus, by forming Fourier spectrum of the optical intensity profile and evaluating the peak height of the spectrum we can evaluate the fringe contrast, and in turn, estimate the oscillation amplitude. The spatial fringe shift due to environmental change in δ_0 does not change the Fourier spectrum as it is not a function of x. An angular misalignment due to environmental disturbance can change the spectrum peak height, but the effect is relatively small (see below).

3. Experimental results and discussion

3.1. Thin film specimens

A pair of platinum-titanium (Pt-Ti) coated silicon (Si) thin film specimens is used in the present experiments. The Si substrate is cut along the [1 0 0] plane and 750 μm in thickness. The Ti layer is coated on the Si substrate and the Pt is coated over the Ti layer. The thickness of the Pt and Ti layers are 100 nm and 10 nm, respectively. In one specimen (the treated specimen) of the pair, the Ti layer is coated after the substrate surface is treated with oxygen-plasma bombardment. This treatment makes the Si surface hydrophilic, and therefore strengthens the Ti-Si bond. In the other specimen (the untreated specimen), the Ti layer is coated without a surface treatment.

3.2. Characterization of Michelson interferometer

(a) Analysis with simple Michelson method

Figure 5 shows the intensity profiles of the reference and signal beams captured by a CCD placed behind the beam splitter. For the measurement of each profile, the other beam is blocked. Thus, they do not include the interference term in Eq. (1). The reference beam is reflected off the mirror (the Y-end mirror in **Figure 2**) and therefore its profile is Gaussian. The signal beam is reflected off the specimen whose surface is not as flat as the mirror. Consequently, its profile is somewhat deformed. When these intensities are subtracted from Eq. (1) for the evaluation of the relative phase in Eq. (2), this causes some part of the interference beam in its cross-sectional area to miss interference. **Figure 6** indicates those non-interfering portion as a dip.

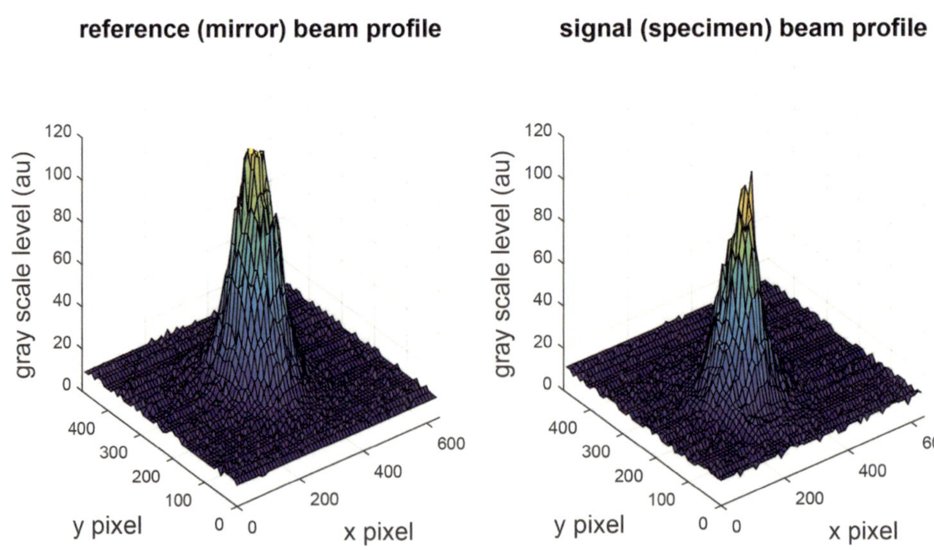

Figure 5. Reference and signal beam profiles.

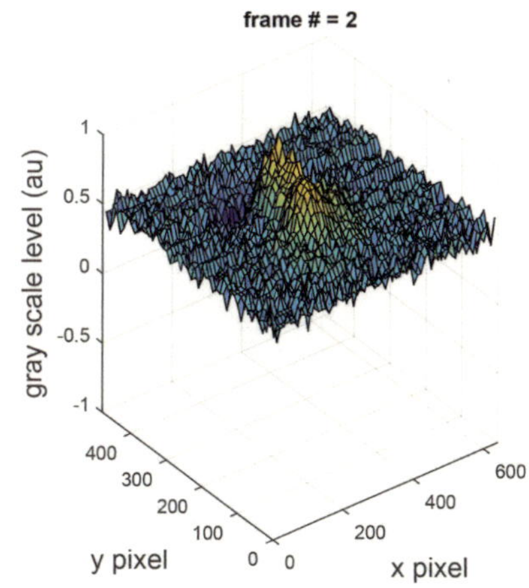

Figure 6. Beam profile of interference term.

The two profiles in **Figure 6** are taken from the same experiment where the specimen is driven by the acoustic transducer under the same condition; the driving frequency and amplitude being the same. The left and right profiles are obviously different. The left profile indicates that the interference term is positive and the left profile indicates that it is more negative. Since the phase change due to the film surface displacement is not large enough to change the sign of Eq. (6) (i.e., it is unlikely that $J_0(\delta)<0$), it is likely that the difference comes from a change in the reference phase difference (δ_0 in Eq. (6)).

Figure 7(a) shows the interference term as a function of time when the acoustic transducer is on and off. Apparently, the fluctuation is significantly higher when the transducer is on. **Figure 7(b)** is the Fourier transform of **Figure 7(a)**. The peak frequency of the disturbance is less than 1 Hz which is lower than the frame rate of the CCD used in the experiment. This indicates that the phase fluctuation affects the approximated expression Eq. (5) when the transducer is on.

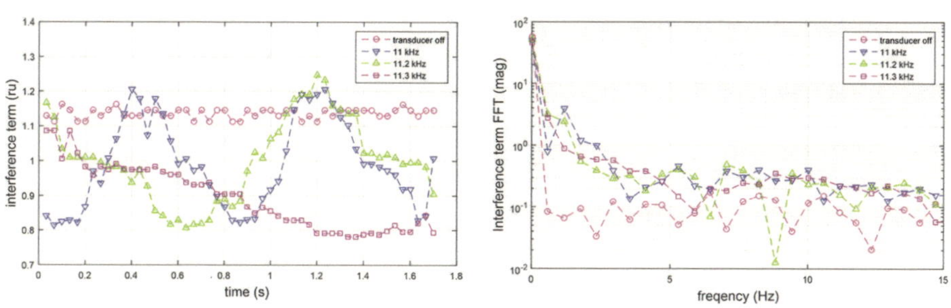

Figure 7. Variation of interference term as a function of time (left) and frequency (right).

It is likely that the temperature rise due to heat emitted from the transducer is one of the causes of the phase fluctuation. An independent temperature measurement indicates that the air temperature easily rises over 0.1°C within 1 s after the transducer is turned on, and that the temperature fluctuates by ±0.1°C approximately every few minutes. In one set of measurement in which the transducer is turned on and off every 3 min, a total temperature rise of 0.4°C is recorded over a period of 30 min. It is suspected that air convection causes the temperature fluctuation.

It is informative to make a rough estimate of the phase change due to the above temperature change. The optical phase change due to the temperature dependence of the refractive index of air can be expressed as follows.

$$d\phi = 2\pi \frac{l}{\lambda} \frac{\partial n}{\partial T} dT \tag{8}$$

Here λ is the wavelength, l is the path length, n is the refractive index of air and dT is the temperature change. The temperature coefficient $\partial n/\partial T$ of air is -0.87×10^{-6} (1/°C) [8]. The arm length of the interferometer used in this experiment is 10 (cm). The wavelength of the laser used in this study is 632.8 nm. So, the phase change due to a temperature change of ±0.1°C over the round trip in the interferometric arm is 20 (cm)/632.8 (nm) $\times0.87\times10^{-6}\times0.1$=2.75% (of the period 2π). Accordingly, the phase error due to the air temperature change of 0.4°C observed over 30 min is 2.75×4 = 11.0% of the period.

The issues of the deformed phase front and the initial phase fluctuation observed in Figures 6 and 7 make it difficult to use Eq. (5) with the total intensity method. In the next section, the carrier fringe method that greatly reduces the influence of the initial phase fluctuation is discussed.

(b) Analysis with carrier fringe method

The fluctuation of the initial phase δ_0 can be evaluated with the carrier fringe method as well. In this case, a change in δ_0 causes a shift in the carrier fringe location. The top row of **Figure 8** shows the spatial shift of carrier fringes at three different driving frequencies. In conducting the measurement at the three driving frequencies, the voltage input to the transducer is adjusted so that the oscillation amplitude of the transducer surface is the same for all the frequencies. It is seen that the shift, i.e., the change in the reference phase δ_0 is rather random, supporting the above speculation that convection of air in the interferometric arm causes the initial phase fluctuation.

The lower row of **Figure 8** is the Fourier spectrum of the spatial intensity variation of the fringe pattern observed at the respective driving frequencies. The first and main spectral peak observed at the spatial frequency of 0.02 (1/pxl) represents the carrier fringe periodicity of the light intensity; cos αx in Eq. (6) or the fringe pattern (see top of **Figure 9**). The Fourier spectrum is computed from the intensity profile over the horizontal span of 600 pxl (across a horizontal line near row 250 in the fringe image in **Figure 9**). Thus, the minimum frequency is 1/600 (1/pxl). This means that the frequency of 0.02 (1/pxl) corresponds to the 0.02/(1/600) =12th harmonics, or the periodicity of 600/12=50 (pxl). The fringe patterns in **Figure 9** indicate this periodicity.

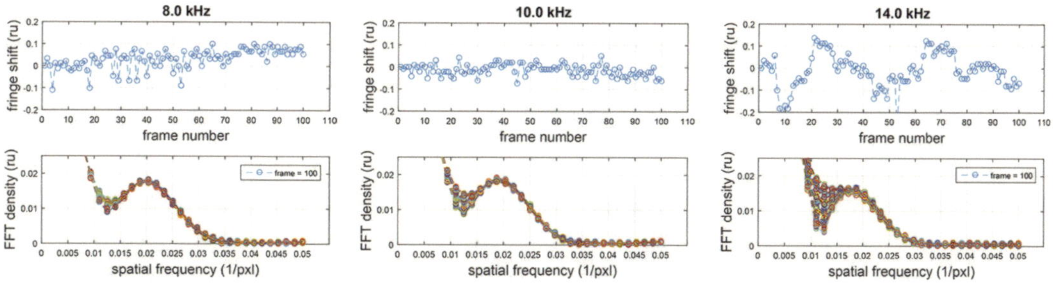

Figure 8. Fluctuation of the initial phase and corresponding change in Fourier spectrum observed with carrier fringe method.

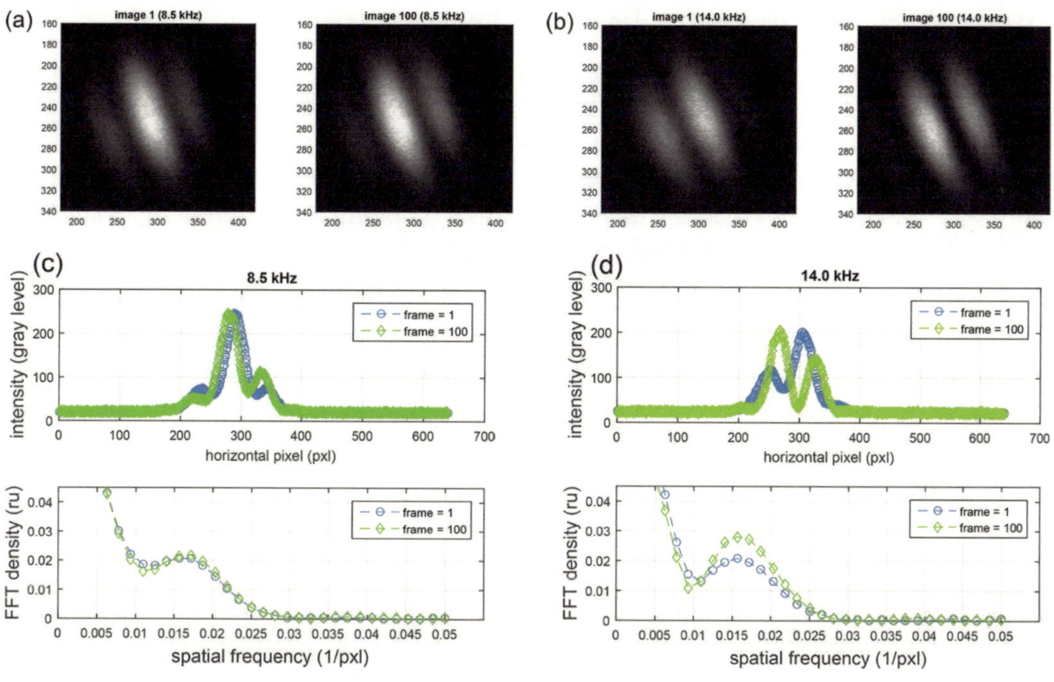

Figure 9. Effect of fringe shift and intensity profile change on Fourier spectrum.

The lower row of **Figure 8** clearly illustrates the advantage of the carrier fringe method. The spectrum plotted for each driving frequency is superposition of 100 frames. Since the CCD frame rate is 30 fps, 100 frames correspond to approximately 3 s in time. Notice that the spectra observed at driving frequency 14.0 KHz show that the spectrum data are scattered in the frequency range left of the peak. This observation indicates that the random variation of the reference phase change is reflected in the low spatial frequency region of the spectrum; as $\cos\delta_0$ changes, the total intensity detected by the CCD fluctuates, and that changes the low spatial frequency component. However, the peak height of the FFT spectrum is unaffected. This is because the peak value corresponds to the spatial dependence $\cos \alpha x$, not $\cos\delta_0$ as a function of time. Thus, it can be said that the peak height difference comes from the oscillation amplitude δ in Eq. (6). Here, the change in the total intensity depends on the initial phase δ_0 ; when δ_0 is closer to $\pi/2$ its change is greater as the cosine function has the greatest slope around $\pi/2$ (**Figure 3**). Since the initial phase changes randomly, we have no control over the total intensity.

Figure 9 illustrates the advantage of the carrier fringe method more explicitly. This figure depicts two representative cases observed at driving frequency of 8.5 and 14.0 KHz. For each frequency, the first and last frames of 100 consecutive frames are shown. **Figure 9(a)** and **(b)** shows the fringe images, the upper graphs of **Figure 9(c)** and **(d)** are the spatial intensity profiles over 640 horizontal pixels, and the lower graphs of **Figure 9(c)** and **(d)** are the corresponding Fourier spectra. In the case of 8.5 KHz driving, the fringe pattern shifts approximately by 10 pxl over the 100 frames, but its profile is unchanged. In this case, the peak value of the Fourier spectrum is the same for the first and last frames.

In the case of 14.0 KHz driving, the fringe shift is similar or less than the 8.5 KHz case but the intensity profiles are changed; the fringe image for the first frame (the left image of **Figure 9(b)**) shows that the right bright fringe is stronger than the left bright fringe in intensity whereas the image of the last frame (the right image of **Figure 9(b)**) shows that the left bright fringe is stronger in intensity. The top graph of **Figure 9(d)** indicates the difference in the intensity patterns between the first and last frames more explicitly. In this case, the peak values of the Fourier spectrum for the first and last frames are different.

As indicated above, the fringe shifts are due to the random change in the initial phase. The change in the intensity profile is most likely caused by angular misalignment of the reference and signal beams. These observations indicate that while the carrier fringe method is not affected by fluctuation of the initial phase, the angular misalignment must be suppressed as much as possible to reduce errors.

3.3. Frequency domain analysis with carrier fringe method

The consistency of the spectrum peak observed in **Figure 8** indicates that it is possible to evaluate the oscillation amplitude quantitatively. Consider the relation between the input voltage to the acoustic transducer and the oscillation amplitude of the film surface. At a given oscillation frequency, the electric power of the transducer is proportional to the square of the applied (input) voltage, and the mechanical power associated with the elastic motion of the film is proportional to the square of the oscillation amplitude. Thus, the input voltage and the film surface oscillation amplitude are proportional to each other. Under the condition that the oscillation frequency is orders of magnitude higher than the CCD's frame rate, an increase in the oscillation amplitude reduces the fringe contrast (because the CCD cannot follow the fast shift of the fringes). Therefore, it is expected that the blurriness of the fringe pattern increases with the input voltage to the transducer. With the reduction of fringe contrast, the peak value of the Fourier spectrum decreases. Further, by fitting the reduction in the fringe contrast with the use of Eq. (6), it is possible to estimate the oscillation amplitude accurately.

http://dx.doi.org/10.5772/66205

Figure 10 compares fringe contrasts as a function of the input voltage. As expected, the fringe becomes blurrier with the increase in the input voltage. **Figure 11** plots the Fourier spectra obtained at the six voltages (including 0 V) indicated in **Figure 10**. With the increase in the input voltage, the spectrum peak decreases.

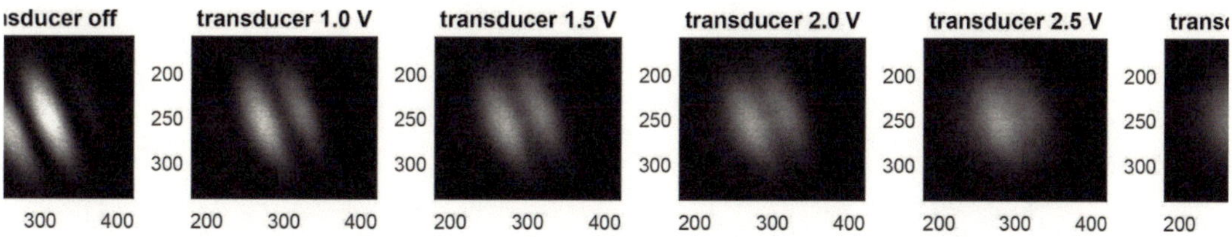

Figure 10. Fringe contrast at various input voltage to transducer.

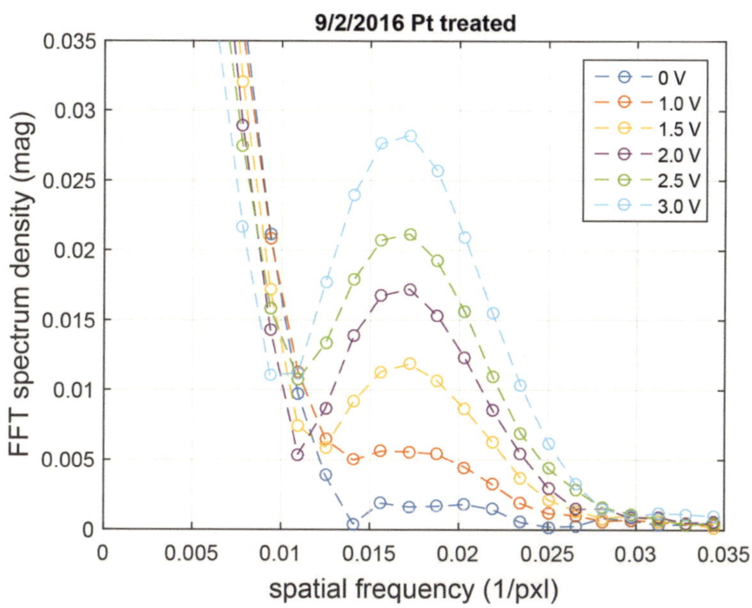

Figure 11. Fourier spectra for six input voltages shown in **Figure 10**.

Based on the proportionality between the input voltage and the oscillation amplitude, it is possible to estimate the oscillation amplitude by fitting the experimental relation between the input voltage and the spectrum peak value to $J_0(\delta)$ (Eq. (5)). When the input voltage is zero hence $\delta = 0$, J_0 takes the maximum value of unity. As the oscillation amplitude increases, $J_0(\delta)$ decreases to the first root at $\delta = 2.4048$. Thus, by evaluating the peak value relative to the case when the transducer is turned off, it is possible to estimate the oscillation amplitude $d = \delta/k$. **Figure 12** plots the peak values shown in **Figure 11** relative to the highest value obtained with the null input voltage using a factor a in $J_0(aV)$ as the fitting parameter. Here V denotes the input voltage. The measured spectrum peak values fit to the Bessel function reasonably well.

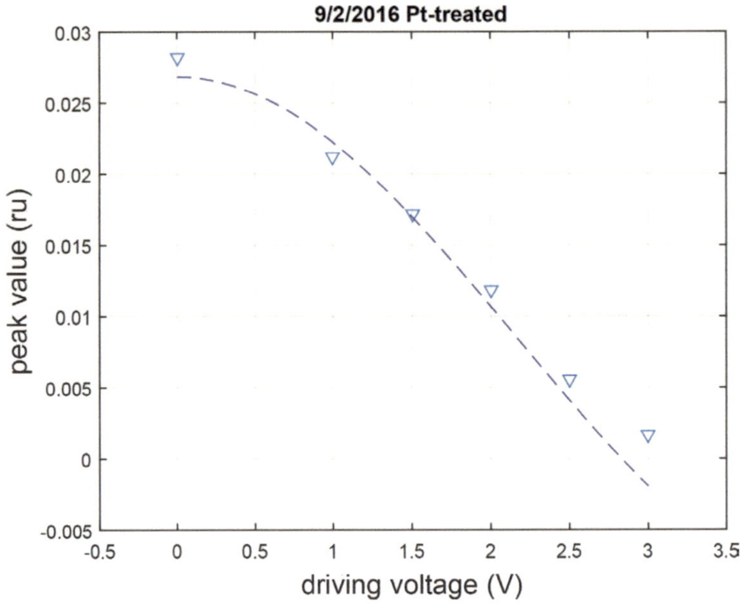

Figure 12. Spectrum peak value as a function of input voltage to transducer.

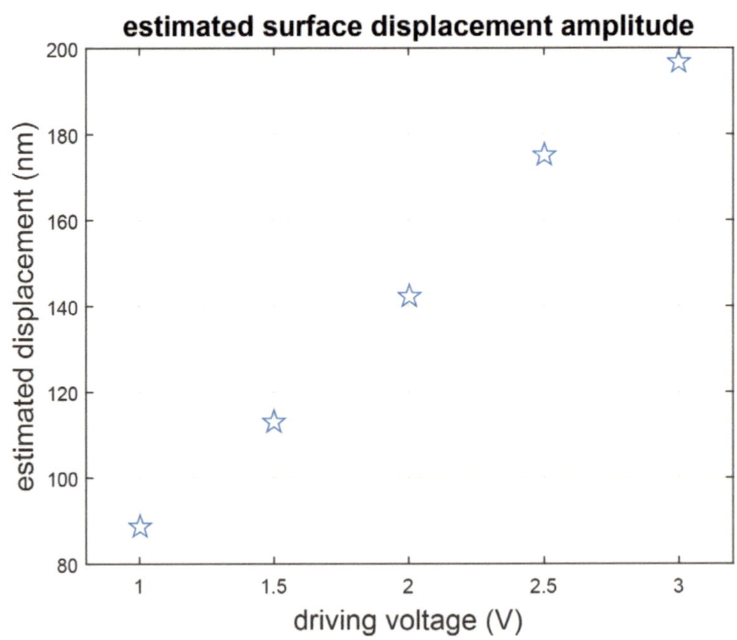

Figure 13. Oscillation amplitude estimated for each voltage input.

http://dx.doi.org/10.5772/66205

Now that the Bessel function-like behavior of the spectrum peaks is confirmed, the value of δ can be estimated from the spectrum peak value obtained for each input voltage relative to the peak value obtained with the transducer turned off (zero input voltage). Subsequently, the oscillation amplitude can be found from $d = \delta/k$. **Figure 13** plots the value of d found in this fashion.

3.4. Driving frequency sweep

By repeating the same measurement as the lower row of **Figure 8**, it is possible to find out the frequency dependence of the film surface. As is the case of **Figure 8**, the input voltage to the acoustic transducer is adjusted so that the oscillation amplitude of the transducer surface is the same for all the driving frequencies tested. **Figure 14** shows the result of such a series of measurement for a driving frequency range of 8–14.5 KHz. Here, the vertical axis is the value of the Fourier spectrum peak obtained from the intensity profile across the fringe pattern averaged over five rows near the vertical center of the fringe images. In accordance with the argument made above, the higher the peak, the smaller the film surface oscillation. Since the transducer surface has the same oscillation amplitude for all the driving frequencies, the frequency dependence of the film surface oscillation observed in this figure represents the transducer surface to the film surface transfer function. Since the transfer function of the substrate and film materials themselves are considered to be unity

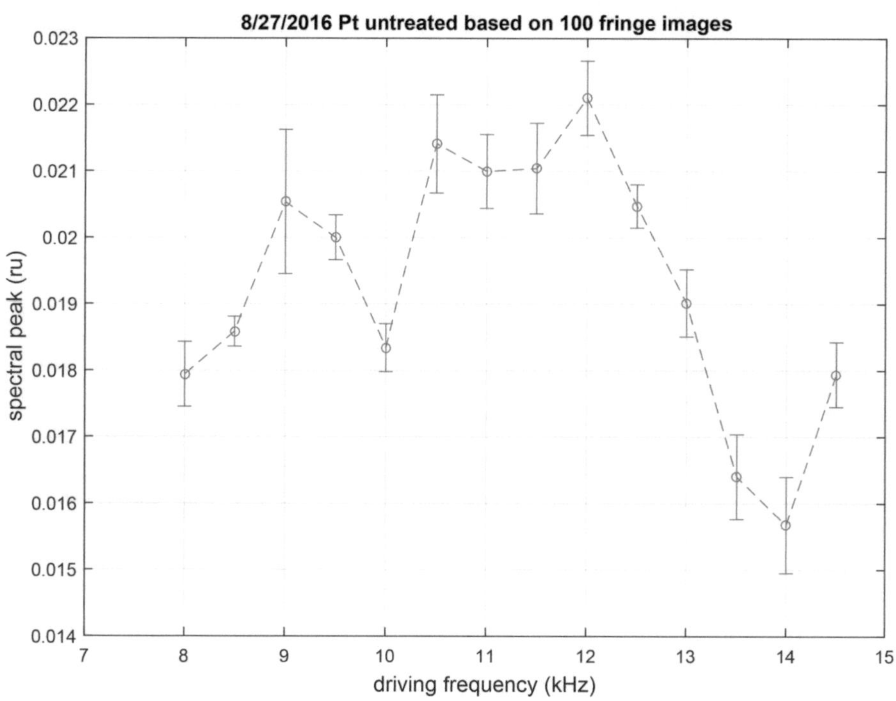

Figure 14. Driving frequency sweep of Fourier spectrum peak value.

in this frequency range, the transfer function represents the elastic property of the film-substrate interface. The lower the spectrum peak the greater the oscillation on the film surface, which can be interpreted as the greater oscillation of the interface.

The transfer function shown in **Figure 14** is obtained for five rows near the vertical center of the fringe image. By repeating the same procedure for other rows, it is possible to draw a map of the transfer function. **Figure 15** shows a three-dimensional map obtained in this fashion for four specimens; the untreated, treated, and bare silicon specimens. Here one horizontal axis is the row number and the other horizontal axis is the driving frequency. The top two plots are the cases when silicon substrates only are used (called the bare silicon specimens), and the bottom two plots are cases when the treated and untreated specimens are used. The treated and untreated specimens are attached to two different acoustic transducers. To eliminate the effect associated with the use of the different transducer, one bare silicon specimen is attached to the same transducer as the treated specimen and the other bare silicon specimen is attached to the other transducer used for the untreated specimen.

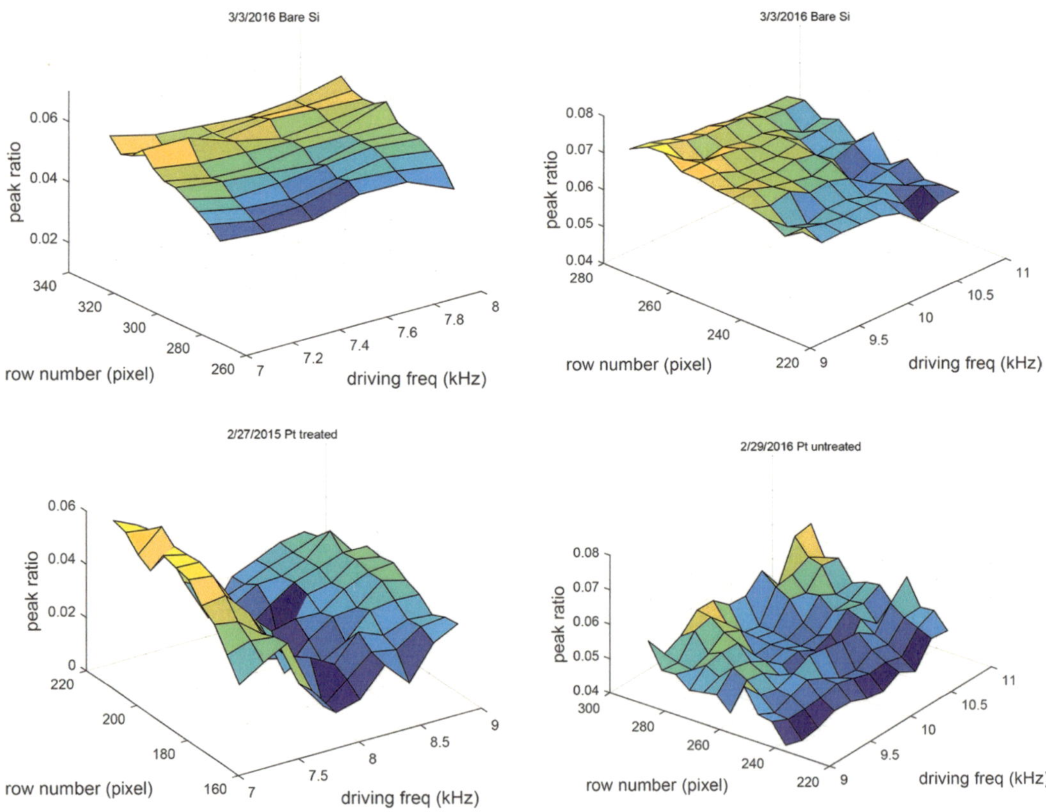

Figure 15. Fourier spectrum peak for several rows as a function of driving frequency.

The two plots for the bare silicon specimens appear to be flat, indicating that the bare silicon specimens do not have clear frequency dependence in the oscillation. On the other hand, the treated and untreated specimens show frequency dependences. As mentioned above, the frequency dependence represents the elastic characteristics of the substrate-film interface.

It is interesting to note that the untreated specimen shows a crater-like pattern around row 250 through 280 near driving frequency of 10 KHz. It is possible to interpret this pattern as representing the so-called blister effect [9–11]; the interface has a weakly adhered spot where the film experiences membrane-like oscillation when the specimen is driven. The treated specimen does not show a crater-like pattern. Instead, there is a valley running through all rows around the driving frequency of 8 KHz. We observed similar patterns in the treated and untreated specimens for a number of times. It is possible that the precoating surface treatment makes the adhesion more uniform so that the chances of the specimen having blisters is lower. These observations particularly interest us because there is no established technique to evaluate the blister effect non-destructively.

4. Conclusion

An optical interferometric method to characterize the elastic behavior of the interface of thin-film systems is discussed. The thin-film specimen is configured as one of the end-mirrors of a Michelson interferometer with the film side facing the beam splitter. The specimen is oscillated sinusoidally with an acoustic transducer. The harmonic response of the film surface to the acoustic oscillation is detected as relative optical phase difference between the two interferometric arms. An algorithm to estimate the amplitude of the film surface oscillation from the relative optical phase measurement is discussed.

Environmental noise that compromises the relative phase measurement is analyzed. The use of a carrier fringe system in conjunction with analysis in the spatial frequency domain is proposed as a method to reduce the influence of environmental noise is discussed. Under some conditions, the effectiveness of the proposed method is demonstrated with experiment.

A sample set of data obtained with Pt-Ti-Si thin-film system is presented. The three-dimensional mapping of the adhesion strength obtained with the carrier fringe method indicates some behavior of the film surface that can be interpreted as representing the so-called blister effect. It is interesting to note that the blister-like behaviors observed in the surface-treated and non-treated specimen are different from each other. This observation is of particular interest to us as non-destructive evaluation of the blister effect is not easy. More investigation is under way.

Acknowledgements

This work was supported by the National Research Foundation of Korea (NRF) grant funded by the Korea government (MSIP), NRF-2013M2A2A9043274, NRF-2011-220-D00002, and the Louisiana Board of Regents, LEQSF(2016-17)-RD-C-13.

Author details

Sanichiro Yoshida[1*], David R. Didie[1], Jong-Sung Kim[2] and Ik-Keun Park[2]

*Address all correspondence to: syoshida@selu.edu

1 Department of Chemistry and Physics, Southeastern Louisiana University, Hammond, Louisiana, USA

2 Department of Mechanical and Automotive Engineering, Seoul National University of Science and Technology, Nowon-gu, Seoul, South Korea

References

[1] Lemons RA, Quate CF: Acoustic microscopy. In: Mason WP, Thurston RN, editors. Physical Acoustics. London: Academic Press; 1979; Vol. XIV, pp. 1–92.

[2] Weglein RD: Acoustic microscopy applied to SAW dispersion and film thickness measurement. IEEE. Trans. Sonics. 1980; 27: 82–86.

[3] Atalar A: An angular-spectrum approach to contrast in reflection acoustic microscopy. J. Appl. Phys. 1978; 49: 1530–1539.

[4] Atalar A: A physical model for acoustic signatures. J. Appl. Phys. 1979; 50: 8237–8239.

[5] Telschow KL, Deason VA, Cottle DL, Larson JD: III: Full-field imaging of gigahertz film bulk acoustic random motion. IEEE. Trans. Ultrason. 2003; 94: 79–88.

[6] Yoshida S, Didie DR, Didie D, Sasaki T, Park HS, Park IK, Gurney D: Opto-acoustic method for the characterization of thin-film adhesion. Appl. Sci. 2016; 6: doi:10.3390/app6060163.

[7] Sciammarella CA, Sciammarella FM: Experimental Mechanics of Solids. Hoboken: Wiley; 2012. ISBN-10: 0470689536.

[8] Effect of Temperature on Refractive Index (dn/dt). Available from: http://www.ohara-gmbh.com/e/katalog/tinfo_2_4.html [Accessed on 2016-09-11].

[9] Bedrossian J, Kohn RV: Blister patterns and energy minimization in compressed thin films on compliant substrates. Commun. Pure Appl. Math. 2015; 68: 472–510.

[10] Dennenberg H: Measurement of adhesion by a blister method. J. Appl. Polym. Sci. 1961; 5: 125–134.

[11] Volinsky AA, Moody NR, Gerberich WW: Interfacial toughness measurements for thin films on substrates. Act Mater. 2002; 50: 441–466.

Chapter 5

Interferometry and its Applications in Surface Metrology

Dahi Ghareab Abdelsalam and Baoli Yao

Additional information is available at the end of the chapter

http://dx.doi.org/10.5772/66275

Abstract

Interferometry has been a time-honored technique for surface topography measurement. Interferometric measurements of surface shape are relative measurement techniques in which the shape of a known surface is compared with that of an unknown surface, and the difference is displayed as a series of interference fringes. Noise attached in the interference fringes can have catastrophic effects on the phase-unwrapping process, so denoising is essential before reconstruction. Some noise may be generated due to vibrations when multiple images over a finite time period are captured for reconstruction by phase-shifting technique. This harmful noise is drastically reduced when fast phase shifting–based single-shot parallel four-step combined with Fizeau interferometer is applied. Measuring the shape of strongly curved surfaces using two-beam interferometry is very complicated due to the higher fringe density. This problem may be solved by multiple-beam interferometry, thanks to the very sharp interference fringes. The experimental results show the feasibility and high precision of multiple-beam interferometry.

Keywords: interferometry, surface topography, optical aberrations, phase shifting, phase unwrapping

1. Introduction

Calibration of surfaces by optical instruments such as interferometers is a necessary step in many applications in engineering and science. The merit of using optical instruments over stylus instruments is that the optical instruments do not physically contact the surface under test and hence protect the surface from damage. In recent years, automatically controlled interferometers were engineered and provided with computer-aided technologies. A combination of moving parts controlled by various computer techniques and sophisticated electronics, and wave front fitting techniques were used to ensure precision and reliability.

However, all two-beam interferometers suffer from the fact that they produce $\cos^2$ intensity distributions. This fact makes two-beam interferometers unpopular to characterize strongly curved surfaces and steep edges because of the too high density of fringes which makes the feature too complex to measure. Multiple-beam interferometers are used to characterize these surfaces successfully thanks to the very sharp fringes. In this chapter, we present new frontiers in both two- and multiple-beam interferometers carried out by the author. As modern interferometers use a laser as the light source, spurious and speckle noises arise in the fringe pattern. Numerical techniques should be applied to the fringe pattern to suppress these spurious and speckle noises. In Section 2, limitations of optical instruments including optical aberrations and denoising and effect of noise on phase unwrapping are explained. In Section 3, fundamentals of interferometry with focus on two- and multiple-beam interferometers and their capabilities in testing film thickness, curvatures of strongly curved surfaces, and parallelism of a standard optical flat are described. It is worth mentioning that the in-line configuration of interferometry can feature finer sample spatial details compared with the off-axis configuration. However, using in-line configuration requires the time-sequent phase-shifting (PS) process to eliminate both zero-order and the twin image. Single-shot parallel phase-shifting technique is proposed for real-time measurement. In Section 4, single-shot parallel four-step phase-shifting Fizeau interferometer for three-dimensional (3-D) surface micro-topography measurement is explained. Section 5 gives concluding discussions and remarks.

2. Limitations of optical instruments

This section briefly discusses some of the limitations of optical instruments. Many optical instruments use a microscope objective to magnify the features on the surface under test. There are two fundamental limitations of the optical instruments utilizing a microscope objective: the first is the numerical aperture (Na) of the object which is given by

$$\mathrm{Na} = n \sin(\alpha) \tag{1}$$

where n is the refractive index of the medium between the objective and the surface and α is the acceptance angle of the aperture as shown in **Figure 1**. For optical instruments based on interference microscopy, a correction factor should be added to the interference pattern due to the impact of the Na. This correction can usually be estimated by well-known methods [1]. The second limitation is the optical resolution of the objective. The resolution determines the minimum distance between two lateral features on a surface that can be measured. The spatial resolution s is approximately given by

$$s = \lambda / 2\mathrm{Na} \tag{2}$$

where λ is the wavelength of the incident radiation [2]. For a theoretically perfect optical system with a filled objective pupil, the optical resolution is given by the Rayleigh criterion, where the

http://dx.doi.org/10.5772/66275

1/2 in Eq. (2) is replaced by 0.61. If the objective is not optically perfect (i.e., aberration-free) or if a part of the beam is blocked (e.g., in a Mirau interference objective, or when a steep edge is measured), the value becomes higher (worse).

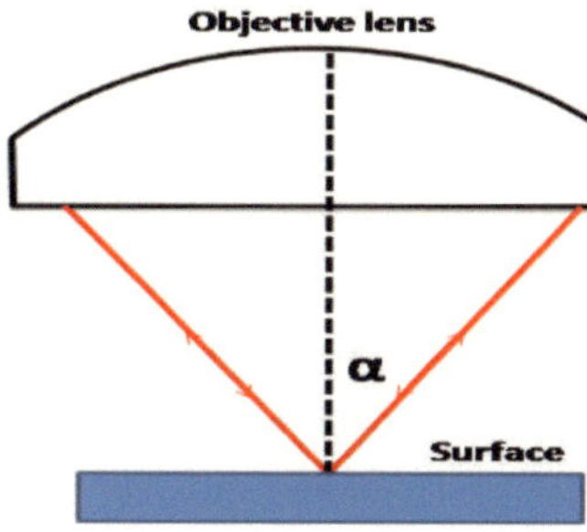

Figure 1. Numerical aperture of a microscope objective lens.

2.1. Optical aberrations

A system with aberrations has a wavefront phase surface that deviates from the ideal spherical wave. Aberrations are found in most practical imaging systems, and their effect reduces image quality. Aberrated systems tend to cause space-variant imaging, where the impulse response is not the same for each image point. **Figure 2** shows the representation of an ideal spherical and aberrated wavefronts. The difference between the ideal spherical wavefront and aberrated wavefront is a wavefront error *W(x,y)*, where *x* and *y* are the coordinates in the pupil plane. It is worth noting that the source of aberrated wavefronts may come from the imperfections in the imaging optics.

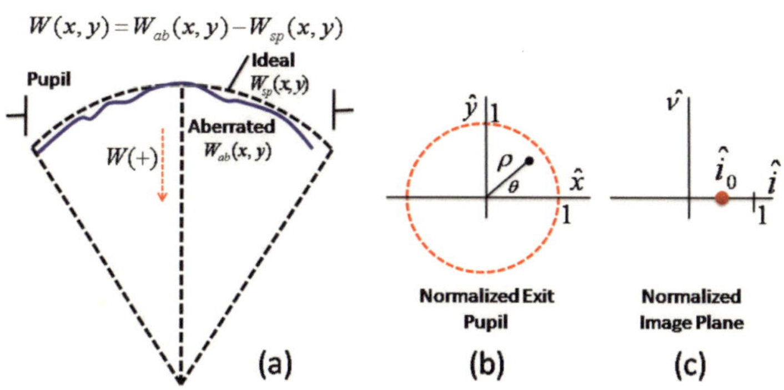

Figure 2. (a) Spherical (sp) and aberrated (ab) wavefronts, (b) Seidel aberration coordinate definitions for the normalized exit pupil, and (c) Seidel aberration coordinate for the normalized image plane.

Wavefront optical path length (OPD) is commonly described by a polynomial series. The Seidel series is used by optical designers because the terms have straightforward mathematical relationships to factors such as lens type and position in the image plane. Another series,

Zernike polynomials, is used in optical testing and applications where the aberrations do not have a simple dependency on the system parameters. Both formulations assume a circular pupil. Seidel polynomials are often used to describe monochromatic aberrations for rotationally symmetric optical systems, such as most lenses and mirrors. A common form that is applied in conventional imaging systems is described by [3]

$$W(\hat{i}_0;\rho,\theta) = \sum_{j,m,n} W_{klm}\hat{i}_0^{\,k}\rho^l \cos^m\theta;\ k = 2j+m,\quad l = 2n+m, \tag{3}$$

where ρ is a normalized radial distance in the exit pupil and θ is the angle in the exit pupil as shown in **Figure 2(b)**. For computational reasons, the angle θ is defined here relative to the x-axis in a counter-clockwise direction. However, note that this angle is often defined relative to the y-axis in traditional aberration treatments. The normalized exit pupil has a radius of 1 where the physical coordinates (x, y) are divided by the exit pupil radius to get normalized coordinates $(\hat{x},\hat{y})$. $\hat{i}_0$ is the normalized image height, defined along the $\hat{i}$ axis in the imaging plane as indicated in **Figure 2(c)**. The normalized image height is the physical height of a given point in the image divided by the maximum image radius being considered. Since the Seidel polynomials assume a rotationally symmetric system, the pupil and image plane coordinate systems are simply rotated to find the wavefront OPD function for an image point that is off the $\hat{i}$ axis. The indices j, m, n, and so forth, in Eq. (3), are a numbering and power scheme. W_{klm} are the wavefront aberration coefficients, and the five primary Seidel aberrations correspond to $k + l = 4$. These primary aberrations are known as spherical aberration, coma, astigmatism, field curvature, and distortion. The coefficients have units of distance (μm), although they are usually discussed relative to the optical wavelength (i.e., so many "waves").

For simulation purposes, it is convenient to convert from polar to Cartesian coordinates. Referring to **Figure 2(b)**,

$$\rho = \sqrt{\hat{x}^2 + \hat{y}^2} \text{ and } \rho\cos\theta = \hat{x}, \tag{4}$$

and the primary aberrations are then written as

$$\begin{aligned} W(\hat{i}_0;\hat{x},\hat{y}) = {} & W_d(\hat{x}^2+\hat{y}^2) + W_{040}(\hat{x}^2+\hat{y}^2)^2 + W_{131}\hat{i}_0(\hat{x}^2+\hat{y}^2)\hat{x} \\ & + W_{222}\hat{i}_0^2\hat{x}^2 + W_{220}\hat{i}_0^2(\hat{x}^2+\hat{y}^2) + W_{311}\hat{i}_0^3\hat{x}. \end{aligned} \tag{5}$$

The first term in this series is not one of the five primary aberrations, but is a defocus term. It is the wavefront OPD that is "created" in moving the image plane along the optical axis from the paraxial focus position. The second, third, fourth, and fifth Seidel aberration terms in Eq. (5) are spherical, coma, astigmatism, field curvature, and distortion, respectively. Simulating the effects of these aberrations by plotting some wavefront OPD surfaces is shown in **Figure 3**. **Figure 3** illustrates that spherical aberration (*W*040) and field curvature (*W*220) are wavefront curvature-like terms that are spherically symmetric with respect to the pupil coordinates. Coma (*W*131) and astigmatism (*W*222) are not spherically symmetric and depend on the image point position.

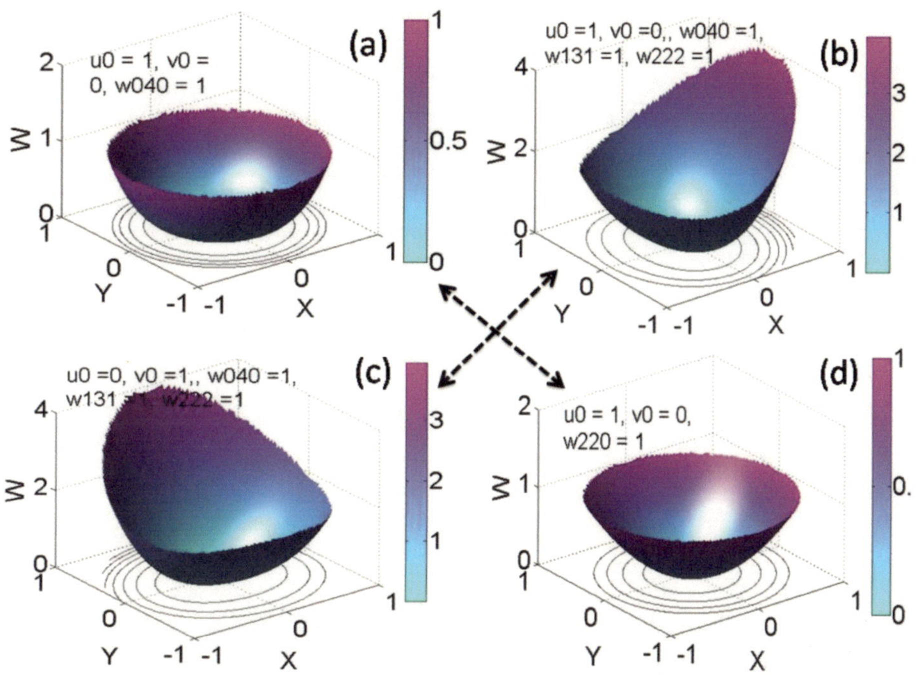

Figure 3. Example wavefront OPD surface and contour plot using Seidel 5. These represent phase surfaces that are applied in the exit pupil.

Because of the great coherence of the laser light, the fringe pattern may be easily obtained. This advantage of laser light makes most of the modern interferometers use a laser as the light source. In fact, this advantage can also be a serious disadvantage, as spurious and speckle noises arise. Special precautions must be taken into account to suppress these spurious and speckle noises. Some of them are practical such as inserting many stops in the optical system and the others are numerical such as applying windowed Fourier transform (WFT) [4] and flat fielding with apodization techniques [5].

2.2. Denoising and effect of noise on phase unwrapping

When an optically rough surface is illuminated by an expanded laser beam, the formed image is a speckle pattern (bright and dark spots). Noise can have catastrophic effects on the phase-unwrapping process. Application of simple filtering techniques in classical image processing to suppress speckle noise tends to do more harm than good, because they blur the image indiscriminately. Alternative techniques must be applied to obtain a clean interference pattern.

Once a clean image is obtained, unwrapping process is applied easily for reconstruction. Let us see how the noise affects on phase unwrapping, suppose that we have a discrete signal whose amplitude exceeds the range $[-\pi,\pi]$ as shown in **Figure 4(a)**. We can wrap the signal $x(n)$ by calculating the sinusoidal and the cosinusoidal values of $x(n)$. The four quadrant arctangent function (atan2) of $\sin(x)$ and $\cos(x)$ is then calculated using the following equation:

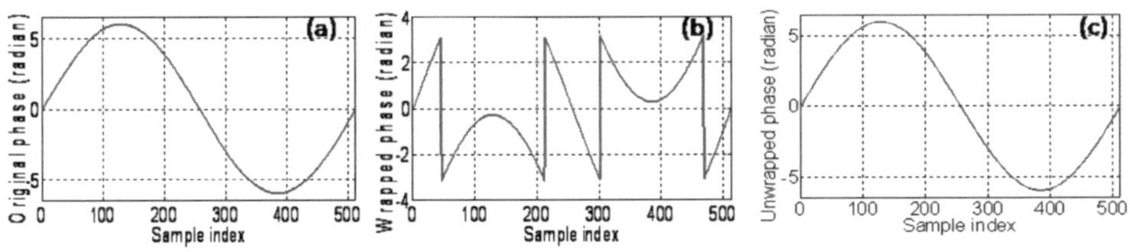

Figure 4. (a) Continuous phase, (b) wrapped phase, and (c) the phase unwrapped signal.

$$a\tan 2(u,v) = \begin{cases} \tan^{-1}(u/v) & 1st.\ quadrant \\ \tan^{-1}(u/v)+\pi & 2nd.\ quadrant \\ \tan^{-1}(u/v)-\pi & 3rd.\ quadrant \\ \tan^{-1}(u/v) & 4th.\ quadrant \end{cases} \tag{6}$$

where u and v are real numbers. We can express the wrapping process mathematically as $x_W(n) = w[x(n)]$. The 2π jumps that are present in the wrapped phase signal that is shown in **Figure 4(b)** must be removed in order to return the phase signal $x_W(n)$ to a continuous form and hence make the phase usable in any analysis or further processing. This process is called phase unwrapping and has the effect of returning a wrapped phase signal to a continuous phase signal that is free from 2π jumps. We can express the unwrapping process mathematically as $x_U(n) = x_W(n) + 2\pi k$, where $x_U(n)$ is the unwrapped phase signal and k is an integer.

The phase-unwrapped signal is shown in **Figure 4(c)**. The wrapped phase signal that is shown in **Figure 4(a)** is a very simple signal to unwrap. This is because $x_W(n)$ is a simulated signal that does not contain any noise, but if the signal is noisy, a fake-phase wrap may be produced by noise in the signal. The existence of a fake wrap will affect the unwrapping of the sample. Let us use a computer simulation to illustrate this. Suppose that we have the discrete signal $x_W(n)$ and then we add white noise to this signal as $x_{\text{noise}}(n) = x(n) + \text{white noise}$. The noise variance is set to a higher value of 0.8. The original noisy signal is shown in **Figure 5(a)**. Wrapping and unwrapping phase signals are shown in **Figure 5(b)** and **(c)**, respectively. As shown in **Figure 5**, the higher noise level has seriously affected the phase-unwrapping process and the phase unwrapping of the signal became a challenging task. This is due to the existence of a fake wrap in the signal.

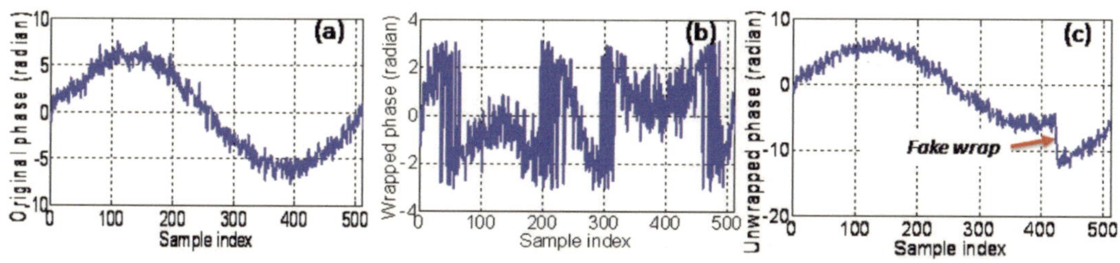

Figure 5. (a) Original noisy signal, (b) the phase wrapped signal, and (c) the phase unwrapped signal. Here, the noise variance has been increased to a value of 0.8.

Some techniques such as windowed Fourier transform and flat fielding with apodization may be used to suppress the noise and solve the problem of fake wrap. **Figure 6(a)** shows a simulated noisy closed fringe, and a clean image is shown in **Figure 7(b)** obtained using WFT technique. Profiles at the middle of **Figure 6(a)** and **(b)** are shown in **Figure 6(c)**. **Figure 7(a)** shows an experimental inline interferogram of a nano-pattern taken by Mach-Zehnder interferometer and a clean image is shown in **Figure 7(b)** obtained using flat fielding with apodization technique. Profiles at the middle of **Figure 7(a)** and **(b)** are shown in **Figure 7(c)** up and down, respectively.

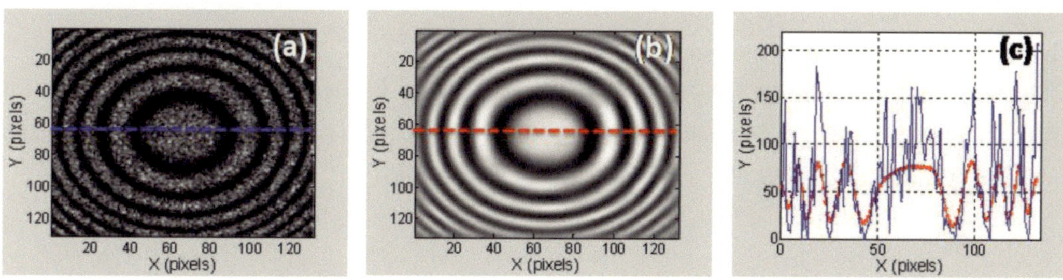

Figure 6. A simulated noisy closed fringes (a), a clean image of **Figure 7(a)** obtained using WFT technique (b), and profiles at the middle of **Figure 7(a)** and **(b)** **(c)**.

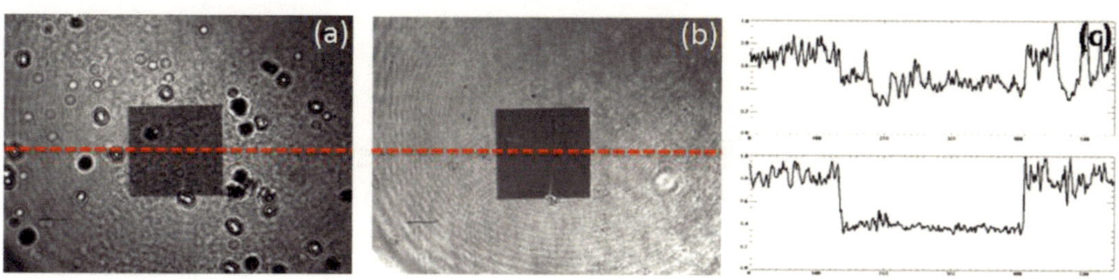

Figure 7. An empirical inline interferogram of a nano-pattern object (a), a clean image of **Figure 7(a)** obtained using flat fielding with apodization technique (b), and profiles at the middle of **Figure 7(a)** and **(b)**, up and down, respectively, (c).

Several basic interferometric configurations are used in optical-testing procedures, but almost all of them are two-beam interferometers. In Section 3, we review fundamentals of interferometry with focus on two- and multiple-beam interferometers and its capability in featuring the topography of surfaces.

3. Fundamentals of interferometry

Interferometry is the technique of superposing two or more waves, to create an output wave that differs from the input waves [6]. An interferometer is an optical instrument that can measure small wavefront deformations with a high accuracy, of the order of a fraction of the wavelength. Two-beam interferometers produce an interferogram by superimposing two wavefronts, one of which is typically a flat reference wavefront and the other a distorted wavefront from the object, whose shape is to be measured. To study the main principles of interferometers, let us consider from Maxwell's equations that the electric field of a plane wave, with speed, c, frequency, f, and wavelength, λ, travelling in the z-direction, is given by

$$E(z,t) = \begin{pmatrix} E_x \\ E_y \end{pmatrix} e^{i(kz-\omega t)} \tag{7}$$

where $\omega = 2\pi f = 2\pi c/\lambda$ is the circular frequency and $k = 2\pi/\lambda$ is the circular wave number. Assume that $E_y = 0$, that is, the light is linearly polarized in the x-direction. At the location $z = 0$, the electric field $E = E_x \cos\omega t$. The intensity is given by the square of the amplitude, thus

$$E(z) = \langle E.E \rangle = (E_x^2)\langle \cos^2 \omega t \rangle. \tag{8}$$

Consider a two-wave interferogram with flat wavefronts $E_1(t)$ and $E_2(t)$ reflected from the two mirrors of Michelson interferometer and combined at the detector. According to the principle of superposition, we can write

$$E(t) = E_1(t) + E_2(t). \tag{9}$$

Combining Eqs. (7)–(9), with some additional assumptions, gives finally

$$I = I_1 + I_2 + 2\sqrt{I_1 I_2}\cos\left(\frac{4\pi\Delta L}{\lambda}\right) \tag{10}$$

Eq. (10) is the essential equation of interference. Depending on the term $4\pi\Delta L/\lambda$, the resultant intensity on a detector can have a minimum or a maximum, and it depends on the path difference or the wavelength. It is evident from Eq. (10) that the intensity has maxima for $4\pi\Delta L/\lambda = 2p\pi$, with $p = 0, \pm 1, \pm 2, ...$, so that $\Delta L = p\lambda/2$ and minima for $\Delta L = (p + 0.5)\lambda/2$. If the intensities I_1 and I_2 are equal, Eq. (10) reduces to

$$I = 2I_1\left(1 + \cos\left(\frac{4\pi\Delta L}{\lambda}\right)\right) = 4I_1\cos\left(\frac{2\pi\Delta L}{\lambda}\right). \tag{11}$$

This means that the minimum intensity is zero and the maximum intensity is $4I_1$. Also, it is clear that if I_1 or I_2 are zero, the interference term vanishes and a constant intensity remains. The relative visibility, V, of the interference can be defined as

$$V = \frac{I_{\max} - I_{\min}}{I_{\max} + I_{\min}} = \frac{2\sqrt{I_1 I_2}}{I_1 + I_2}. \tag{12}$$

If the magnitude of the optical path length between the two beams is greater than the temporal coherence length of the light source of the two beams, fringes will not be observed. As the OPD returns to zero, fringe visibility reaches a maximum. Temporal coherence $L_c = \lambda_c^2/\Delta\lambda$ is inversely proportional to the spectral bandwidth of the light source of the two beams, where λ_c is the center wavelength and $\Delta\lambda$ is the spectral bandwidth, measured at the full-width half maximum (FWHM) at short coherent length. It is worth mentioning that temporal coherence goes as the Fourier transform of the spectral distribution of the source. **Figure 8** shows two types of temporal coherence length: the left side is the long coherent length emitted from a laser source (spectral bandwidth very small). The Fourier transform of a zero bandwidth source is a constant, so the temporal coherence is infinite. The right side is the short coherent length emitted from a femtosecond laser with FWHM around 30 μm.

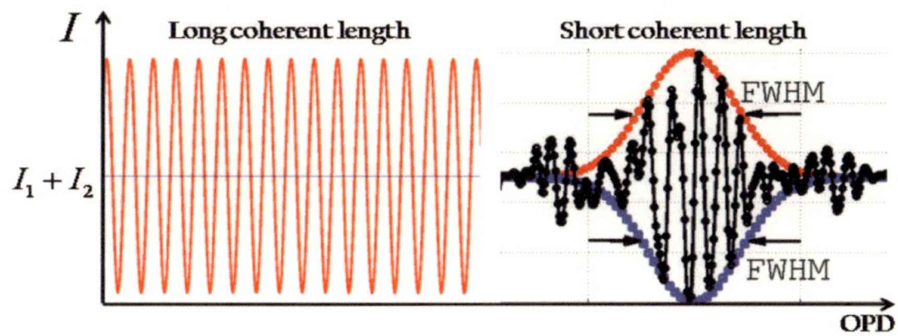

Figure 8. Fringe visibility degradation due to temporal coherence can be improved by varying the OPD between the two beams.

3.1. Two-beam interferometry

It has been noted that measuring the feature height of asphere surfaces and strongly curved surfaces using two-beam interferometers is very complicated due to the higher fringe density [7], so it has been shown that it is possible to measure the asphere form using multiple beam fringes. This is because two-beam interferometers suffer from the fact that they produce $\cos^2$ intensity distributions. **Figure 9(a)** and **(b)** compare precontour fringes of asphere surface obtained by two-beam Fizeau interferometry and multiple-beam Fizeau interferometry, respectively. As seen from **Figure 9(b)**, the multiple-beam Fizeau interferometry has the ability to resolve very small irregularities compared to common two-beam Fizeau interferometry. In this chapter, we review some of two- and multiple-beam interferometers for surface micro-topography measurement.

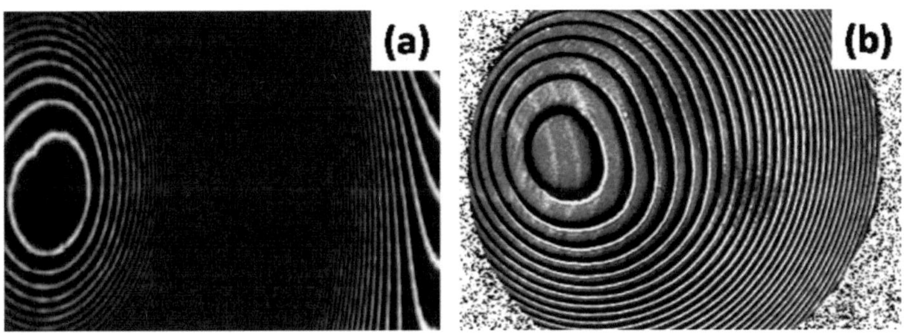

Figure 9. Precontour fringes of asphere surface obtained by (a) two-beam Fizeau interferometry and (b) multiple-beam Fizeau interferometry.

3.1.1. Twyman-Green interferometer

The light source used in a Twyman-Green interferometer is a quasi-monochromatic point source that is collimated by a collimating lens. This collimated light is incident on a beamsplitter which divides the beam into two copies: a reference beam and a test beam. The interferometer is used here for testing a spherical optical flat. The reference beam is incident on the known reference optical flat and returns to the beamsplitter. The test beam is incident on the unknown test part and also returns to the beamsplitter. The beams from the reference and the object interfere at the beamsplitter and constitute an interferogram relayed by an imaging lens to the observation plane. **Figure 10** shows a Twyman-Green interferometer for testing a curved surface in reflection [8, 9]. The reference is an optical flat of 1 inch in size and flatness of $\lambda/20$ nm. The object being tested is a curved surface of radius of curvature around. The sample is mounted carefully and four inline interferograms (no tilt between the reference and the object) are captured with phase shift between images of $\pi/2$ as shown in **Figure 11**. The intensities in the four fringe patterns can be expressed as follows:

http://dx.doi.org/10.5772/66275

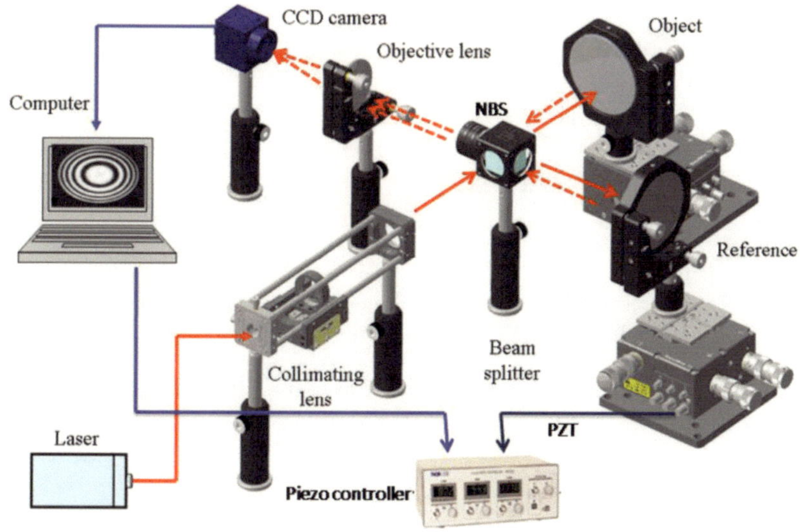

Figure 10. Optical schematic of the Twyman-Green interferometer.

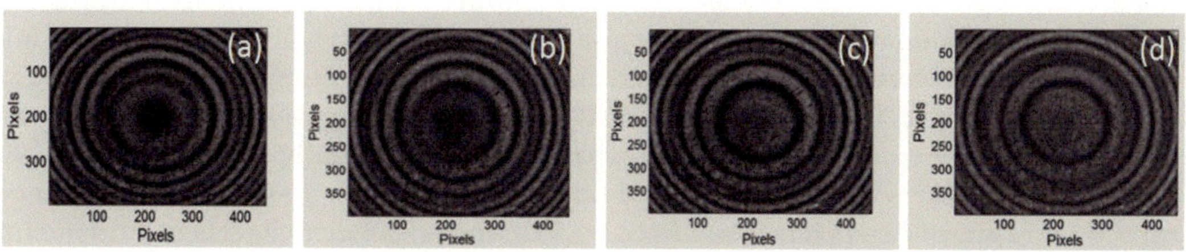

Figure 11. Intensity images of a curved object with a phase shift of 0π, 0.5π, 1π, and 1.5π, for (a–d), respectively.

$$I_j(x,y) = I_O + I_R + 2\sqrt{I_O I_R}\cos(\varphi + (j-1)\pi / 2), \tag{13}$$

where I_O and I_R are the intensities of the object and the reference waves, respectively, φ is the phase encoded in the intensity distribution, and $j = 1, 2, 3$ and 4 (four frames) is the number of the phase-shifted frames.

Using the four-phase step algorithm, the phase distribution φ can be expressed as follows:

$$\varphi = \tan^{-1}\left(\frac{(I_4 - I_2)}{(I_1 - I_3)}\right). \tag{14}$$

The evaluated phase is wrapped between $-\pi$ and π due to arctangent function. The wrapped phase map resulted from the four frames in **Figure 11** is shown in **Figure 12(a)**.

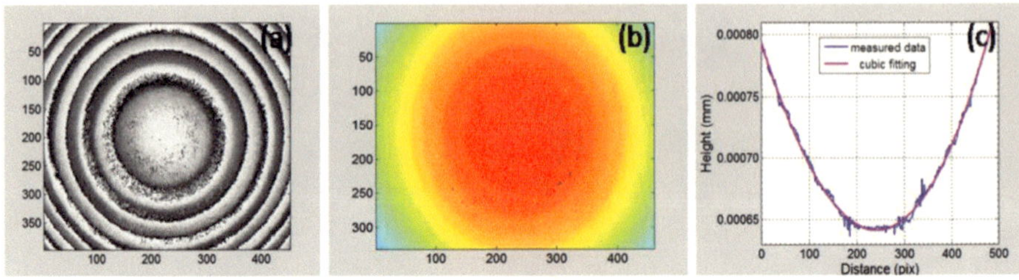

Figure 12. (a) Wrapped phase map resulted from the four frames of **Figure 10**; (b) 3-D unwrapped phase map of (a); and (c) two-dimensional height at the middle of (b) in the *x*-direction.

The wrapped phase map is then unwrapped to remove the 2π ambiguity and the unwrapped phase map is shown in **Figure 12(b)** and profile along **Figure 12(b)** is shown in **Figure 12(c)**.

3.1.2. Fizeau interferometers

Fizeau interferometers are most commonly used for testing surface figure, flatness, and parallelism of optical components. **Figure 13** shows a schematic diagram of a commercial laser phase-shifting Fizeau interferometer equipped with a tunable laser. Two-beam Zygo interferometer of type VeriFireMST and wavelength 632.467 nm at PTB, Germany, was used to measure the flatness of a high-flatness optical flat of size 60 mm (photograph of the optical flat being tested is shown in **Figure 13**). The reference of the interferometer is a transmission optical flat of 100 mm in size and flatness of $\lambda/500$ nm. Zygo's advanced phase analysis methods are coupled with fast Fourier transformation to separate each of these individual frequencies [10]. When the optical flat is carefully positioned on the interferometer, four different frequency patterns are obtained due to reflections from the surfaces as shown in the schematic diagram in **Figure 13**. The reflections are from the transmission flat (S_1), plate front surface (S_2), plate back surface (S_3), and reference surface (S_4).

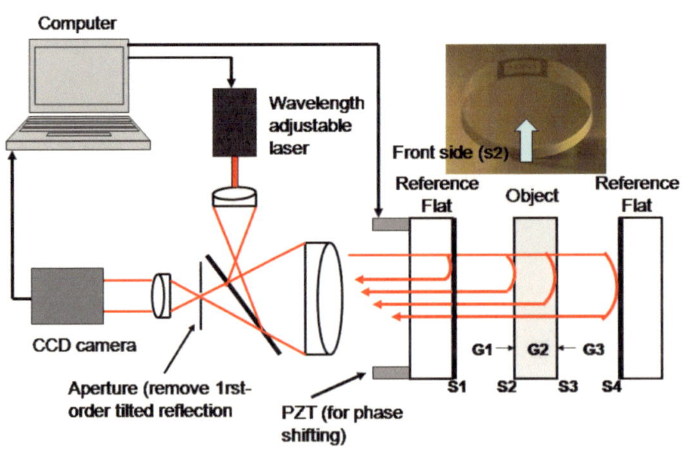

Figure 13. Familiar transmitted wavefront test geometry using a commercial laser Fizeau interferometer.

http://dx.doi.org/10.5772/66275

Because the object (optical flat) is relatively thin ($G_2 = 15$ mm), the gap G_1 is the intermediate thickness of 100 mm, and G_3 the thickest (115 mm) as shown from the three peaks of the OPD spectrum plot in **Figure 14(a)**. Each peak in the spectrum corresponds to the OPL of a particular elemental cavity. The spatial phase variation for the interferometer cavity is calculated according to the OPL. The interferogram in a four-surface cavity is shown in **Figure 14(b)**. The horizontal fringes correspond to the interference of the front side S_2 of the optical flat and the reference, while the second inclined dark fringes correspond to the interference of the back surface of the object S_3 and the reference. The software analyzes each fringe pattern and the surface from of the optical flat was obtained. **Figure 14(c)** shows the phase map of the front side of the sample, and profiles through *x*- and *y*-directions are shown in **Figure 15**, respectively. The measurement has been calculated at a temperature of 20.5°C with uncertainty in the measurement of 15 nm.

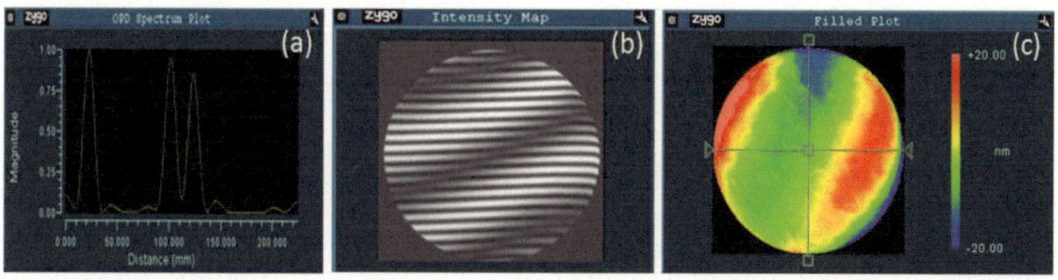

Figure 14. (a) Peaks of the OPD spectrum; (b) interferogram in a four-surface cavity; and (c) phase map of the front side of the sample.

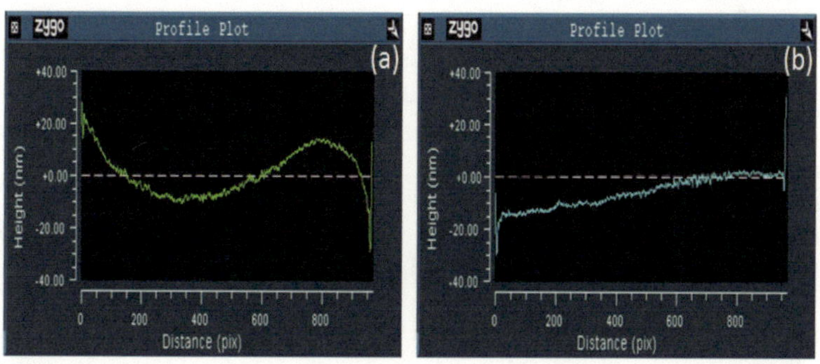

Figure 15. (a) 2-D surface height of **Figure 14c** along *x*-direction; and (b) along *y*-direction.

Another type of plano-concave surface of radius of curvature of 12744.1 mm at a focal length of −6000 mm has been tested using two-beam Zygo interferometer of type VeriFireMST and a wavelength of 632.467 nm. Circular fringes of the curved surface being tested are shown in **Figure 16(a)**; as shown from **Figure 16(a)**, the number of fringes/12 mm is around 15 fringes, which means the surface is nearly strong. The phase map of **Figure 16(a)** is shown in **Figure 16(b)**. Two-dimensional surface height of **Figure 16(b)** along *X*- and *Y*-directions is shown in **Figure 16(c)**.

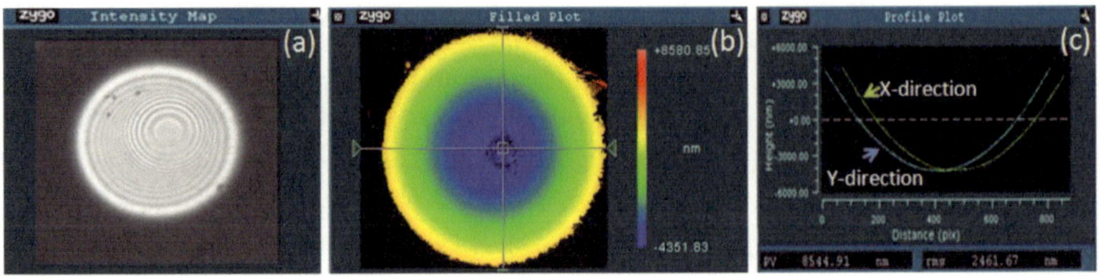

Figure 16. (a) Interferogram of the plano-concave surface; (b) phase map of (a); and (c) 2-D surface height of **Figure 16(b)** along X- and Y-directions.

3.2. Multiple-beam interferometry

Multiple beam fringes are extremely sharp. Simple measurements with such fringes can reveal surface micro-topography with a precession close to ($\lambda/500$). This advantage made multiple-beam interferometers so popular in revealing strongly curved surfaces and steep abrupt edges. Sharp fringes are obtained when the surfaces forming the cavity are coated with higher reflectivity film. The theory of the intensity distribution of Fabry-Perot fringes at reflection from an infinite number of beams collected was dealt with by [11]. Recently, Abdelsalam [12] modified an analytical equation correlate intensity distribution at reflection with a number of beams collected. It is found that 30 number of beams collected produce the same intensity profile as infinity number of beams are collected. In this section, we review multiple-beam Fizeau interferometry for thin film and curved strongly surfaces measurements.

3.2.1. Multiple-beam Fizeau interferometer for film thickness measurement

The schematic diagram of the Fizeau interferometer for film thickness measurement is illustrated in **Figure 17(a)**. Details of the measurement technique are explained by the author in [13]. The fringe pattern is digitized into the computer and then thinned to get the maximum or minimum of each individual fringe by a written program.

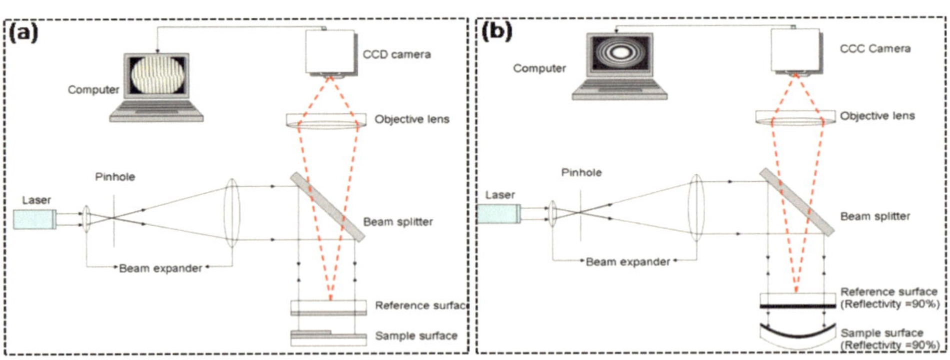

Figure 17. Schematic diagram of multiple-beam Fizeau interferometer for measurement of (a) film thickness and (b) curved surfaces.

Three different scan directions, *x*-scan, *y*-scan, and *x*-*y*-scan, were tested to obtain the best measurement. The program scans the image row by row for the *x*-scan, or by column in the *y*-scan, or in mixture for the *x*-*y*-scan. In **Figure 18(b)**, the *x*-scan, which scans row by row, was used. At each row, the program finds the pixel coordinate of the minimum intensity of that row. Selected fringes in the middle of **Figure 17(a)** were filtered and processed in the program until these fringes were thinned. The thinning of the selected fringes is shown in **Figure 18(b)**. The isometric plot shown in **Figure 18(c)** shows the average of the measured step height h. The average step height value of the film determined from at least 10 readings was 30.6 nm. It was found that the average step height value is very close to the nominal value (31 ± 3 nm).

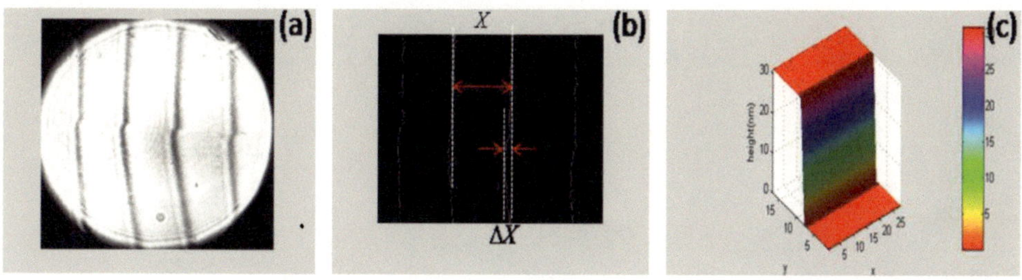

Figure 18. (a) Multiple-beam reflection fringes captured from Fizeau-Tolansky interferometer, (b) thinning of selected fringes in the middle of **Figure 2**, and (c) an isometric plot of (b).

3.2.2. Multiple-beam Fizeau interferometer for curved surfaces measurement

The schematic diagram of the Fizeau interferometer for curved surfaces measurement is illustrated in **Figure 17(b)**. Three curved surfaces of 25.4 mm in size and different radius of curvatures were coated with silver film of reflectivity nearly 90% and mounted parallel and close with the calibrated reference of nominally $\lambda/50$ flatness [7]. The reflectivity of the reference should be the same with the reflectivity of the object to obtain good contrast. **Table 1** shows types of curved surfaces being tested using **Figure 17(b)**.

Types of spherical smooth surfaces	Nominal value	Measured value
Large radius of curvature	$R \cong 38000.0$ mm	$R \cong 37715.8$ mm
Intermediate radius of curvature	$R \cong 18000.0$ mm	$R \cong 17980.2$ mm
Short radius of curvature	$R \cong 8000.0$ mm	$R \cong 8314.7$ mm

Table 1. The types of curved surfaces with nominal radius of curvatures and the measurement values by Zernike polynomial fitting method.

The three curved surfaces one by one were inserted in the interferometer and adjusted carefully until the inline interferogram is captured. **Figure 19(a–c)** shows the three inline interferograms of the corresponding three curved surfaces, large, intermediate, and short radius of curvatures, respectively, after correction with flat fielding. The interferograms were reconstructed by Zernike polynomial fitting to extract the 3-D surface height as shown in **Figure 19(d–f)**. In Zernike polynomials fitting, the surface height function $Z_r(x_r, y_r)$ can be represented by a linear combination of M polynomials $F(x_r, y_r)$ and their weighting coefficients G [7]:

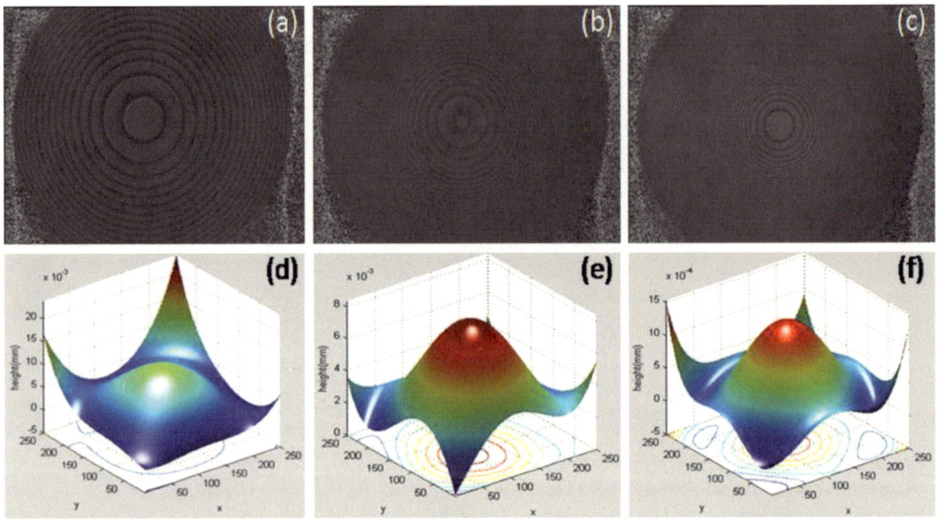

Figure 19. (a–c) captured interferograms after correction with flat fielding of the three curved surfaces, large, intermediate, and short radius of curvatures, (d–f) the corresponding 3-D surface height.

$$Z_r(x_r, y_r) = \sum_{j=1}^{M} F_j(x, y) G_j \tag{15}$$

where r is the sample index, so it is important to calculate the coefficients to represent the surface.

3.2.3. *Testing parallelism degree on standard optical flat using Fizeau interferometer*

Testing parallelism on standard optical flat of 25 mm in size using Fizeau interferometer is shown in **Figure 1(a)**. The optical flat is positioned on the front side as shown in **Figure 20(c)**, the left one. The interference pattern between the reference and the object is obtained and captured, as shown in **Figure 21(a)**, by color charge-coupled device (CCD) camera of frame rate of 15 fps, and pixel area of 2456 (y) × 2058 (x) μm^2 with a pixel size of 3.45 μm. The number of fringes of **Figure 21(a)** over 13 mm is found around 31 fringes or $\delta_1 = 31\lambda/2$. The optical flat is then positioned on the back side as shown in **Figure 20(c)**, the right one. The number of fringes of **Figure 21(b)** over 12 mm is found to be 32 fringes or $\delta_2 = 32\lambda/2$. The number of fringes can be accounted for manually or automatically by writing a small program.

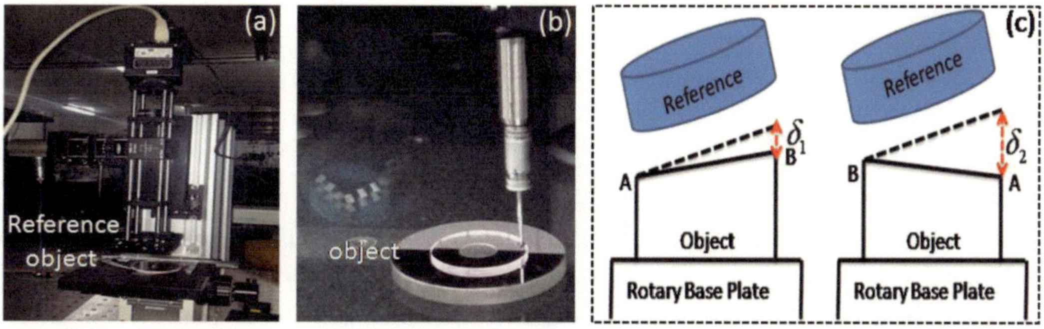

Figure 20. Testing parallelism degree on standard optical flats over 12 mm in length using (a) Fizeau interferometer and (b) coordinate measuring machine (CMM). Schematic diagram shows the locations of front and back surfaces of the sample in the interferometer (c).

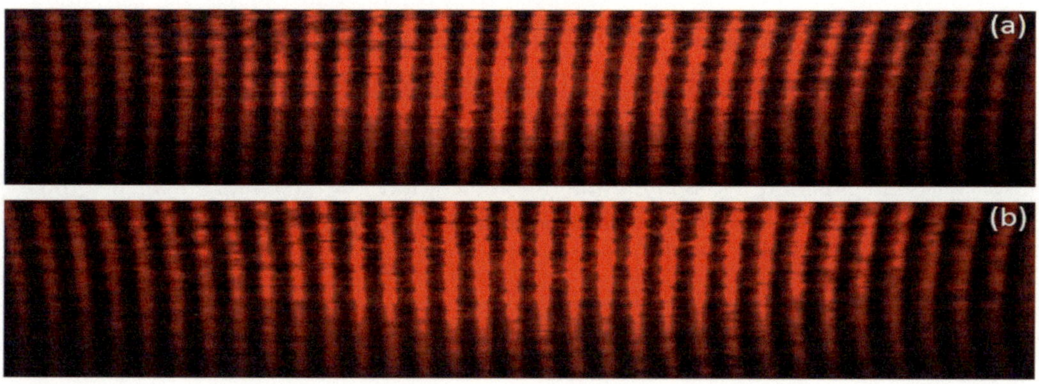

Figure 21. Fringe pattern produced at 12 mm on the (a) front side and (b) back side of the standard optical flat.

The change in the angular relationship is $\delta_2 - \delta_1 = 1\lambda/2$. But due to the rotation through 180°, there is a doubling effect. Therefore, the error in parallelism $=(\delta_2 - \delta_1)/2 = \lambda/4 \cong 160$ nm. The wavelength used in the experiment is a He-Ne laser of 632.8 nm, thus the optical flat has an error in parallelism of 160 nm over 12 mm at the middle of 25 mm.

The same size of front and back surfaces of the optical flat has been tested with coordinate measuring machine (CMM) with a suitable tip as shown in **Figure 20(b)** and the average difference between the two surface sides is calculated to be 200 nm.

3.3. Displacement interferometry

Displacement interferometry is usually based on the Michelson configuration or some variant of that basic design. Displacement measurement is defined simply a change in length. It is usually carried out by counting the number of fringes when either the object being measured or the reference surface is displaced. The fringes are counted by photodetectors and digital electronics and the fraction is estimated by electronically sub-dividing the fringe [14, 15]. **Figure 22(a)** shows a configuration of homodyne interferometer. The homodyne interferometer uses a single frequency, f_1, laser beam. The beam from the reference is returned to the non-polarized beamsplitter (NPBS) with a frequency f_1, but the beam from the moving measurement path is returned with a Doppler-shifted frequency of $f_1 \pm \delta f$. These beams interfere in the NPBS and enter the photodetector. **Figure 22(b)** shows a heterodyne interferometer configuration. The output beam from a dual-frequency laser source contains two orthogonal polarizations, one with a frequency of f_1 and the other with a frequency of f_2 (separated by about 3 MHz using the Zeeman effect). A polarizing beamsplitter (PBS) reflects the light with frequency f_1 into the reference path. Light with frequency f_2 passes through the beamsplitter into the measurement path where it strikes the moving retro-reflector causing the frequency of the reflected beam to be Doppler shifted by $\pm\delta f$. This reflected beam is then combined with the reference light in the PBS and returned to a photodetector with a beat frequency of $f_2 - f_1 \pm \delta f$. This signal is mixed with the reference signal that continuously monitors the frequency difference, $f_2 - f_1$. With a typical reference beat of around 3 MHz, it is possible to monitor δf values up to 3 MHz before introducing ambiguities due to the beat crossing through zero. The displacement being measured for both homodyne and heterodyne is calculated from this equation $d = \lambda N/2$, where N is a fringe count and λ is the wavelength of the incident radiation. Homodyne interferometers have an advantage over heterodyne interferometers because the reference and measurement beams are split at the interferometer and not inside the laser.

Figure 22. Homodyne interferometer configuration (a), and heterodyne interferometer configuration (b).

4. Fast phase-shifting interferometry

Phase shifting is an attractive and very robust technique for the analysis of fringe patterns. Since PS takes multiple images over a finite time period, it is sensitive to the time-dependent phase shifts due to vibrations. These vibrations are difficult to correct since the optimum algorithm depends on the frequency and the phase of the vibration. For a given vibration

http://dx.doi.org/10.5772/66275

amplitude, the phase error is a function of the ratio of the vibration frequency relative to the frame capture rate. Several methods have been used to try to get around the vibration problem in phase-shifting interferometry (PSI). One of those methods is speeding up the data collection process by capturing all frames simultaneously to be in single shot. Single-shot algorithm is reliable, fast, and less sensitive to vibration and turbulence in surface micro-topography measurement. In this section, we review one technique for fast phase-shifting-based single-shot parallel four step combined with Fizeau interferometer.

4.1. Single-shot parallel four-step phase-shifting Fizeau interferometer

In this section, the common path Fizeau interferometer is combined with a parallel four-step phase-shifting mechanism, thus real-time measurement is achieved [16]. By simultaneously capturing all four interferograms, this system is insensitive to vibration. The schematic diagram of the common path Fizeau interferometer combined with parallel four-step phase-shifting is shown in **Figure 23**. A helium-neon laser beam with vertical polarization passes through a collimating lens was expanded by the beam expander (BE). The collimated beam of the laser light falls upon the beamsplitter and are split into two copies. The transmitted copy from the beamsplitter is incident on the interferometer (the reference and the object) and then reflected from the interferometer with reference wave and object wave carrying the information of the tested curved surface. Note that the sample being tested was mounted as an object, and the quarter-wave plate of $\lambda/10$ flatness was mounted as a reference in the interferometer. The reflected reference and objects waves are introduced into another quarter-wave plate, whose fast axis is inclined at an angle of 45° relative to the polarization direction of the original reference wave. Thus, the object and reference waves are converted into the perpendicularly circularly polarized lights.

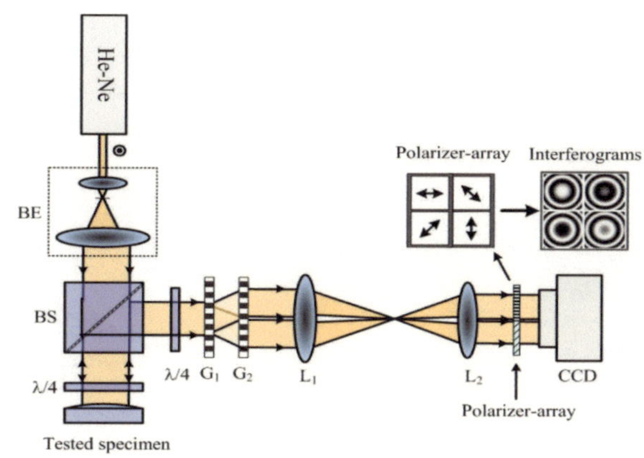

Figure 23. Configuration for single-shot parallel phase-shifting Fizeau interferometry, *BE*, beam expander; L_1-L_2, achromatic lenses with focal lengths $f_1 = 300$ mm, $f_2 = 150$ mm.

Two Ronchi phase gratings, G_1 and G_2, are located on the paths of the object and the reference waves. The axial distance between the two gratings is d, and their grating vectors are perpendicular with each other. After passing through the gratings G_1 and G_2, both the object and the reference waves are diffracted into different orders. A polarizer array is mounted in front of the CCD camera to perform the polarization phase shifting. Thus, we can achieve the single-shot recording of the interferogram containing the information of the four phase-shifted interferograms whose phases of the reference (a quarter wave plate) wave are constant and the phases of the object were shifted. **Figure 24(a)** shows the four phase-shifted interferograms with $\pi/2$ rad generated from the proposed setup. Using the four-phase step algorithm [17], the phase distribution is wrapped between $-\pi$ and π due to arctangent function. The wrapped phase map is shown in **Figure 24(b)**. The wrapped phase map is then unwrapped [18] to remove the 2π ambiguity and the unwrapped phase map is shown in **Figure 24(c)**.

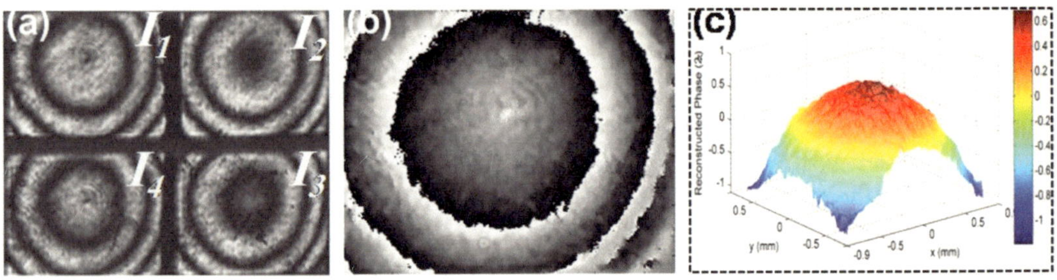

Figure 24. Experimental results of the single-shot, four-step phase-shifting using on-axis Fizeau interferometer; (a) intensity images of a spherical object with phase shift of 0, $\pi/2$, π, and $3\pi/2$; (b) wrapped phase map resulted from the four frames of (a); (c) 3-D unwrapped phase map of (b).

Another sample of step height has been tested using this technique. **Figure 25(a)** shows the four phase-shifted interferograms with $\pi/2$ rad generated from the proposed setup. The 3-D phase map is shown in **Figure 25(b)** and two-dimensional profile along the middle of **Figure 25(b)** is shown in **Figure 25(c)**. The step height has been measured again as shown in **Figure 23(c)** to confirm our method. As shown from **Figure 25(c)**, the proposed method is efficient, more robust, and highly accurate.

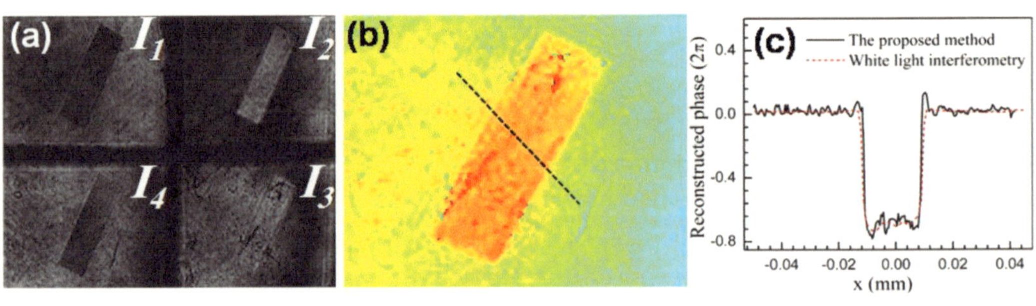

Figure 25. Experimental results of the single-shot, four-step phase-shifting using on-axis Fizeau interferometer; (a) intensity images of a step height object with phase shift of 0, $\pi/2$, π, and $3\pi/2$; (b) 3-D phase map of (a); (c) 2-D profile along the black line of (b).

http://dx.doi.org/10.5772/66275

5. Conclusion

In conclusion, we have presented new frontiers in interferometry carried out by the author for surface characterization. In this chapter, the fundamentals of interferometry and its ability to investigate the shape of surfaces with focus on denoising and impact of noise on phase unwrapping are presented. Also, limitations of optical instruments and optical aberrations measurement are discussed. Finally, we have described a fast phase-shifting technique, namely single-shot parallel four-step phase-shifting Fizeau interferometer for surface characterization. Experimental results are presented to verify the principles.

Author details

Dahi Ghareab Abdelsalam[1*] and Baoli Yao[2]

*Address all correspondence to: shosho_dahi@yahoo.com

1 Engineering and Surface Metrology Laboratory, National Institute for Standards, Egypt

2 State Key Laboratory of Transient Optics and Photonics, Xi'an Institute of Optics and Precision Mechanics, Chinese Academy of Sciences, Xi'an, China

References

[1] Greve, M.; & Sehm, K.: Direct determination of the numerical aperture correction factor of interference microscopes. Proceedings of the XI International Colloquium on Surfaces, Chemnitz, Germany, (2004), 156–163.

[2] Hecht, E.: Optics. Pearson Education, 5th edition. (2016).

[3] Wyant, J; & Creath, K.: Basic wavefront aberration theory for optical metrology. in Applied Optics and Optical Engineering. (1992); Vol. XI, R. R. Shannon and J. C. Wyant (eds.), Academic, New York, NY.

[4] Kemao, Q.: Two-dimensional windowed Fourier transform for fringe pattern analysis: principles, applications and implementations. Optics and Lasers in Engineering. (2007); 45, 304–317.

[5] Abdelsalam, D.; & Kim, D.: Coherent noise suppression in digital holography based on flat fielding with apodized apertures. Optics Express. (2011); 19, 17951–17959.

[6] Born, M.; Wolf, E.: Principles of Optics. Cambridge University Press. England (1980), 459–490.

[7] Abdelsalam, D.; Shaalan, M.; Eloker, M.; & Kim, D.: Radius of curvature measurement of spherical smooth surfaces by multiple-beam interferometry in reflection. Optics and Lasers in Engineering. (2010); 48, 643–649.

[8] Abdelsalam, D.; Baek, B & Kim, D.: Curvature measurement using phase shifting in-line interferometry, single shot off-axis geometry and Zernike's polynomial fitting. Optik-International Journal for Light and Electron Optics. (2012); 123, 422–427.

[9] Abdelsalam, D.; Min, J; Kim, D; & Yao, B.: Digital holographic shape measurement using Fizeau microscopy. Chinese Optics Letters. (2015); 13, 100701–1 to 100701–5

[10] Deck, L.L.: Fourier-Transform phase-shifting interferometry. Applied Optics. (2003); 42, 2354–2365.

[11] Holden, J.: Multiple-beam interferometer-intensity distribution in the reflected system. Proceedings of the Physical Society. (1949); 62, 405–417.

[12] Abdelsalam, D.: Numerical evaluation of the intensity distribution of a multiple-beam Fizeau fringe in reflection at finite number of beams collected. Optik- International Journal for Light and Electron Optics. (2012); 123, 422–427.

[13] Abdelsalam, D.; Baek, B; Abdel-Aziz, F; & Kim, D.: Highly accurate film thickness measurement based on automatic fringe analysis. Optik- International Journal for Light and Electron Optics. (2012); 123, 1444–1449.

[14] Birch, P.: Optical fringe sub-division with nanometric accuracy. Precision Engineering. (1990); 12, 195–198.

[15] Peggs, N; & Yacoot, A.: A review of recent work in sub-nanometre displacement measurement using optical and x-ray interferometry. Philosophical Transactions of the Royal Society of London A. (2002); 260, 953–968.

[16] Abdelsalam, D.; Yao, B.; Gao, P.; Min, J.; & Guo, R.: Single-shot parallel four-step phase shifting using on-axis Fizeau interferometry. Applied Optics. (2012); 51, 4891–4895.

[17] Kumar, U.; Bhaduri, B.; Kothiyal, M.; & Mohan Krishna.: Two wavelength micro-interferometry for 3-D surface profiling. Optics and Lasers in Engineering. (2009); 47, 223–229.

[18] Ghiglia, D.; & Pritt, M.: Two-Dimensional Phase Unwrapping: Theory, Algorithm, and Software, Wiley, New York, NY, USA (1998).

Chapter 6

Dynamic Speckle Interferometry of Thin Biological Objects: Theory, Experiments, and Practical Perspectives

Alexander P. Vladimirov and Alexey A. Bakharev

Additional information is available at the end of the chapter

http://dx.doi.org/10.5772/66712

Abstract

Relation between the phase dynamics of the waves sounding thin biological object and the dynamics of the speckles in the object image plane was theoretically detected using a model dealing with interference of multiple waves with random phases. Formulas determining the dependence of time-average intensity $\tilde{I}$ and temporal autocorrelation function $\eta = \eta(t)$ of this intensity at a point of the image plane with mean value $\langle x \rangle$, mean square deviation σ_u, and correlation time τ_0 of the difference between the optical paths Δu of the wave pairs in the neighborhood of a conjugate point of the object plane were obtained. A relation between a normalized temporal spectral function of stationary process $\Delta u(t)$ and a temporal spectral radiation intensity fluctuation function was substantiated. An optical device relevant to the model used in the theory was developed. Good quantitative coincidence between the theory and the experiment was shown by means of dosed random variation of path difference Δu. The calibration procedure for the device determining σ_u was developed; errors and the sensitivity limit of the technique were assessed. Application of value σ_u as a cell activity parameter on biological objects, namely, a monolayer of live cells on a transparent substrate in a thin cuvette with the nutrient solution was substantiated. It was demonstrated that the technique allows determination of herpes virus in the cells as early as 10 min from the experiment start. A necessity to continue upgrading of the technique was pointed out as well as its prospects for studying the cell reaction to toxic substances, bacteria, and viruses considered.

Keywords: interference, speckle, speckle dynamics, phase object, live cells, cell activity, viruses

1. Introduction

If a rough object is illuminated at some angle or a transparent object is illuminated by coherent radiation via matte glass, an inhomogeneous or a speckle is generated in the image plane. Speckles of random brightness and size are generated as a result of mutual interference of multiple waves with random amplitudes and phases. A random radiation intensity value at any point of the image plane can be regarded as resulting from superposition of multiple waves arriving from the area in the neighborhood of the conjugate point in the object plane. Minimum transverse speckle size $2b_s$ is related to the linear resolution of the lens $2a_s$, with formula $a_s = b_s/m$, where m is magnification generated by the lens.

At present, the statistic properties of stationary speckles have been well studied, and they can be checked for in various publications [1–4].

If the phases of sounding waves vary due to the processes occurring on the surface of the reflecting object or inside a transparent body, the speckle pattern will vary. Speckle dynamics manifests itself as a speckle shift and (or) a change of their structure. At present, there are numerous papers on application of this phenomenon and interference of speckle fields during studies of solid, liquid, or gaseous mediums [5–10].

The speckles generated by biological objects have been called biospeckles in the literature. Some researchers mean that biospeckles are speckle dynamics due to the processes occurring in biological objects. Below, we are going to regard biospeckles as speckles generated by biological objects.

There is vast literature on the properties of biospeckles generated by various objects: seeds, fruit, vegetables, plant leaves, bacteria, skin, patients' extremities, etc. Numerous examples of such studies can, for example, be found in a monograph [11]. Despite a large number of publications on biospeckles, there is just one technique implemented in clinical practice. This technique was originally proposed in Ref. [12], and it permits determination of the blood flow velocity in the blood vessels of the patients' retinas or extremities. The latest publications aiming at upgrading of the technique can be checked for in Ref. [13]. Successful application of this technique might be related to the fact that authors developed its theory at a certain stage of the research [14].

When live cells are sounded by coherent waves, the intracellular processes can alter the amplitudes and phases of the waves, thus changing the speckle image of an object. This provides an opportunity to study intracellular processes by the features of biospeckle dynamics in principle so that, for example, the problem of therapeutic drug management could be solved. That is why establishing of the relation between the physical-chemical intracellular processes and the biospeckle dynamics is a task vital both from the scientific and practical point of view. The authors explain absence of notable progress in solution of this problem by absence of a theory establishing a relation between the parameters that characterize the physical-chemical processes influencing the variation of the cell-sounding wave phases and the parameters characterizing speckle dynamics. Recently, one of the authors of this paper developed a similar theory for thin biological object so that this paper aims at familiarizing the readers with the adopted model, the results of the theory test on the model objects, and with some perspectives of its application.

2. Interference of two and multiple waves

2.1. Complex amplitude of light wave and radiation intensity

The electromagnetic nature of the light waves was theoretically substantiated by British physicist Maxwell in his paper [15]. Ruling out the currents from his equation system, he obtained an equation that describes the propagation of electromagnetic disturbances:

$$\Delta\vec{f} - \frac{\frac{1}{c^2}d^2\vec{f}}{dt^2} = 0, \tag{1}$$

where $\vec{f}$ is the electric $\vec{E}$ or magnetic $\vec{H}$ field vector tension, and c is light velocity.

Note that this equation exactly coincides with the equation of disturbance motion in elastic no compressive medium.

Further, let us target on plane monochrome waves that are a special case of a solution to Eq. (1). These are the waves wherein the electric and magnetic field vary according to the cosine law and vector $\vec{f}$ is a function of a single coordinate and time. Let us restrict ourselves to discussing the tension of the electric-filed vector. In a form independent from the origin selection for a plane monochrome wave, we have [16]:

$$\vec{E} = Re\left\{\vec{E}_0 e^{+i(\vec{k}\vec{r}-\omega t)}\right\}, \tag{2}$$

where $\vec{E}_0$ is some constant complex vector, $\vec{k}$ is a wave vector equaling $(\omega/c)\ \vec{n} = (2\pi/\lambda)\ \vec{n}$, $\vec{r}$ is a radius vector of a point of space, ω is the wave frequency, λ is the wavelength, and $\vec{n}$ is a single vector coinciding with the direction of the light wave propagation.

Let us further accept that all the waves have the same direction of the electric field vectors, then, when the waves impose their amplitudes, they can be added as scalar values. Besides, further in Eq. (2), we shall omit symbol *Re* and operate exponents instead of cosines. It is possible because in the problems that we are discussing below the final result will differ by an insufficient factor. Instead of Eq. (2), we have:

$$E(x, y, z, t) = A(x, y, z)\exp(-i\omega t), \tag{3}$$

where expression $A(x, y, z) = |A(x, y, z)|exp[i\theta(x, y, z)]$ was named complex amplitude in optics, $|A(x, y, z)|$ is the module of the complex amplitude, and $\theta(x, y, z)$ is the wave phase at the observation point.

Now let us introduce the notion of light intensity as a value proportionate to volumetric density of radiation energy averaged by the time interval substantially exceeding the wave oscillation period:

$$I(x,y,z) = \lim_{T\to\infty}\frac{1}{T}\int_{-T/2}^{T/2}|E(x,y,z,t)|^2dt = |A(x,y,z)|^2. \tag{4}$$

Thus, the radiation intensity at a point of space equals the squared complex amplitude module.

2.2. Interference of two waves

Let us discuss the light intensity distribution in superposition of two monochrome waves. Suppose that two waves of the same length λ were emitted by one point source in various directions, then two plane waves 1 and 2 crossing at angle θ were shaped by the optical systems. Let us take some point in the area of beam superposition. For certainty, let the wave amplitudes be the same equaling A_0, but their initial phases φ differ. In compliance with Eq. (4), we have:

$$\begin{aligned} I = A\times A^* &= \left[A_0e^{i(\vec{k}_1\vec{r}+\varphi_1)} + A_0e^{i(\vec{k}_2\vec{r}+\varphi_2)}\right]\times\left[A_0e^{-i(\vec{k}_1\vec{r}+\varphi_1)} + A_0e^{-i(\vec{k}_2\vec{r}+\varphi_2)}\right] \\ &= 2I_0 + 2I_0\cos\left[(\vec{k}_1-\vec{k}_2)\,\vec{r} + \varphi_1 + \varphi_2\right], \end{aligned} \tag{5}$$

where $I_0 = A_0^2$. Eq. (5) describes periodic light intensity distribution in the neighborhood of point $\vec{r}$, which was called light interference by T. Jung. Elementary calculations can demonstrate [17, 18] that minimum distance Δ between neighboring intensity maximums or minimums called bandwidth or period of the interference fringes is determined by formula (Eq. (6)):

$$\Delta = \frac{\lambda}{2\sin\frac{\theta}{2}}. \tag{6}$$

It follows from Eq. (6) that if θ tends to zero, Δ tends to infinity, which corresponds to tuning of the interferometer to "endless" band. If angle θ between vectors $\vec{k}_1$ and $\vec{k}_2$ equals 180°, then $\Delta = \lambda/2$, which corresponds to the wave interference in colliding beams. For θ = 60° value, $\Delta = \lambda$.

Now let us discuss the contrast of the interference fringes γ introduced by Michelson and determined by formula $\gamma = (I_{max}-I_{min})/(I_{max} + I_{min})$, where I_{min} and I_{max} are the minimum and the maximum intensity values, respectively. From Eq. (5), it follows that in the case of a point light source discussed here and constant wavelength λ, contrast $\gamma = 1$. Experience shows that if the light source is not point and (or) it emits light in some wavelength interval, the fringes contrast is less than 1.

It is commonly believed that case $\gamma = 0$ corresponds to completely incoherent light; if $0 < \gamma < 1$, the light is partially coherent, and coherence is the ability of waves to interfere. Interference of partially coherent light can be studied in Ref. [19]. In the text below, we will suppose that the light waves discussed here are completely coherent, i.e., two waves of the same amplitude generate an interference pattern with the contrast equal to 1. We can suppose with practical precision that similar waves are generated by laser light sources.

2.3. Interference of multiple waves with random amplitudes and phases

The creation of laser and discovery of the speckle structure of scattered radiation were immediately followed by a number of theoretical papers on statistical properties of speckles in free space [1] and in the image area of scattering surface [2]. **Figure 1** presents a typical speckle pattern observed in the image plane of a rough surface. The picture was obtained by the authors of this manuscript during its preparation. In this section, we discuss some main features of speckle fields obtained by Goodman [3, 4] on a simple model describing interference of multiple waves.

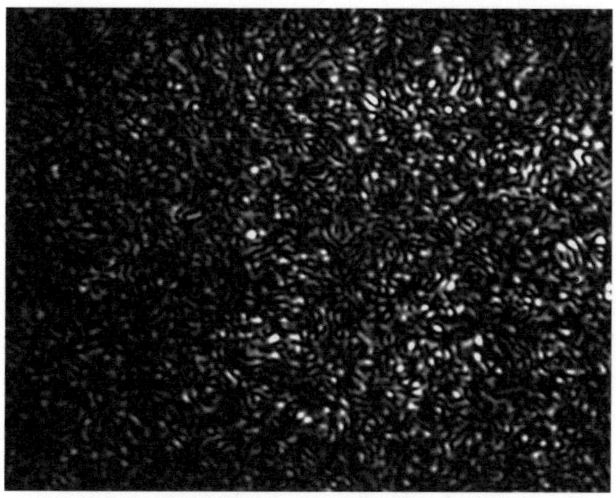

Figure 1. Speckles in the image plane of a rough surface.

According to Goodman's model, the waves that arrived at an arbitrary point of the free space from elementary areas of surface can be regarded as plane monochrome waves with random amplitudes $a_j/\sqrt{N}$ and phases φ_j, where j is the wave number, $j = 1, 2, \ldots N$. It was supposed that the amplitude and phase of the same wave and the amplitudes and phases of different waves are independent, and the values of a_j^2 averaged by the object ensemble are nonzero. It was considered that phases φ_j were homogeneously distributed in the area from $-\pi$ to $+\pi$. The presence of the object ensemble means the presence of numerous macroscopically identical scattering objects, each object generating N plane and monochrome waves with random amplitudes and phases. Any value averaged by the object ensemble is found by means of fixation for every object of the ensemble with subsequent calculation of its mean value. In the text below, we will denote the ensemble-average with angular parenthesis. Note that mathematically, object ensemble-averaged value of some function f of random arguments $x_1, x_2, \ldots, x_m$, is determined in the following way:

$$\langle f(x_1,x_2,...,x_m)\rangle = \int_{-\infty}^{+\infty}\cdots\int_{-\infty}^{+\infty} f(x_1,x_2,...,x_m)\rho(x_1,x_2,...,x_m)dx_1dx_2...dx_m, \tag{7}$$

where $\rho(x_1,x_2,...,x_m)$ is joint probability density of random values $x_1,x_2,...,x_m$. If these values are independent, $\rho(x_1,x_2,...,x_m) = \rho(x_1)\rho(x_2)...\rho(x_m)$, and the calculation of integral (Eq. (7)) may simplify substantially.

If all the waves are linearly polarized in the same mode, then, according to J. Goodman, the following ratio holds for total complex amplitude A at some point $\vec{q}$ of the free space:

$$A = ae^{-i\theta} = A^r + iA^i = \frac{1}{\sqrt{N}}\sum_{j=1}^{N} a_j e^{i\varphi_j}, \tag{8}$$

where A^r and A^i are real and imaginary parts of the total complex amplitude, respectively. Eqs. (9)–(11) obtained by Goodman on the basis of the discussed model that characterize the statistical properties of speckles are cited as follows:

$$\langle A^r\rangle = \langle A^i\rangle = \langle A^rA^i\rangle = 0, \tag{9}$$

$$\langle A^rA^r\rangle = \langle A^iA^i\rangle = \frac{1}{2N}\sum_{j=1}^{N}\langle a_j^2\rangle, \tag{10}$$

$$\rho(A^rA^i) = \frac{1}{2\pi\sigma^2}e^{-\frac{(A^r)^2+(A^i)^2}{2\sigma^2}}, \tag{11}$$

$$\rho(I,\theta) = \rho(I)\times\rho(\theta) = \frac{1}{\langle I\rangle}e^{-\frac{I}{\langle I\rangle}}\times\frac{1}{2\pi}, \quad I\geq 0, \ -\pi\leq\theta\leq\pi, \ 2\sigma^2 = \langle I\rangle, \tag{12}$$

$$\langle I^n\rangle = n!\langle I\rangle^n. \tag{13}$$

From Eqs. (9)–(13), it follows that at an arbitrary point of a free field, the real and imaginary parts of total complex amplitudes are independent, uncorrelated, and distributed according to the Gauss' law. Radiation intensity I and resulting phase θ are independent, value θ is homogeneously distributed in the range from $-\pi$ to $+\pi$. From Eq. (12), it follows that probability P_I exceeding some threshold equal to I by the light intensity that is given by Eq. (14):

$$P_I = e^{-\frac{I}{\langle I\rangle}} \tag{14}$$

Thus, in a speckle field, the most probable intensity value is value I equal to zero. With increasing intensity, the probability of its detection decreases exponentially. From Eq. (13), it also follows that speckle contrast C equal to the ratio of mean square deviation intensity to the mean intensity equals 1.

Experience shows that experimental dependence $P_I(I)$ agrees well with theoretical dependence Eq. (14) for scattering surfaces that lack the mirror constituent of scattered radiation, and whose height of heterogeneity of the surface relief is comparable with wavelength λ. The statistical properties of the speckles corresponding to other models of rough surface can be studied, for example, in Refs. [20, 21].

The model proposed by Goodman was further developed in Ref. [22] to obtain the formula allowing determination of three-dimensional speckle sizes. It was supposed that point scattering centers were located in some three-dimensional area transparent for radiation. A formula was obtained that allowed determination of three-dimensional speckle sizes for an area of an arbitrary shape with random location of the radiation source, the object, and the observation site by the width of a spatial autocorrelation function of intensity of scattered radiation. The formulas determining the transverse and longitudinal speckle sizes for two objects of a simple shape are given below. Let us examine a transparent area shaped like a right-angle parallelepiped of size 2X and 2Y on *ox* and *oy* axes, respectively, and of size 2Z on *oz* axis. Point scatterers are located within the area. Let the coordinate origin be located in the center of the area. Then, if the direction of illumination is arbitrary at distance ρ_q on *oz* axis, speckles with minimum sizes $\Delta\tilde{q}_x$, $\Delta\tilde{q}_y$, $\Delta\tilde{q}_z$ are generated as *xd, yd,* and *zd,* respectively:

$$\Delta\tilde{q}_x = \frac{\lambda\rho_q}{2X}, \ \Delta\tilde{q}_y = \frac{\lambda\rho_q}{2Y}\ , \ \Delta\tilde{q}_z = \frac{5\lambda\rho_q^2}{\pi(X^4 + Y^4)^{1/2}}. \tag{15}$$

If the object is cylindrical, axis *oz* coincides with the axis of the cylinder, and the coordinate origin is located in the center of the object, then in similar observation and illuminating conditions

$$\Delta\tilde{r} = 1.22\frac{\lambda\rho_q}{D}\ , \ \Delta\tilde{q}_z = 2\frac{\lambda\rho_q^2}{R^2}\ , \ \Delta r = (\Delta q_x^2 + \Delta q_y^2)^{1/2}\ , \tag{16}$$

where *R* is the radius of the cylinder, D =2R. In the literature [5, 23], it was shown that the mean *xd, yd,* and *zd* speckle sizes are threefold compared with the minimum.

In the preceding text above, we confined ourselves to the main features of speckle fields in a free field. The speckles generated in the scattering image plane have very similar properties. For the speckles in the area of the object images in Eqs. (15) and (16), values 2X, 2Y, and *R* equal the size of a diaphragm of the relevant shape located near the lens.

3. Theory of dynamic speckle interferometry of thin phase objects

To study various properties of transparent objects, a variety of interference, shadow and speckle techniques are conventionally used [24–26]. As a rule, these techniques are oriented toward the analysis of macroscopic processes. With that, the logic of advancement in speckle optics and the practical needs pose the problem of studying microscopic processes occurring at the structure level. In particular, in biology, the problem of therapeutic drug management poses the problem of studying the processes in cells and their membranes. As at the structure level, the properties of biological media are random, when waves pass through various parts of the cell, their phase randomly varies in time. Therefore, a variation of the radiation intensity at the observation point is also a random process. The complexity of theoretical analysis of similar phenomena is that in the general case, there is necessity for dispersive ratios characterizing the wave phase variation in space and time.

In practice, there is an option when random values of the wave phases varying in space and time are independent. In particular, this option may be generated if the area of spatial correlation of a physical value causing the wave phase variations is less than the wavelength of light λ. In this case, the solution to the problem of establishing a relation between the wave phase dynamics in a thin transparent (phase) object and the dynamics of the light intensity in its image plane simplifies considerably. The solutions to this problem attempted for studying the properties of particular objects are found in the literature [27, 28]. In the para below, we are giving the general solution to the problem obtained by the authors of this paper.

3.1. Model of the object

At the first stage, the aim of the theoretical analysis is to obtain the expression for radiation intensity $I(\vec{q})$ at some point $\vec{q}$ of the observation plane, and then for temporal autocorrelation function of a random process $I = I(t)$. We will obtain the expression for value $I(\vec{q})$ using the model of a three-dimensional diffuser published in the abovementioned paper [22]. Let a point source of coherent radiation with wavelength λ located at point 1 illuminate point scattering centers randomly located in thin diffuser 2 near *(xoy)* plane as is shown in **Figure 2**. Let the position of the point source be given by radius vector $\vec{s}$. To simplify the transformations, let us admit that the refraction indexes of the medium inside and outside the diffuser are the same and equal 1. At distance L_0 from *(xoy)* plane, in plane $(\beta_x 0 \beta_y)$ there is a thin lens with focal distance f and diaphragm diameter D. Planes *(xoy)* and $(q_x 0 q_y)$ are conjugate. We consider all the waves discussed linearly polarized in the same direction. Let us admit that phase φ_j of the wave scattered by the jth center is random, and the waves from all the scattering centers arrive at an arbitrary point of $(\beta_x 0 \beta_y)$ plane.

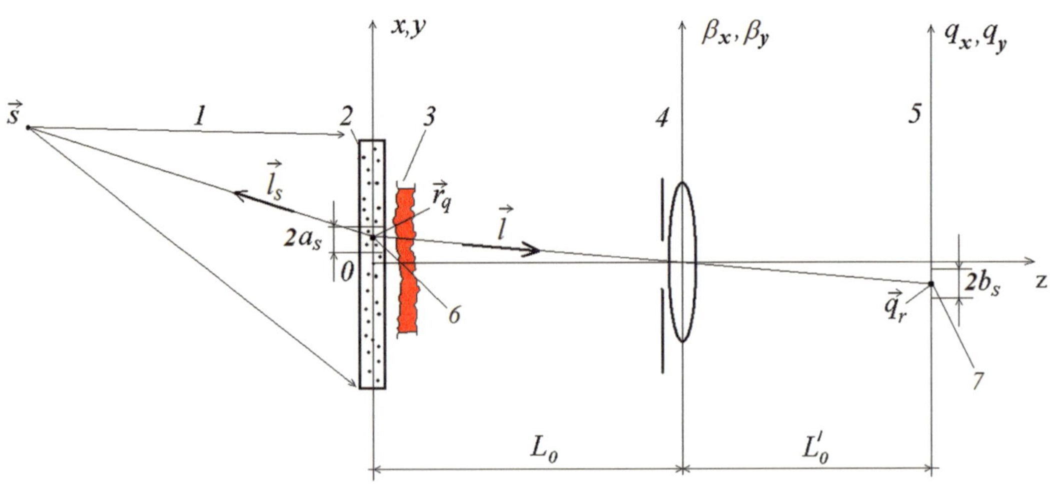

Figure 2. Optical system taken in the theory: (1) light source, (2) diffuser, (3) thin transparent object, (4) the lens with diaphragm, (5) the image plane, (6,7) conjugate points.

Let thin phase object 3 whose refraction index varies in time (**Figure 2**) be located near the diffuser to its right. Let us admit that the longitudinal resolution of the lens exceeds the sum of the diffuser thickness, the object thickness, and the distance from the object to the diffuser. We also suggest that the point scattering centers are fairly rare, so the random phases of the waves that have passed through the object are independent.

3.2. Radiation intensity

First, let us obtain the expression for radiation intensity $I(\vec{q})$ at some point $\vec{q}$ of plane (q_x0q_y) in the absence of the phase object. We suppose that the optical system does not permit separate scattering centers, and that the number of the scattering centers is fairly large in the area of the transverse lens resolution. For total complex amplitude $A(\vec{\beta})$ at arbitrary point $\vec{\beta}$ of plane $(\beta_x0\beta_y)$ we have:

$$A(\vec{\beta}) = \sum_{j=1}^{M} a_j(\vec{\beta}), \tag{17}$$

where M is the count of scattering centers, $a_j(\vec{\beta})$ is the complex amplitude of the jth wave at point $\vec{\beta}$. We will obtain the complex amplitude of light $A(\vec{q})$ at point $\vec{q}$ adding the amplitudes of waves that arrived from the points of plane $(\beta_x0\beta_y)$ to point $\vec{q}$, taking amplitude $P(\vec{\beta})$ and phase $exp\ ik|\vec{\beta}|/2f$ lens transmission into consideration [7]:

$$A(\vec{q}) = \int_{-\infty}^{+\infty}\int P(\vec{\beta})e^{\frac{ik|\vec{\beta}|}{2f}}e^{ik|\vec{L}_q(\vec{\beta})|}\sum_{j=1}^{M} a_j(\vec{\beta})d\beta_x d\beta_y, \tag{18}$$

where i is an imaginary unit, $k = 2\pi/\lambda$ is the wave number, and $\vec{L}_q(\vec{\beta})$ is the vector connecting points $\vec{\beta}$ and $\vec{q}$. Henceforward, the inferior index of the vector denotes the position of the vector head.

Let us take the relation between the complex amplitude of light in proximity to point $\vec{r}_j$ and at point $\vec{\beta}$ in the same form as in Ref. [7]:

$$a_j(\vec{\beta}) = \sqrt{I_0(\vec{r}_j)\xi(\vec{r}_j)}e^{i\left\{k[|\vec{L}_s(\vec{r}_j)|+|\vec{L}_\beta(\vec{r}_j)|]+\varphi_j\right\}}, \tag{19}$$

where $I_0 = I_0(\vec{r})$ is a distribution of the illuminating radiation intensity, $\xi = \xi(\vec{r})$ in the general case is a complex coefficient accounting the share of the radiation going from point $\vec{r}$ to point $\vec{\beta}$, $\vec{L}_s(\vec{r})$ is the vector connecting points $\vec{r}$ and $\vec{s}$, and $\vec{L}_\beta(\vec{r})$ is the vector connecting points $\vec{r}$ and $\vec{\beta}$.

Let us take arbitrary point 6 in plane (xoy) and its conjugate point 7 in plane (q_x0q_y) that are given by radius vectors $\vec{r}_q$ and $\vec{q}_r$, respectively (**figure 2**). It is known that the wave going from point 6 generates an Airy pattern with the center in point 7 as the result of light diffraction on the diaphragm of diameter D. The radius of the central spot b_s of the pattern equals $1.22\lambda L_0^{'}/D$, where $L_0^{'}$ is the distance from the lens to plane (q_x0q_y). The areas of radius b_s in plane (xoy) correspond to the area of radius $a_s = b_s/m$, where m is the magnification generated by the lens. It is known that 85% of the energy of the wave that passed through the lens falls on the central speckle of the Airy pattern. We are going to neglect the energy of the waves beyond the area of radius b_s. This in turn means that we suppose that the waves only from the scattering centers in the area of radius a_s with the center at point 6 arrive at point 7. Let N be the number of these centers.

Then, supposing that the area of radius a_s, the thickness of the diffuser and value D are small compared to the distances from the object to the radiation source and to the lens, and also from the lens to the image plane, we can obtain the expression for complex amplitude $A(\vec{q})$:

$$A(\vec{q}) = \sqrt{I_0}e^{i\psi}\sum_{j=1}^{N} e^{i\theta_j}, \tag{20}$$

where I_0, ψ are constants, $\theta_j = k\vec{r}_j(\vec{l}_s + \vec{l}) + \varphi_j$, $\vec{l}_s = \vec{l}_s(l_{sx}, l_{sy}, l_{sz})$ and $\vec{l} = \vec{l}\,(l_x, l_y, l_z)$ are single vectors directed from point $\vec{r}_q$ toward the radiation center and to the observer, respectively, complex amplitude $\sqrt{I_0}e^{i\psi}$ determines the complex expression preceding the summation sign. A detailed output of Eq. (20) can be found in Ref. [29].

Let us insert a thin phase object between the diffuser and the lens, as shown in **Figure 2**. Let us suppose that the object will alter only the phase of the jth wave, and there is no light refraction. In this case value, θ_j will change by value ζ_j, where

$$\zeta_j = \frac{2\pi}{\lambda}\left\{\int_{l_j}[n_j(l) - n_0]dl\right\} = \frac{2\pi}{\lambda}u_j, \tag{21}$$

$n_j(l)$ is a distribution of the refraction index in the phase object along the path of the jth wave, l_j is the path length of the jth wave in the object; integrals are found along the wave path, u_j is the optical difference of the jth wave travel path in the phase object.

So instead of Eq. (20), we have:

$$A(\vec{q}) = \sqrt{I_0}e^{i\psi}\sum_{j=1}^{N} e^{i(\zeta_j + \theta_j)} \tag{22}$$

For radiation intensity at point $\vec{q}$ we have the following:

$$I(\vec{q}\,) = A(\vec{q})A^*(\vec{q}) = I_0\sum_{j=1}^{N}\sum_{m=1}^{N} e^{i[k(u_j - u_m) + \theta_j - \theta_m]} = I_0N + 2I_0\sum_{\kappa=1}^{K}\cos\left[k\Delta u_\kappa + \Delta\theta_\kappa\right], \tag{23}$$

where Δu_κ is the relative optical difference of the travel path of the κth pair of scattering centers, $\Delta\theta_\kappa = \theta_j - \theta_m$, $j \neq m$, $\kappa = 1, 2 \ldots K$, $K = N(N-1)/2$.

3.3. Temporal autocorrelation function

First, let us obtain the expression for temporal autocorrelation function of the radiation intensity at point $\vec{q}$, i. e., Eq. (24):

$$R_{1,2}(t_1,t_2) = \langle [I_1 - \langle I_1\rangle] \times [I_2 - \langle I_2\rangle]\rangle = \langle I_1 I_2\rangle - \langle I_1\rangle\langle I_2\rangle, \tag{24}$$

where inferior indexes 1 and 2 denote time points t_1 and t_2, angle parentheses denote averaging by the object (model) ensemble. Let us suggest that at different κ random values Δu_κ are independent, and at the same κ, their time correlation occurs. Suppose also that joint probability density $\rho(\Delta u_{\kappa 1}, \Delta u_{\kappa 2})$ is a two-dimensional Gaussian function that is the same for different κ. So further we are going to omit inferior index κ in expressions Δu_κ. Using the suggestions made in Ref. [29], we obtained the expression for $R_{1,2}(t_1,t_2)$:

$$R_{1,2}(t_1,t_2) = I_0^2 N(N-1) cos[(\langle x_2\rangle - \langle x_1\rangle)] \times e^{-\frac{1}{2}k_{11} - \frac{1}{2}k_{22} + k_{12}}, \tag{25}$$

where $\langle x_1\rangle$ and $\langle x_2\rangle$ are the object ensemble-averaged values $x = k\Delta u$ at time points t_1 and t_2, respectively, k_{11} and k_{22} are dispersions of value x at time moments t_1 and t_2, respectively, $k_{12} = \langle (x_1 - \langle x_1\rangle)(x_2 - \langle x_2\rangle)\rangle$. For the normalized autocorrelation function $\eta_{12} = R_{12}(t_1,t_2)/R_{12}(t_1,t_1)$ we have the following:

$$\eta_{12} = cos[(\langle x_2\rangle - \langle x_1\rangle)] \times e^{-\frac{1}{2}k_{11} - \frac{1}{2}k_{22} + k_{12}}. \tag{26}$$

Let process $x = x(t)$ be stationary. Then $\langle x_1\rangle = \langle x_2\rangle$, $k_{11} = k_{22}$ and, therefore,

$$\eta(\tau) = e^{-k_{11} + k_{11}\rho_{12}(\tau)}, \tag{27}$$

where $\tau = t_2 - t_1$, $\rho_{12}(\tau)$ is a normalized temporal correlation function of random value $k\Delta u$. Let $\rho_{12}(\tau) \to 0$, $\tau \to \infty$. For example, this is a feature of normalized Lorentzian and Gaussian correlation functions. Then function $\eta(\tau)$ levels off to η^* equal to $exp(-k_{11})$. So by value η leveling off with time, we can determine dispersion k_{11} of phase differences varying in time and variation $\sigma_u = \frac{\lambda}{2\pi}\sqrt{k_{11}}$ of value Δu. We used this fact in experiments studying the processes occurring in live cells. These experiments will be discussed in Sections 5 and 6.

3.4. Temporal spectral function

Subtracting constant component η^* from Eq. (27) and renormalizing it, we obtain a new temporal autocorrelation function of radiation intensity fluctuation $\eta'(\tau)$:

$$\eta'(\tau) = \frac{\eta(\tau) - \eta^*}{1 - \eta^*}. \tag{28}$$

Let $\Delta u \ll \lambda$. Then it easy to demonstrate that $\eta'(\tau) \cong \rho_{12}(\tau)$. Therefore, the temporal autocorrelation function of intensity fluctuations corresponds to the temporal autocorrelation function of the wave pair optical path differences. Let us further suppose that random process

$\Delta u = \Delta u(t)$ is not only stationary but also ergodic. As the normalized temporal energetic spectrum for these processes is Fourier's transformation from the normal autocorrelation function, the corresponding normalized temporal spectral functions of intensity fluctuations $g_{\Delta I}(\omega)$ and optical path differences $g_{\Delta u}(\omega)$ are equal.

It was also demonstrated in Ref. [29] that if $\Delta u \geq \lambda$ and $\rho_{12}(\tau)$ is a Gaussian function, functions $g_{\Delta u}(\omega)$ and $g_{\Delta I}(\omega)$ are also Gaussian functions, but spectrum width $g_{\Delta I}(\omega)$ is k_{11}-fold spectrum width $g_{\Delta u}(\omega)$. So at k_{11} increasing spectrum $g_{\Delta I}(\omega)$ widens, and at $\Delta u \geq \lambda$ it widens k_{11}-fold.

3.5. Time averaging technique

A disadvantage of the theory presented in the Section 3.4 is the difficulty of application in the case when the wave phase variations in time happen due to existence of various processes occurring simultaneously at different scale levels. For example, when the target of research is a cell, sounding wave phase variation can occur due to passage of ions via the membrane, to capture large molecules by endocytosis (local variation of the cell shape), due to chemical processes during protein synthesis in the cytoplasm and nucleus of the cell, so in Ref. [30] the technique was upgraded to overcome this disadvantage. The idea was in the application of time-averaging procedure for speckle dynamics. If characteristic time τ_0 of wave phase variation corresponding to the most rapid process is known, averaging time T of the recorded optical signals can be taken as a value exceeding τ_0. In this case, the speckle dynamics will result from slower processes, and interpretation of the experimental data can get simplified. In the para below the results obtained in Ref. [30] are presented in brief.

Using the model discussed in Section 3.1, we obtained expressions for time-average intensity $\tilde{I}(\vec{q})$ at point $\vec{q}_r$ (**Figure 2**). Having substantiated the possibility of discussing continuous function $\tilde{I} = \tilde{I}(t)$ at point $(\vec{q})$, we obtained the expression for temporal autocorrelation function $R_{1,2}(t_1, t_2)$ of time-average intensity:

$$\tilde{I}(\vec{q}) = I_1 + I_2 e^{-k^2\sigma^2/2} \cos(k\mu + \alpha), \tag{29}$$

$$R_{1,2}(t_1, t_2) = I^2 N(N-1) C_0^2 cos[\langle x_2 \rangle - \langle x_1 \rangle] \times e^{-\frac{1}{2}k_{11} - \frac{1}{2}k_{22} + k_{12}}. \tag{30}$$

In Eq. (29) I_1, I_2, α are constants, μ and σ^2 are mean value and dispersion of variable Δu obtained by time-averaging and averaging by a region of radius a_s (see **Figure 2**). Eq. (30) coincides with Eq. (25) to a precision of insufficient coefficients I^2 and C_0^2. But now $x = k\mu$, so the arguments of the cosine and the exponent contain mean values, dispersions, and the correlation moment of a new value $x = k\mu$.

The peculiarity of Eq. (30) is that if averaging time T of the radiation intensity exceeds the correlation time of value μ, normalized function, Eq. (30) takes on the following appearance:

$$\eta(\vec{q}\ ,t_1,t_2) = \frac{R_{1,2}(\vec{q},t_1,t_2)}{R_{1,2}(\vec{q},t_1\ ,t_2)} = cos[\langle x_2\rangle - \langle x_1\rangle]e^{-k_{22}/2+k_{11}/2}. \tag{31}$$

Supposing in Eq. (31) that values k_{11}, k_{22}, and $k\mu$ are small compared with 1, let us decompose Eq. (31) into Taylor's series in the neighborhood of points k_{22} and $k\mu$ equal to zero, having retained the first-order derivatives. We obtain

$$\eta(t) = 1 - \frac{(k_{22}(t) - k_{11})}{2}, \tag{32}$$

where $t = t_2 - t_1$. It is seen from the formula that if averaging time T exceeds the correlation time of random value $k\mu$, the relation between η and k_{22} is linear.

Now among N waves let us have two wave groups with random optical wave path variations $u_n = u_n(t)$ occurring homogeneously in the statistical sense. Let count n of such waves in groups 1 and 2, respectively, equal N_1 and N_2. In practice, groups 1 and 2, for example, lie inside and outside a live cell. At large magnifications, such groups can lie within the cell nucleus and in its cytoplasm. In Ref. [30], it was shown that in this case time-average radiation intensity at conjugate point $\vec{q}_r$ is determined by Eq. (33):

$$\tilde{I}(\vec{q}) = \tilde{I}_1 + \tilde{I}_2 + 2\tilde{I}_{12}cos[k\Delta\mu + \theta], \tag{33}$$

where $\tilde{I}_1$ and $\tilde{I}_2$ are time-average intensities generated by groups 1 and 2 individually, $\tilde{I}_{12} = I_3 e^{-\sigma_1^2/2 - \sigma_2^2/2}$ and I_3, θ are constants, $\Delta\mu$ is the difference of time-average values u_n in groups 1 and 2, σ_1^2 and σ_2^2 are dispersions of values u_n in groups 1 and 2, respectively.

3.6. Relation between the object features and the parameters of speckle dynamics

We used the results of the theory presented in Section 3 to conduct experiments with live cells cultured or precipitated on a transparent substrate. To determine the value η, we took segments of diameter $2a_s$ in the object plane. We regarded a region containing a large number of such segments as an object ensemble. The corresponding segments of the speckle image in the conjugate region were recorded at time points t_1 and t_2, and then they were used to determine correlation coefficient η of digital speckle images.

Analysis of the formulas obtained demonstrates that experimentally obtained dependences $\eta(t)$ and (or) $\tilde{I}(t)$ can in principle be used to determine the mean value, variation, and correlation time of the medium refraction index in small regions of the transparent object. In turn, the refraction index is related to medium density ρ and its specific refractivity $\hat{r}$ via Lorentz-Lorenz formula for liquids. For multicomponent media, the latter is equal to the sum of the products $\hat{r}$ of single molecules on their relative concentration.

It is known that the density of liquids depends on their temperature. Therefore, if the time range where the composition of the medium can be considered constant is selected, under certain conditions the spectrum of intensity fluctuations can be regarded as the energetic spectrum of chemical reactions occurring in the cell areas under study. Similarly, if a time range or the object segments with the temperature (density) that can be considered invariable is selected, the processes of mass transfer in live cells can be studied by variation of correlation coefficient η or average intensity $\tilde{I}(t)$.

4. Cultured cells as research target

4.1. Features and advantages of cultured cells

Cell cultures have been playing a more important and notable role in toxicological, pharmacological, and other investigations. That said, the sphere of their application has been widening, and the technique of in vitro culture has been getting upgraded and automated. Cell cultures are single cell groups grown in invariable conditions. Moreover, the researcher is allowed to vary these conditions within certain limits enabling themself to assess the effect of various factors such as pH, temperature, and amino acid concentration on cell growth. Cell growth can be assessed in a short time period or by increase of the cell count or size, or by inclusion of radioactive precursors into cellular DNA. These real advantages compared with investigations on intact animals put cell cultures on a par with cultured microorganisms as an experimental system.

When working with cell cultures, we can obtain significant results only on a fairly small number of cells. Experiments requiring 100 rats or 1000 humans for clearing up some matter can be conducted using 100 cultures on cover glasses with equal statistical significance. So if every cell is regarded as an independent object of the experiment, one culture on a cover glass can give an answer as reliable as a clinic full of patients can. This is a significant advantage when it concerns humans; besides, it removes a number of ethical problems from the agenda when it is necessary to use a large group of animals for an experiment.

4.2. Cell life cycles

Cell culture monolayers are populations of cells having certain species and tissue origin growing on the surface of a carrier made of plastic, glass etc. A complete cell monolayer may cover more than 90% of the surface, with the cell membranes connected. In such conditions, an average cell size is 20–30 μm at 5.5-μm thickness.

Cell cultures may be roughly divided into two main groups: (1) continuous cultures that are capable of unrestrictedly long existence *in vitro*; (2) diploid ones obtained from normal body tissues retaining many features of the original tissue and capable of restricted (up to 50 divisions) growth in an artificial medium.

In turn, the first group is divided into two subgroups:

(1) high-transformed ones, derived, as a rule, from various tumors and capable of existing in artificial conditions for an uncertain time, (2) low-transformed ones derived from normal

tissues whose transformation source is often unclear, also capable of unrestricted growth but closer to normal body tissues in many respects.

In compliance with the experimental terms, all the three cell culture types were used. The selection of the cultures was chiefly due to by their sensitivity to herpes simplex virus type.

All the cell cultures of the endotherm have similar cell cycle duration—19 to 24 h. The cell cycle consists of the following phases:

1. The presynthetic phase (G1) (phase formula *2n2c*, where *n* is the chromosome count, and *c* is the molecule count in the cell). It starts immediately after the cell division. There is no DNA synthesis yet. The cell actively increases in the dimensions, increases supplies of the substances necessary for division: proteins (histones, structural proteins, and enzymes), RNA, and ATP molecules. Division of mitochondria and chloroplasts (i.e., structures capable of reduplication) occurs. The features of interphase cell organization are being restored after the previous division.

2. The synthetic phase (S) (*2n4c*). The genetic material duplicates by DNA replication. It occurs in a semiconservative way when the DNA double spiral separates into two chains, and a complementary chain is synthesized on each one. This results in the formation of two identical DNA spirals, each one consisting of one old and one new DNA chain. The amount of genetic material doubles. Besides, RNA and protein synthesis continues. A small part of mitochondrial DNA also undergoes replication (most of it replicates in G2 period).

3. The postsynthetic phase (G2) (*2n4c*). DNA is not synthesized anymore, but the synthesis flaws of S period are corrected (reparation). Also, energy and nutrients are accumulated; synthesis of RNA and proteins (mainly nuclear ones) continues.

4. The quiescent phase (G_0) wherein the cell only consumes the survival minimum of nutrients without preparing for another multiplication cycle.

Each phase is characterized by a different intensity level of nutrient absorption/exchange/release, and these processes can localize in various parts of the cells (the nucleus, cytoplasm, and organelles) depending on the cell cycle phase.

It is also noteworthy that in an actively growing continuous cell culture more than 90% cells enter the cycle, while in a diploid cell culture at most 20% cells (more often less than that) do so.

At a certain cycle phase, the percentage of cells in an actively growing cell culture approximately complies with the phase duration ratio; cells in G_0 phase prevail in a closely packed cell monolayer.

4.3. L41, Vero, and HLE-3 cell lines

In compliance with the passport of L41 CD/84 cell line Ref. [32], a strain of continuous cells G-96 derived from the blood of a patient with monocytic leukemia is known, which was used in 1966 by Solovyov et al. to derive a subline (G-41) specifically resistant to Coxsackie B3 virus by triple treatment with large doses of the virus.

By its morphological features, L41 cell line is an even monolayer of distinct epithelium-like polygonal or roundish cells; there is a constant 4–6% share of giant cells. The cytoplasm is fine-granular. The nuclei are roundish and contain 2–4 nucleoli. There are up to 6% abnormal mitosis forms. The share of cells with irregularly shaped nuclei is 8%. The monolayer was generated on day 3–5 from planting into a medium consisting of equal parts of Eagle medium and 199 medium with 10% bovine embryo serum. The cell maintenance medium contains necessary amino acids, salts, and glucose.

The culture is highly sensitive to poliomyelitis, Coxsackie B, ECHO-19, human adenovirus, and measles viruses.

Vero cell line was derived from normal simian renal cells (those of an adult African green monkey). The number of generations and passages: over 120 passages before the test start. The line has been used in a laboratory research since 1962. The monolayer forms on days 3–5 from the planting moment. The multiplying factor is 6–7 on day 5.

The morphological features: epithelium-like cells, polygonal, with notable vacuolization, and distinctly oriented growth zones.

The karyological characteristics: the cells correspond to the monkey karyotype by their structure—44 diploid cells, 3 hyperploid cells, and 53% hypoploid cells.

Species origin: monkey, confirmed karyologically. Data on contamination: no bacteria, fungi, or *Mycoplasma* detected. The cell line is maintained in the growth medium + 10% glycerin in liquid nitrogen. About 80–85% cells restore on defrosting. The culture is highly sensitive to poliomyelitis viruses and arboviruses.

HLE-3 cell line was derived by Yekaterinburg Institute of Viral Infections (YRIVI) staff from normal human lung tissue. The technique for obtaining this cell line was developed in the cell line laboratory of YRIVI by Glinskikh et al. in 1980 from the lung of a 12-week human embryo from a healthy female whose genelogy was free from malignant or hereditary diseases.

The number of generations and passages: 20 passages at most before the test start.

The monolayer forms on days 4 and 5 from the planting moment. The morphological features: fibroblastic cells with distinct edges.

The karyological characteristics: the cells correspond to the human karyotype by their structure. The modal class contains 87% cells with normal diploid human chromosome set.

Species origin: human, confirmed karyologically with an isoenzyme technique.

Data on contamination: no bacteria, fungi, or *Mycoplasma* detected. The cell line is maintained in the growth medium +10% glycerin in liquid nitrogen. About 70% of the cells restore on defrosting.

Sensitivity to viruses: the culture is highly sensitive to polioviruses 1, 2, and 3, Coxsackie B3, BArinsECHO 3, 6, 11, 13, 19, 20, 24, 28, RS viruses, and herpes simplex virus.

5. Procedure and technique of experimental research

We used the theory presented in Section 3 to study the processes occurring in a thin layer of live cells. In the text below, a description of the experiment, its errors, and the measurement calibration is given. The latter procedure by default stemmed from the experiment conducted to check one of the tenants of the theory.

5.1. Thermostats and temperature control

Precise maintenance of the temperature in the medium for the cells plays an important role in the experiments. After several attempts to create a small thermal chamber to maintain the temperature in the small region, we decided on placing the entire optical system into a thermostat of suitable dimensions. We used three thermostats: (a) a self-made laboratory thermostat, (b) a liquid thermostat of ЗЦ-1125М type, and (c) an air bath of ТСЭ-200 type.

We used the laboratory thermostat in the cases when we had to place the substrate with cells in a horizontal position. The photo of the thermostat is shown in **Figure 3**.

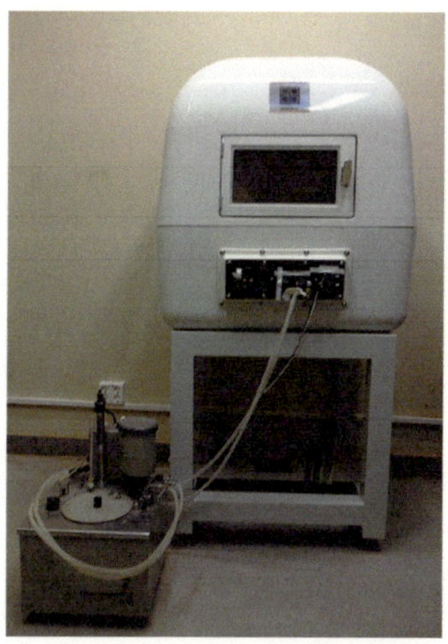

Figure 3. Photograph of laboratory thermostat.

We used a dustproof chamber of a scanning atomic-force microscope. To heat the air inside the chamber, we used a liquid ultrathermostat U10 connected to the radiator with pipes which is common for Russia. The radiator was in the lower part of the chamber under the table for the microscope: it was blown over by six small ventilators. A metallographic microscope of Axio 40 MAT type that we used to generate speckle images of the cells was placed on the table.

The temperature of the cuvette in all the thermostats was determined by a temperature sensor of DS18B20 type of precision to ±0.1°C. The signals from the sensor entered the computer; the temperature values were displayed in the monitor or recorded into the computer memory in the preset time.

The laboratory thermostat maintained the temperature of the cuvette to ±0.1°C precision for several hours, and the liquid thermostat 3Ц-1125M did so for several days. An air bath was used to study the cell reaction to temperature.

5.2. Optical and television systems

When the laboratory thermostat was used, the transparent substrate with cells was placed at the bottom of a special cuvette. In turn, the cuvette was placed on the stage of the microscope. The transparent lid of the cuvette was placed so that (1) the nutrient solution about 1 mm thick was placed above the cells and (2) there was no free liquid surface. Above the lid, there was matte glass that was illuminated by a diverging beam from the semiconductor laser module. The speckle image of the cells was entered into the computer using a monochrome TV camera of Videoscan—415/P/C-USB type. The camera had a photosensor array of 6.5 × 4.8 mm size with 780 × 572 cells (pixels) of 8.3 × 8.3 μm size. The frame input frequency was up to 25 Hz. The signals from the TV camera entered the laptop computer of Aspire 3692 WLMi 8 type by Acer firm via a USB port. A semiconductor laser module of KLM-D532-20-5 type with wavelength $\lambda = 0.532$ μm and 20 mW power was used as the light source. Note that the above TV camera and laser were used in all of our optical systems.

In the liquid thermostat and in the air bath, optical systems with the upright position of the substrate with cells in the optical cuvette were used. The logic chart of the optical systems is given in **Figure 4**, and its photograph is presented in **Figure 5**.

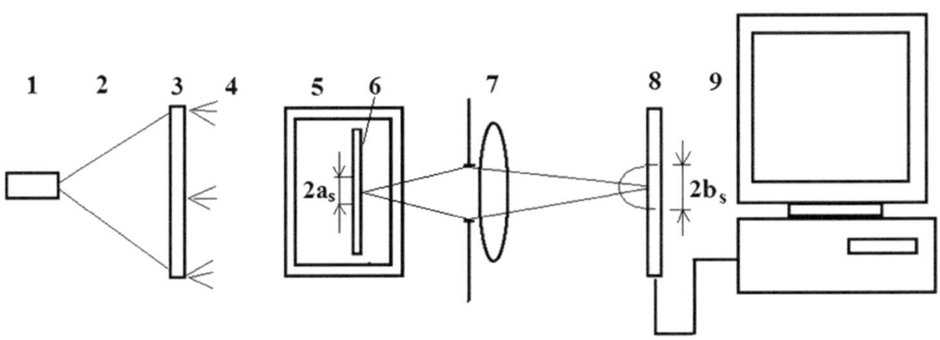

Figure 4. Logic chart of optical device: (1) laser module, (2) illuminating beam, (3) matte glass, (4) scattered radiation, (5) cuvette, (6) substrate with cells, (7) lens with diagram, (8) photosensor array, (9) computer.

5.3. Software

Input of the image frames into the computer was made using the software coming standard with camera Videoscan—415/P/C-USB. The frames entered into the computer were processed to obtain dependences $\eta(t)$ and $\tilde{I}(t)$ using two original computer programs. The first program was intended to process the frames already entered into the computer. The operator displayed the first frame of the speckle image onto the monitor and selected the fragment of the frame to determine η using the mouse or the keyboard. Then it gave the second frame or a mass of frames. In the first case, the program displayed value η using Eq. (34) and digital value $\tilde{I}$, in the second case, it displayed dependences $\eta(t)$ and $\tilde{I}(t)$ onto the monitor. The first program also allowed to obtain the distribution of values η. The operator selected two frames corresponding to time moments t_1 and t_2. Then the operator selected a segment on the displayed image and divided it into subsegments. The program digitized the fragment and determined value η in every subsegment using Eq. (34). The obtained values were recorded into the computer memory as a matrix in text format to be processed.

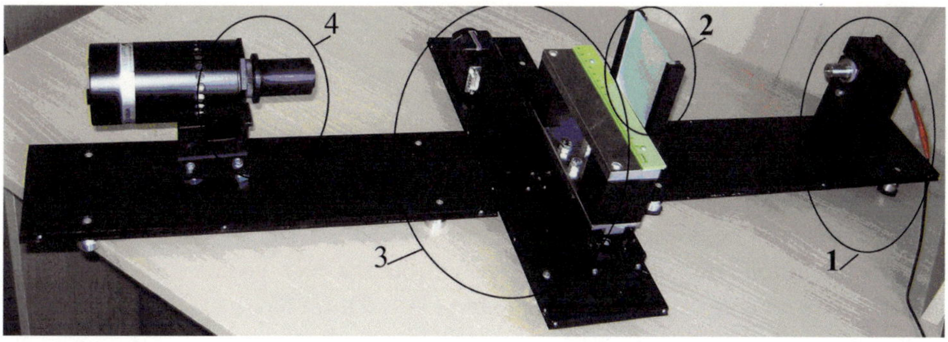

Figure 5. Photography of the optical device: (1) laser module with microobjective, (2) matte glass, (3) lock of the object on the platform of the motorized translator, (4) camera with lens.

Value η was determined by Eq. (34):

$$\eta = \frac{\frac{1}{m\times n}\sum_{i=0}^{m-1}\sum_{j=0}^{n-1}[A_{i,j}-\overline{A}][B_{i,j}-\overline{B}]}{\sqrt{\frac{1}{m\times n}\sum_{i=0}^{m-1}\sum_{j=0}^{n-1}[A_{i,j}-\overline{A}]^2}\times\sqrt{\frac{1}{m\times n}\sum_{i=0}^{m-1}\sum_{j=0}^{n-1}[B_{i,j}-\overline{B}]^2}}, \tag{34}$$

where $A_{i,j}$ are digitized signals at a segment of m × n-pixel size at initial time point t_1, $B_{i,j}$ are the signals in the same segment at a different time point t_2, i, and j are the segment pixel numbers xd and yd, respectively, $\overline{A}$ is the mean signal value in the segment at the start time, and $\overline{B}$ is the mean signal value in the segment at time moment t_2.

The second program in DOS medium permitted real-time determination of digital values η and $\tilde{I}$. First, the operator set x and y coordinates of the pixels (up to 40 pieces) on the program interface. By the operator's command, the program determined digital values of mean intensity $\tilde{I}$ in the above pixels, and it obtained η values in the neighborhood of the pixels using Eq. (34). Masses η and $\tilde{I}$ were saved in *txt* format in preset files. The program could work for several days continuously.

5.4. Errors and calibration of optical measurements

We assessed the error of value η determination by Eq. (34). In compliance with the indirect measurement error assessment techniques recommended in Russia [31], the mean square deviation of random error $S(\tilde{\eta})$ in the indirect measurement result is determined by Eq. (35):

$$S(\tilde{\eta}) = \sqrt{\sum_{i}^{m} \left(\frac{\partial \eta}{\partial a_i}\right)^2 \times S^2(\tilde{a}_i)}, \quad (35)$$

where $\tilde{\eta}$ is experimentally found value of η, $a_i (i = 1, \ldots, m)$ are values $A_{i,j}$ and $B_{i,j}$ featuring Eq. (34), $\frac{\partial \eta}{\partial a_i}$ is the first derivative of function η by argument a_i, calculated at point $\tilde{a}_1, \ldots, \tilde{a}_m$, $\tilde{a}_i$ is the result of measuring value a_i, and $S(\tilde{a}_i)$ the mean square deviation of random errors in the result of measuring the a_i-th argument.

We performed the transformations by Eq. (35) and assessed the error of value η determination in a typical experiment. We selected the variant with 8-bit digitization of the radiation intensity averaging-out half the dynamic range, the minimum speckle size slightly exceeding the TV camera pixel size, and $S(\tilde{a}_i) = 0.7$. Calculations showed that value $\partial\eta/\partial a_i$ featuring Eq. (35) consists of sum m of random 10^{-2}-order values of different signs. The random sign value appears due to randomness of intensity deviation from the average value in the speckle field. In the model experiment, a reflecting rough object in the form of a metal plate was used, and value η varied due to its shift. For a 10 × 10-pixel fragment of the speckle image, we obtained that $S(\tilde{\eta})$ decreases steadily with increase of η in the range from 0.3 to 0.999, and the relative error of η determination does not exceed 1%.

Essentially, the calibration technique for the optical systems intended to determine the optical path dispersion value was developed in Ref. [29]. In this technique, batched random variations of the wave phase difference were set by means of shifting a 1-mm thick transparent plate. The plate shift u_x was performed with a 0.12-µm pitch. The plate roughness was prechecked with a WYKO NT-1100 optical profilometre. **Figure 6** shows experimental and theoretical dependences of η on the plate shift u_x.

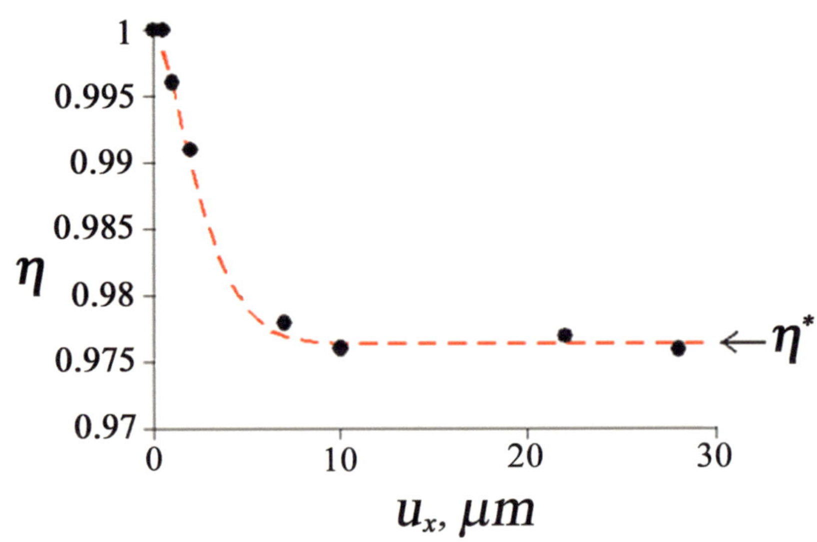

Figure 6. Theoretical (---) and experimental dependences $\eta(u_x)$.

As is seen from the graphs, when the plate is shifted for a value exceeding the characteristic surface roughness size, the correlation coefficient levels off to η^*. The theoretical curve was obtained for the Gaussian function by Eq. (27). The difference between the theory and the experiment was in the range of 2.5%. Roughness parameter difference R_a obtained by level η^* from that measured by the profilometre was in the range of 5%. The experiment details and the digital derivations can be found in Ref. [29]. This experimental technique can be used to calibrate the equipment used to determine value k_{22}. As for the calibration of the device for determination of value $\langle x_2 \rangle$, further research is needed to perform this procedure.

Good coincidence of the data obtained by the speckle dynamics and with the optical profilometre (**Figure 6**) is to a great extent determined by small errors of determining the speckle image correlation coefficient η and with high sensitivity of the technique. Let us assess the sensitivity limits of the equipment for the determination of values $\langle x_2 \rangle$ and σ_u. Let us admit that in Eq. (26) values $\langle x_1 \rangle$, k_{11}, k_{22}, and k_{12} are equal to zero. Typical values η_{12} equaled 0.99 in the absence of the object. Then for wavelength $\lambda = 0.532$ μm, we obtain that $\Delta u = (\lambda/2\pi) arccos 0.99 = 12$ nm. Hence, it follows that the limit sensitivity of the device related to the optical path difference generated in the range of the linear resolution of the lens equals 12 nm. Let us admit that the object thickness is invariable, and the optical paths vary due to variation of the mean refraction index. For instance, for the 10-μm cell thickness, we obtain that the refraction index will vary by 1.2×10^{-3}. Now let us find the limit sensitivity of the equipment in determination of mean square deviation $\sigma = [\lambda/(2\pi)]\sqrt{k_{11}}$ of value Δu by value η^*. Supposing again in formula $\eta^* = exp(-k_{11})$ that $\eta^* = 0.99$, and $\lambda = 0.532$ μm, we obtain that $\sigma = [\lambda/(2\pi)]\sqrt{-ln 0.99} = 8$nm.

It is noteworthy that values η_{12} and η^* equal to 0.99 appeared due to application of medium-quality equipment. If limit values η_{12} and η^* equal to 0.999 are reached due to noise decrease, then the sensitivity to the mean value and dispersion of value Δu will equal 4 and 3 nm, respectively.

6. Experiments

6.1. Studying speckle dynamics in the image plane of a cultured cell monolayer

We studied the features of speckle dynamics caused by an activity of cultured cells on L41 cells discussed above in Section 4.3. After the formation of a monolayer, the substrate with the cells was placed into an optical cuvette filled with nutrient solution. A cell-free substrate of similar thickness was placed near. Next, the cuvette was fixated on the optical device shown in **Figure 5** that was placed in liquid thermostat 3Ц-1125M. The typical magnification by the optical system was 0.25, and the typical linear resolution of the lens was 60 μm.

Figure 7 presents a typical speckle pattern recorded in the image plane of the cuvette with substrates.

After equalization of the temperatures of the cuvette and the thermostat the mechanical system was checked for stability. To do so, we used the above software, selected the segment of the frame near the image of the cuvettes. Using two speckle images of the segment recorded in 1 or 2 min, we determined the value of correlation coefficient η of their digital images. If value η equalled 0.99 or 0.98, the system was considered mechanically stable. Further, we recorded the film of the speckle dynamics lasting 20–40 s with 25 Hz frequency. Using the software discussed in Section 5.3, we obtained dependences of the speckle image fragment correlation coefficient η on the time using the frames of the film. In **Figure 7**, typical selected fragments are shown by white frames.

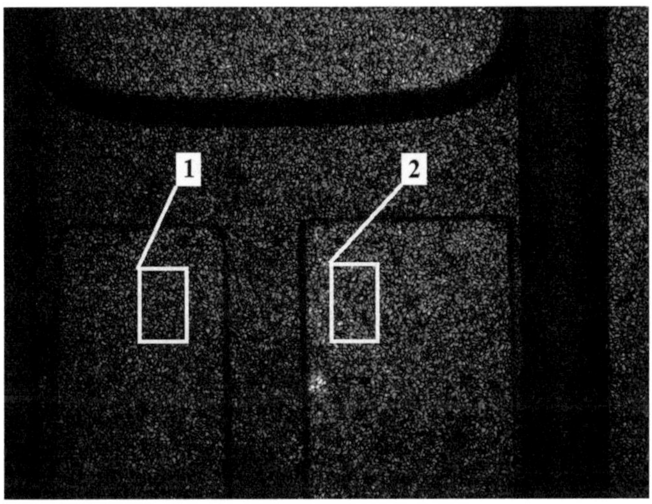

Figure 7. Typical speckle image of a cuvette with transparent substrates: the substrate with cells is on the left, and the cell-free one is on the right. The selected image fragments are denoted by numbers.

Figure 8 shows typical dependences of $\eta(\tau)$ for the cells in the nutrient solution (dependence 1) and for the nutrient solution (dependence 2) obtained after processing of the film. The dotted lines show theoretical dependences obtained using Eq. (27) for normalized Lorentzian function $k_{12}(\tau)$. Analysis of the experimental dependences obtained in different segments of the substrate with cells within its image showed that the deviation of the theory from the experiment was in the range of 8–11%.

As seen from the graphs in **Figure 8**, in about 5 s dependences $\eta(\tau)$ level off. The mean square deviation of the last four points from their mean value does not exceed 1%. In compliance with the theory discussed above in Section 3, levelling off dependence $\eta(\tau)$ means that random process $\Delta u = \Delta u(t)$ is stationary in time. Value Δu is typical (mean) optical wave path difference within the region with cells of 60-μm diameter. As fragments of about 1-mm size correspond to the selected segments of the speckle images in the object plane (**Figure 7**), random process $\Delta u = \Delta u(t)$ can also be regarded as homogeneous in this fragment.

As was pointed out in Section 3, by values of variable $\eta = \eta^*$ we can determine the corresponding dispersions of the wave pair optical paths σ_1^2 and σ_2^2 in the horizontal segment of dependences 1 and 2. Supposing that optical wave path variations in the cells and in the nutrient solution are uncorrelated, we can show that the optical path dispersion in cells can be determined using Eq. (36):

$$\sigma_u^2 = \sigma_1^2 - \sigma_2^2. \tag{36}$$

Mean square deviation σ_u of values Δu obtained by Eq. (36) came to 14 nm.

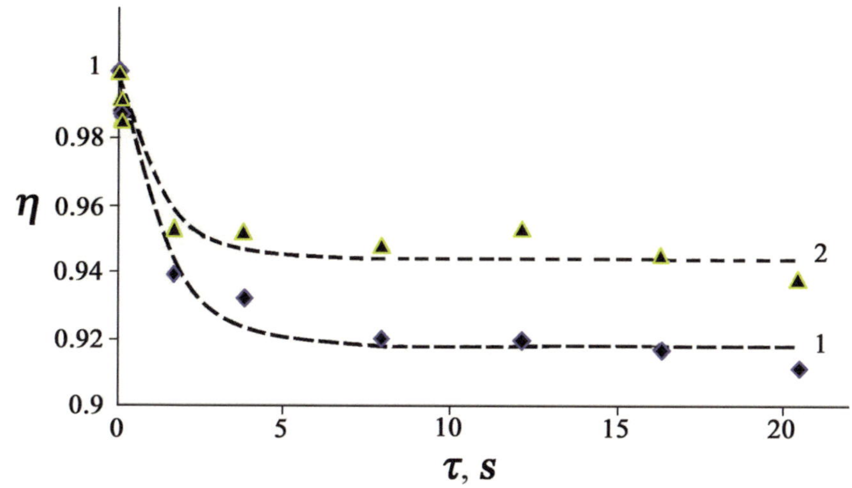

Figure 8. Joint dependences of η on the time for the cells in the nutrient solution (1) and for the nutrient solution (2).

The homogeneity and stationarity of process $\Delta u = \Delta u(t)$ in a segment of a 1-mm order can be explained by the fact that in a monolayer, the cells are closely packed, so there is no variation of their shape due to translation and division. In these conditions local deviations of the medium refraction index from its mean value are possible at the structural level. As was discussed above in Section 3, chemical reactions and phenomena of mass transfer can be the reasons for refraction index variation.

6.2. Defrosted cells and speckle dynamics

We conducted an experiment with L41 cells precipitated onto a transparent substrate immediately after defrosting. The interest in similar experiment was caused by the fact that, as distinct from a monolayer of cultured cells, after defrosting the cells do not attach to the substrate immediately, being in motion. When a cell is moved for a distance comparable to its size, the sounding wave pair path difference can vary by a value comparable to wavelength of radiation λ and exceeding it. Therefore, in compliance with Eq. (26) variation of speckle image fragment correlation coefficient can be caused both by the cosine argument variation and by variation of the values in the exponent. As a cosine can be both positive and negative, appearance of negative values of the variable close to 1 would speak for correctness of Eq. (26) and our theory.

Besides, the objective of the conducted research was to study the possibility to apply the technique at large magnifications to analyze the processes occurring in different parts of one cell.

The experiments were conducted on a laboratory device with a horizontal position of the substrate with cells. When the temperature reached the value near 36°C a glass 0.1-mm thick was placed into a cuvette with nutrient solution, and poured frozen cells from Dewar vessel onto it. After the temperature stabilization in 30–60 min, recording of the speckle dynamics film was started. The optical magnification was about ×8, the exposure time equaled 9 s, and the frames were recorded for several hours.

A typical speckle pattern of cells precipitated on a substrate is shown in **Figure 9**. Viewing of the films showed that the cells contact other cells being in continuous random motion. The typical shift of a cell in one direction was comparable to its dimensions. There were cells making shifts for a larger distance, and there were also cells that could be visually regarded as stationary.

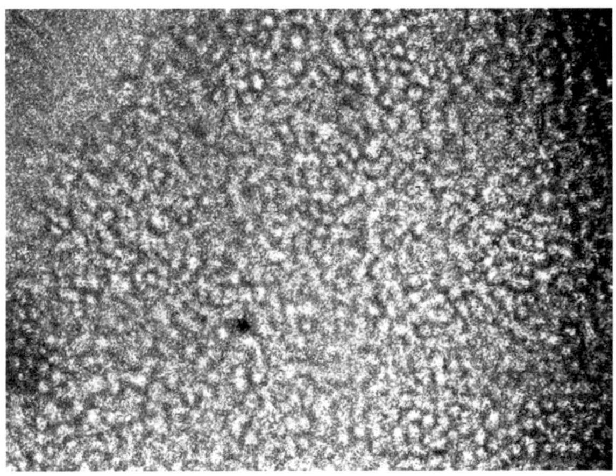

Figure 9. Speckle image of defrosted cells.

Figures 10–13 demonstrate typical dependences $\eta = \eta(\tau)$ obtained for different sizes of speckle image segments. The graph in **Figure 10** corresponded to a 4 × 4-pixel segment or the cell fragment size of about 4 × 4 μm. Originally, the segment was in the center of the cell image. The rest of the dependences were obtained by means of data averaging in segments containing from 4 to 200 cells.

As seen from **Figure 10**, for a randomly moving cell value η randomly varies in the range from -1 to +1 around zero. The obtained result qualitatively confirms correctness of Eq. (26) containing dependence of value η on the optical wave path difference by the cosine law. Variation of η from 1 to -1 means that the positive image has changed to a negative one, or vice versa. This is possible, for example, if in all the 16 pixels radiation intensity varies by the cosine law with the same period (the same Δu) but with the different initial phase. In **Figure 10**, value η does not reach +1 and -1 again. It points out that either during the cell motion its shape changing in a 4×4-μm segment is inhomogeneous, or value η varies not only by the cosine law.

If the data averaging region covers plenty of cells (see **Figures 11–13**), values of variable η are positive and reach a horizontal segment in about 0.5 h. The obtained data can be explained by the fact that at a fixed time point values x included in the cosine argument in Eq. (26) can reach large values with different signs in different cells. But in averaging by a large count of cells (the

http://dx.doi.org/10.5772/66712

objects of the ensemble) variables $\langle x_1 \rangle$ and $\langle x_2 \rangle$ have values close to zero. Then in Eqs. (26) and (31) the dependence on the cosine disappears, but the dependence on k_{22} remains.

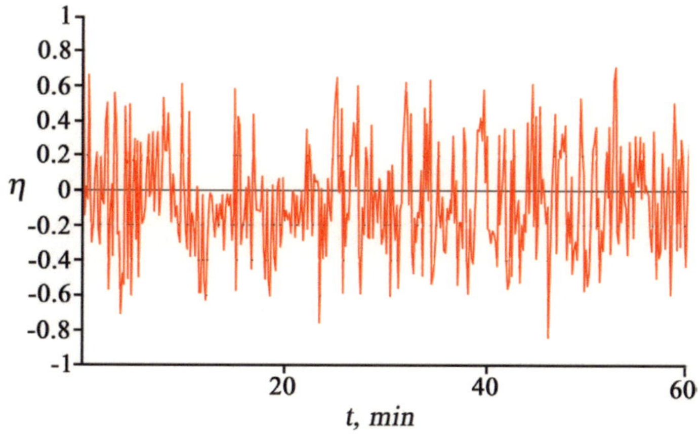

Figure 10. Dependence $\eta = \eta(t)$ for fragment inside the cell image.

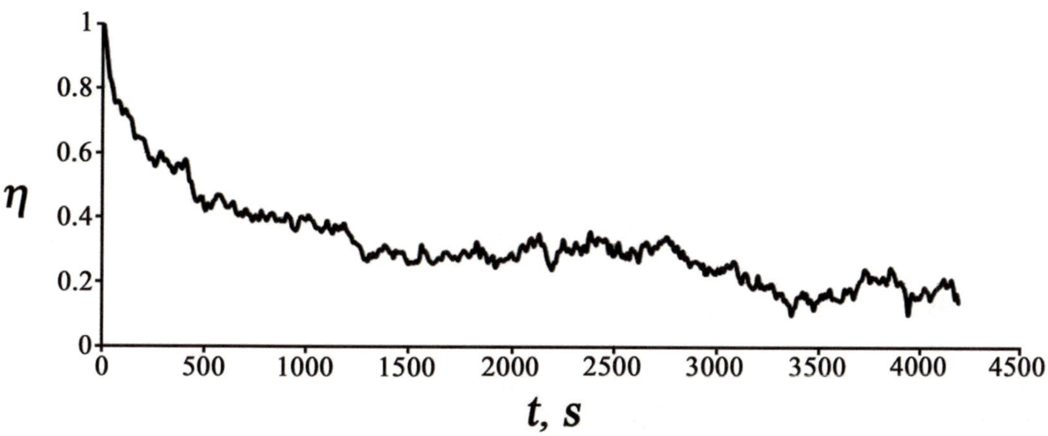

Figure 11. Dependence $\eta = \eta(t)$ corresponding to four cells.

Dependences $\eta(t)$ shown in **Figures 11–13** can be interpreted in two ways. Supposing that the 9-s speckle averaging time exceeds the correlation time of the fastest processes in the cells, one can suppose that by Eq. (31) dependence $\eta(t)$ corresponds to an unstable process wherein value k_{22} first increases continuously and then levels off. On the other hand, we can suppose that graphs of $\eta(t)$ correspond to stationary process $\Delta u(t)$ with the correlation time about 30 min.

To clear up this matter, we selected different segments of the view-field containing about 100 cells. For each of these segments, four dependences $\eta(t)$ lasting about 1 h were built for different time intervals. The form of dependences was well reproducible, the multiple correlation coefficients of the four masses was in the range from 0.86 to 0.96. We came to the conclusion that in fragments containing hundreds of defrosted cells random processes $\Delta u(t)$ can be regarded as homogeneous in space and stationary in time for several hours.

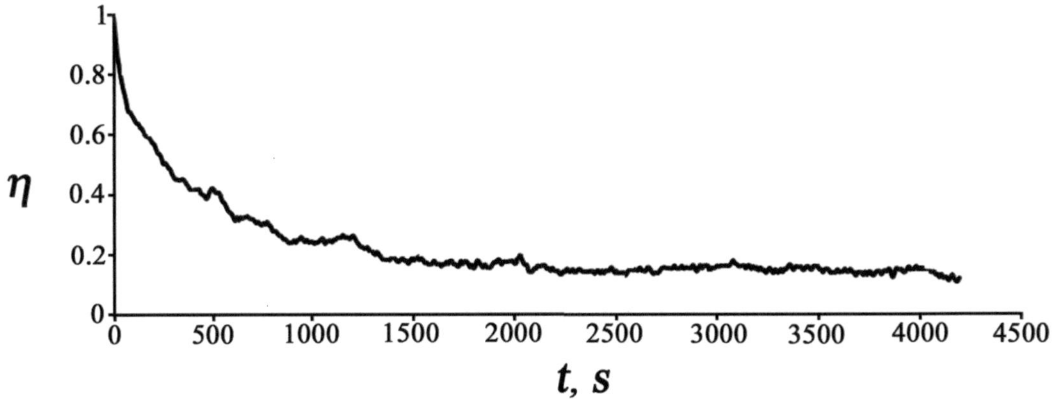

Figure 12. Dependence $\eta = \eta(t)$ corresponding to 60 cells.

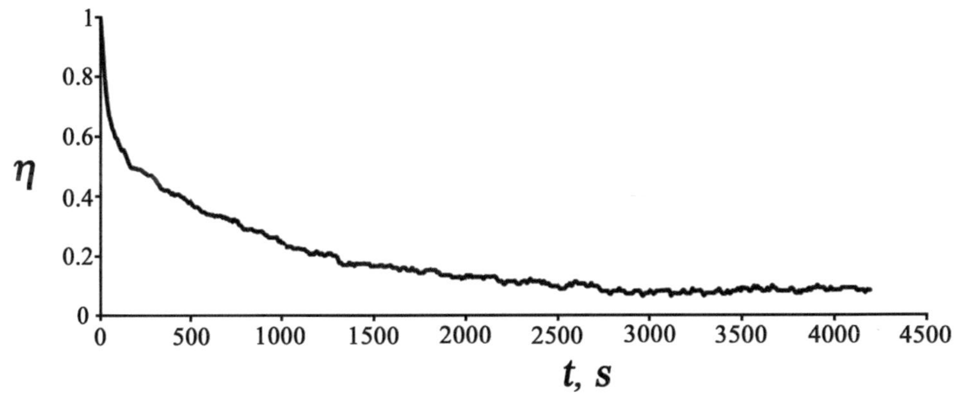

Figure 13. Dependence $\eta = \eta(t)$ corresponding to 200 cells.

6.3. Application of speckle dynamics for studying the reaction of a cell mass and fragments of individual cells to temperature variations

The theoretically detected and experimentally confirmed relation between the correlation coefficient of speckle images η and the dispersion of the optical wave pair path difference σ_u^2 was immediately used in our first experiments studying dependence of σ_u on temperature T. The details of the experiment can be found in Ref. [33]. A segment of L41 cell monolayer containing hundreds of cells was the averaging region. Value σ_u^2 corresponding to the cells was obtained as the difference of values σ_u^2 corresponding to the cells in the nutrient solution and to the nutrient solution. The cuvette with cells was first heated to a temperature around 40°C. Then the heating was stopped, and when the cuvette cooled to room temperature, speckle dynamics films lasting 20–40 s were recorded with 25-Hz frame rate. The optical system presented in **Figure 5** was used. **Figure 14** presents dependence of σ_u on temperature T obtained experimentally. As seen from the given graph, an approximately linear relation between σ_u and T is found.

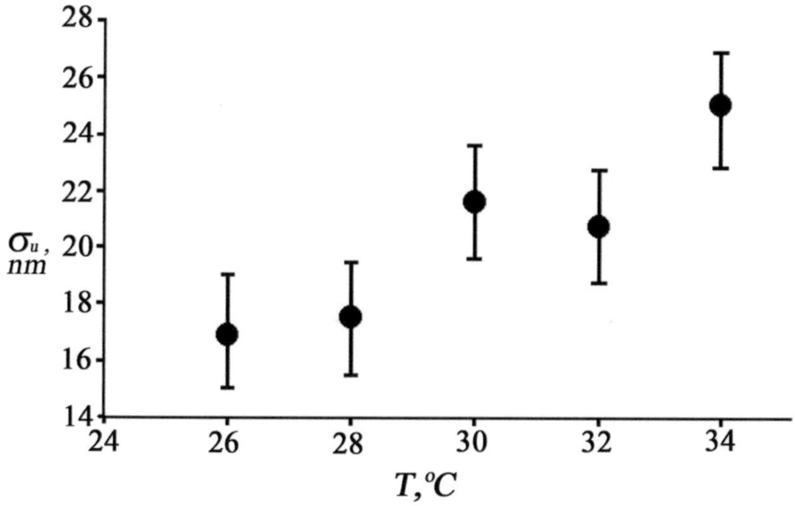

Figure 14. Dependence of σ_u on T for L41 cells.

The purpose of the next experiment was studying the reaction of small fragments inside a cell and a small group of cells to temperature variation. As distinct from the previous experiment, the frames were recorded continuously with heating of the thermostat from the room temperature to 43°C in about 2 h. The averaging time (the frame exposure time) equaled 9 s. We used an air bath of TCЭ-200 type and an optical system with an upright position of the substrate. An L41 cell monolayer was the research objective. For the experiment, we selected a segment that contained at least a small cell-free area (**Figure 15**) in the view-field. The magnification was ×8, and about 10–30 pixels of the TV camera matrix fell onto an image of an individual cell.

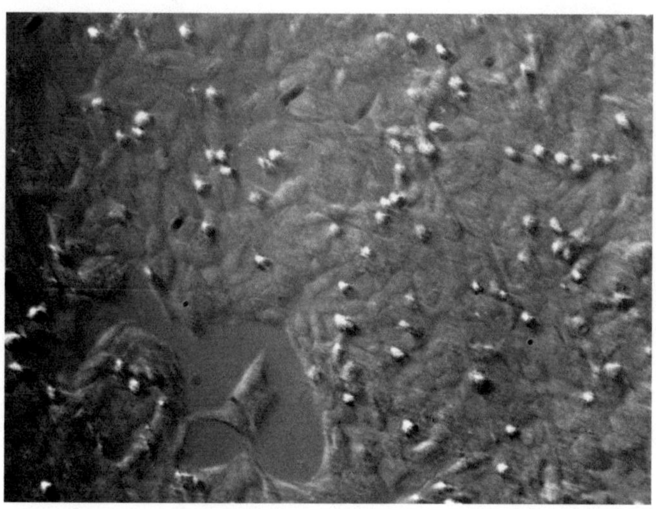

Figure 15. Photographs of cells in white light. A cell-free fragment is visible in the lower part of the frame.

Next in **Figure 16**, there is distribution of value η obtained at the temperature of 30°C by 2 speckle cell images shown in **Figure 15**. The time interval between the frames equaled 18 s. Values η were found in segments of 10 × 10-pixel size.

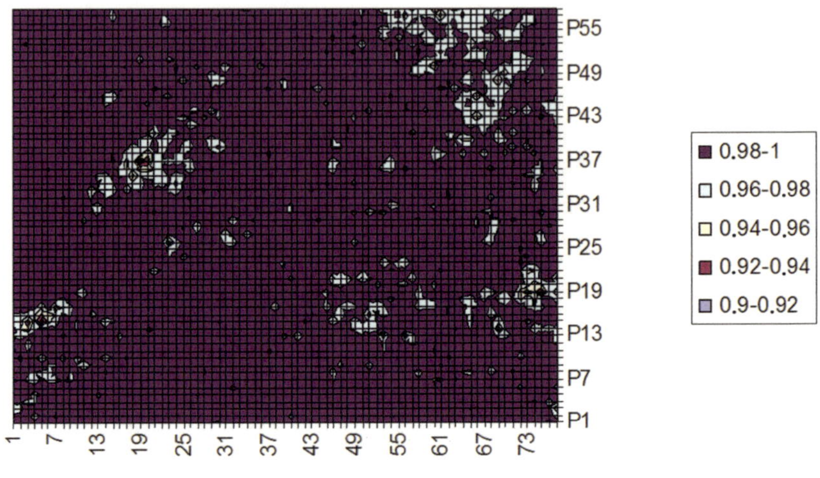

Figure 16. Distribution of value η obtained by two speckle images.

Figures 17, 18 and **19** show typical dependence σ_u on temperature and joint dependences of σ_u and temperature T on time. Value σ_u was obtained using two dependences $\eta(t)$ corresponding to the cells in the nutrient solution and to the nutrient solution. Segments containing 60 cells (**Figure 18**) and small regions inside the cells were averaging regions (**Figure 17**). As is seen from the pictures, considerable fluctuations of σ_u are observed with temperature increase. Fluctuations of value σ_u differ from one cell part to another. If the temperature gets stabilized in 30 min, σ_u also stabilizes (**Figure 19**). The correlation coefficient of masses σ_u and T shown in **Figure 19** equals 0.88.

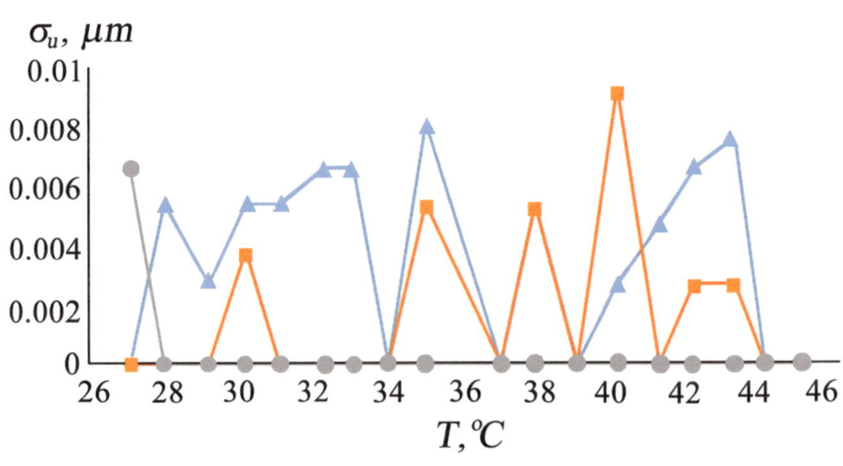

Figure 17. Dependences of σ_u on T for three segments inside one cell. Red colour – cell edge, gray colour – cytoplasm, blue – cell center.

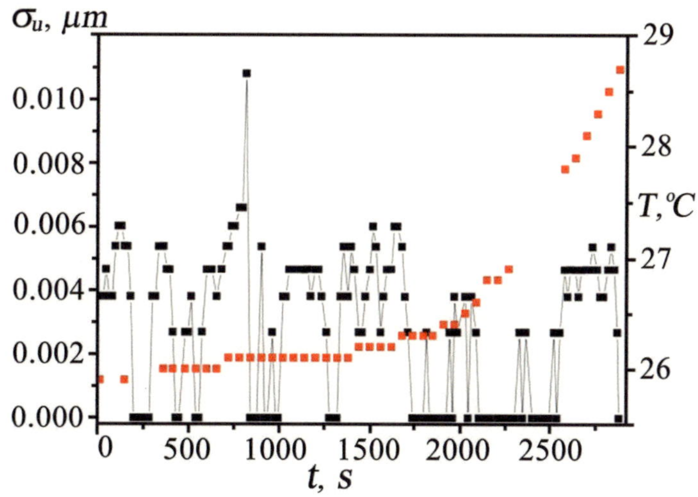

Figure 18. Joint dependences σ_u on time and those of temperature on time.

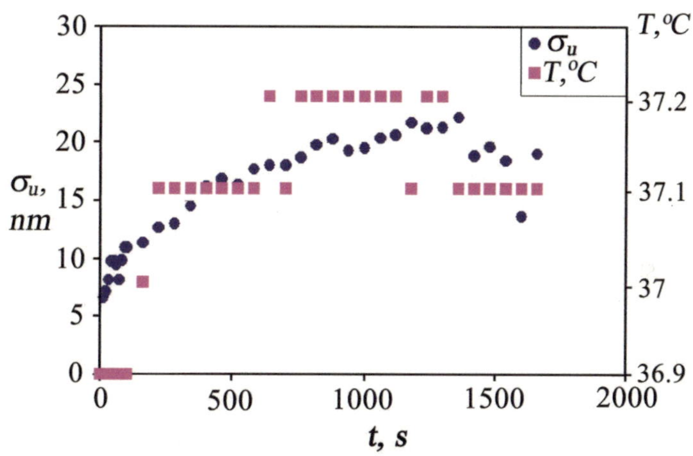

Figure 19. Joint dependences σ_u and temperature on time at small heating rates.

So on the basis of the conducted research, we can conclude that with temperature increase from 25° to 43°C at the rate of about 0.5° a minute, there are fluctuations of value σ_u in space and time. With decreasing variation rate of temperature T by an order variations of σ_u stabilize, and the dependence of σ_u on T becomes linear. To study the dependence of σ_u on T in small segments inside the cells in detail, further research is needed.

6.4. Comparison of theory and experiment: cell activity parameters

Our experiments on cultured and defrosted cells showed qualitative coincidence of theory and experiment. So in random cell motions on the bases of a 1-μm order, the mean difference in the optical paths of two waves can reach and exceed wavelength λ. Then, in compliance with the theory, the value of the cosine and also value η must accept not only positive but also negative values in a random way. Experimental confirmation of this supposition speaks for correctness of the model applied and the calculations made.

We detected good correlation of dependences $\eta(t)$ corresponding to hundreds of defrosted cells obtained in different time intervals. Absence of dependence of a random process on selection of the counting origin means its stationarity. According to the theory, levelling off dependence $\eta(t)$ speaks for stationarity of processes $\tilde{I}(t)$ and $\Delta u(t)$. Homogeneity and stationarity of the intracellular processes in defrosted cells detected using two methods can find practical application. In particular, studying the reaction of hundreds of defrosted cells to the effect of viruses, bacteria, and searching the optimum drugs that prevent their development can be promising. Cultured cells can serve this purpose as well. The advantage of cultured cells is their immobility. That is why above-noted studies can be conducted on a small cell number and on individual cells. The advantage to application of defrosted cells is simplicity of the research target preparation.

At present we suggest that it is value σ_u, or mean square deviation of wave pair optical paths obtained by way of averaging by some region that can be regarded as a cell activity parameter.

This selection is well substantiated from the viewpoint of physics. Indeed, if some processes do not occur in the cells, there is no optical path variation, so $\sigma_u = 0$. If the processes connected with small energy absorption or emission and with transfer of small amounts of substances arise in the cells, small random deviations of the refraction index and the cell shape from the mean value appear. Therefore, values σ_u will be small as well. With intensification of physical-chemical processes in the cells, the values of σ_u will increase.

The selection of parameter σ_u is justified from the viewpoint of biophysics as well. It is known that at room temperature, the metabolic processes in cultured cells are weakly manifested. The culture techniques have shown that with increase of temperature T, the metabolic processes become more distinctly manifested and reach their maximum at 34–37°C. The increase of value σ_u with increase of T (**Figure 14**) in a relatively wide range for hundreds of cells and good linear correlation between σ_u and T in the range of 0.4°C (**Figure 19**) for tens of cells speaks good reason behind applying σ_u as a cell activity parameter.

That said, it is not clear yet what constituents of cell metabolism affect σ_u. Further, research is needed to clarify the matter.

7. Rapid speckle control of cell reaction to herpes simplex virus

The previous sections discussed the theoretical and experimental research that allowed for substantiating the application of speckle dynamics for studies of a thin cell layer activity. This section presents the research results aimed at the study of opportunities for application of the speckle technique for detection of viruses in cells and, in perspective, for therapeutic drug management. The features of speckle dynamics generated by the effect of herpes simplex virus on a monolayer of cultured cells were studied as the first step toward this goal. Initially, it was necessary to understand if the technique can distinguish between the processes in virus-free and virus-infected cells.

7.1. Features of virus development in the cell

Herpes (from Greek herpes—fever) is a viral infection remaining long in the body, predominantly in a latent form Ref. [34]. The most reliable data on herpes were obtained after the virus of this disease was isolated. W. Grüter (1912) first observed development of keratoconjunctivitis in rabbits after administration of the liquid from a human herpetic blister into the scarified sclera. Later, the author made effective successive passages of the viruses on the rabbit eye cornea.

Due to using the negative drug contrasting technique, it was discovered that the external envelope of HSV-1 virus (capsid) has a cubic symmetry type and is an icosahedron. The capsomers composing the capsid are hollow bodies, penta- and hexagonal in the cross-section. Typical viral particles (virions) consist of three main components: a nucleoid located in the central part of the capsid covering the nucleoid and composed from capsomers, and the envelope surrounding these structures. The envelope of the particles has diverse shapes. Sometimes, it repeats a hexagonal capsid projection. Its diameter varies from 170 to 210 nm.

On the basis of on a number of generally agreeing observations, it was established that HSV-1 attachment to the cells is rather slow, and the time is difficult to measure in such cases. The virus attaches to the cells insecurely, so about 50% of the viruses can be detached in various ways.

At the adsorption site, the cell wall forms kind of a "pocket" that transforms into a vacuole, and thus the virus finds itself in a cytoplasm. Then the process of virus disintegration follows resulting in release of the nucleic acid from the proteins of the external envelopes. It is known that the perinuclear space is connected with the extracellular channel system of the endoplasmic reticulum. It is considered that in the period of cell infection, with the centripetal motion of the virus, the channels can serve as natural "passages"for the causative agent, and the structures described by various authors and called vacuoles are nothing but individual sections of the endoplasmic reticulum channels in the cross-section. In 10–12 h postinfection, the characteristic signs of virus formation are undetectable. At the same time, structural viral proteins and nuclear acids are generated to be later used to arrange the nucleic acid and the capsomers into a single structure conventionally called a nucleocapsid or a virinucleon.

The development of herpes is accompanied by the formation of intranuclear inclusions. These formations are considered as assembly sites of viral particles.

HSV-1 causes gradual suppression of macromolecular syntheses in the cell. Suppression of the cell DNA synthesis starts approximately 2 h from the onset of the disease and completes by 7 h. In HSV-1-infected HEp-2 cells, the bulk of the viral DNA is synthesized between the 3rd and the 7th hours from the onset of the disease. It was shown that the DNA of the herpes virus family is reduced in a semiconservative way. To initiate synthesis of the viral DNA, synthesis of early proteins is necessary. After the start of viral DNA synthesis, it continues in the absence of accompanying protein synthesis.

The intranuclear cycle of virus reproduction is completed with its "maturation". Morphologically, it is expressed by covering the capsid with an envelope that further probably has a protective function.

Mass release of virus particles from the cell occurs from 15 to 18 h and is accompanied by the formation of numerous structures of a platy type. At late stages of the disease, various types of viral particles at different stages of formation get beyond the cell limits during its destruction.

7.2. Speckle control procedures

The experiments were conducted in a liquid thermostat of ЗЦ-1125М type with the optical device shown in **Figure 5** inside. Monolayers of cultured virus-free and herpes simplex virus-infected cells were the research targets. Two identical optical cuvettes were prepared for the study. There were two identical substrates in each cuvette, one with the cells and the other cell-free with nutrient solution. The experiments used HSV-1 infectious titre 4.5 lg TCD 50/ml (tissue cytopathic doses) in a dilution of 10^{-3}. Speckle dynamics films of frequency 25 Hz lasting 20–40 s were recorded in the first experiments for 18–20 h at half-hour intervals. Cells of L41 line were the research target. Typical joint dependences $\eta(t)$ for the nutrient solution and virus-free and virus-infected cells are shown in **Figure 20**.

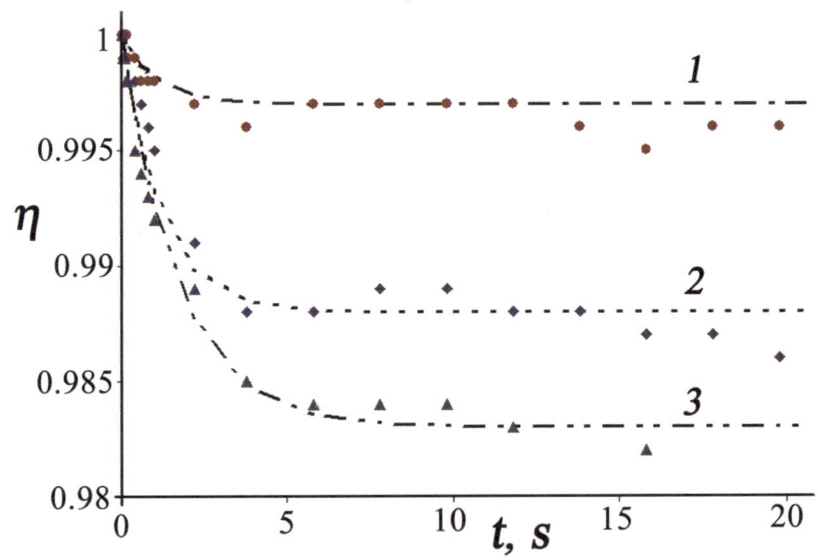

Figure 20. Typical joint dependences $\eta(t)$ for the nutrient solution (1), virus-free cells (2), and virus-infected cells (3).

Analysis of dependences of $\eta(t)$ as well as $\sigma_u(t)$ for virus-free and virus-infected cells shows that they have features agreeing with some phases of virus development in cultured cells, but they are reproduced in about 50% of the cases. The result obtained was probably related to two considerations. First, while dependences $\eta(t)$ were being recorded, the initial frame, starting from the second film, did not correspond to the experiment start. Second, the optical wave path variation could have been caused by several factors with relaxation times of the same

http://dx.doi.org/10.5772/66712

order. Appearance and disappearance of these factors could have occurred in an unpredictable mode. To eliminate the detected flaw, we altered the experimental technique. In the new technique, the program for real-time recording of dependences $\tilde{I}(t)$ and $\eta(t)$ was started 1 h after the administration of the virus. Values $\tilde{I}$ were determined in preselected pixels, and value η was obtained by a segment of 10 × 10-pixel size in the neighborhood of the selected pixel. In 18–20 h, the program was switched off. The frame exposure time was taken as a value exceeding the radiation intensity correlation time found by the graphs in **Figure 20** and equalled 9 s.

7.3. Studies using L41, Vero, and HLE cells

The studies using the upgraded technique were conducted using three cell types: L41, Vero, and HLE-3. Dependences $\tilde{I}(t)$ and $\eta(t)$ of the nutrient solution, virus-free and virus-infected cells were considerably different for all the cell types and were well reproduced in a qualitative sense.

Figure 21 shows typical dependences $\tilde{I}(t)$ for the monolayer of HLE-3 cells.

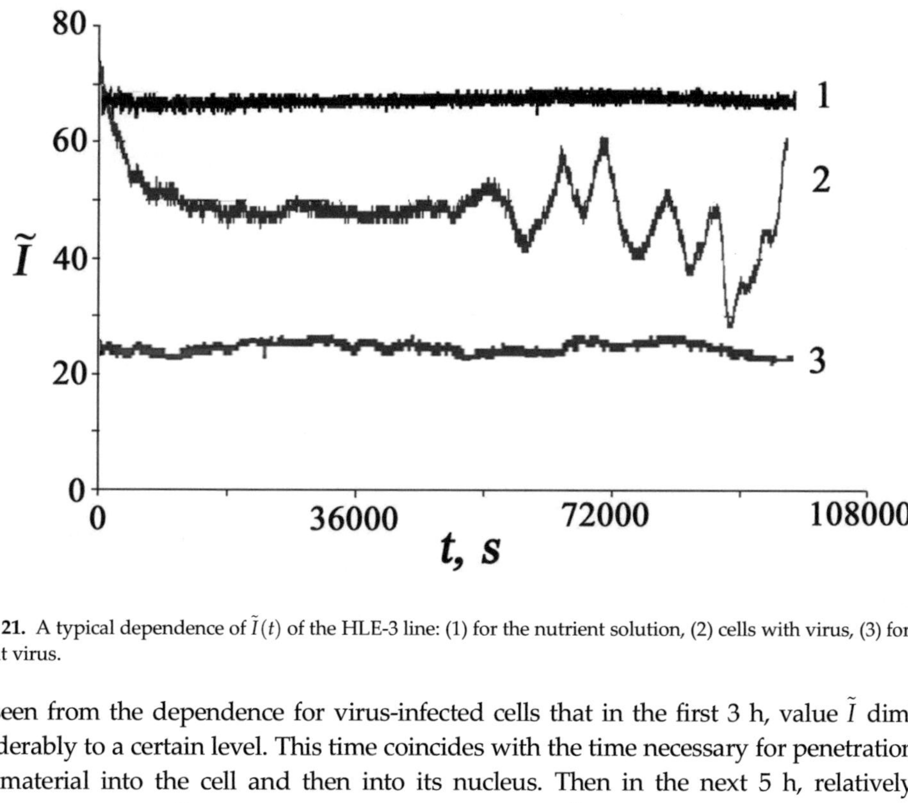

Figure 21. A typical dependence of $\tilde{I}(t)$ of the HLE-3 line: (1) for the nutrient solution, (2) cells with virus, (3) for the cells without virus.

It is seen from the dependence for virus-infected cells that in the first 3 h, value $\tilde{I}$ diminishes considerably to a certain level. This time coincides with the time necessary for penetration of the viral material into the cell and then into its nucleus. Then in the next 5 h, relatively weak fluctuations of value $\tilde{I}$ compared with the variations of $\tilde{I}$ occurring in the virus-free cell take place. This time interval coincides with the time interval during which the cell produces proteins necessary for appearance of new viruses. Then relatively strong quasiperiodic variations of value $\tilde{I}$ reappear. It is known that in this time interval, a capsule with new virions grows.

In **Figures 22–24** typical dependences $\eta(t)$ for L41, Vero, and HLE-3 cells are shown.

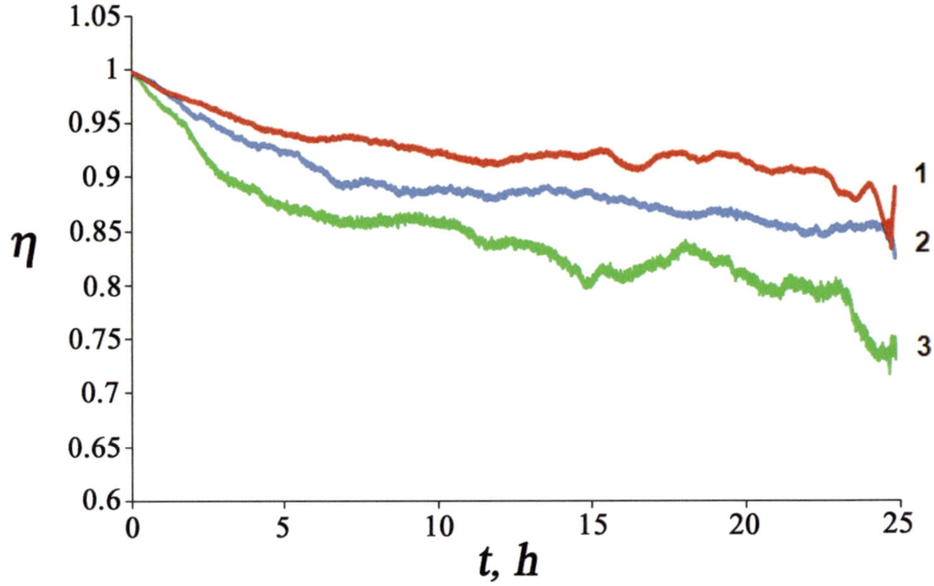

Figure 22. Typical dependences $\eta(t)$ for L41 line: (1) nutrient solution, (2) cells without virus, (3) cells with virus.

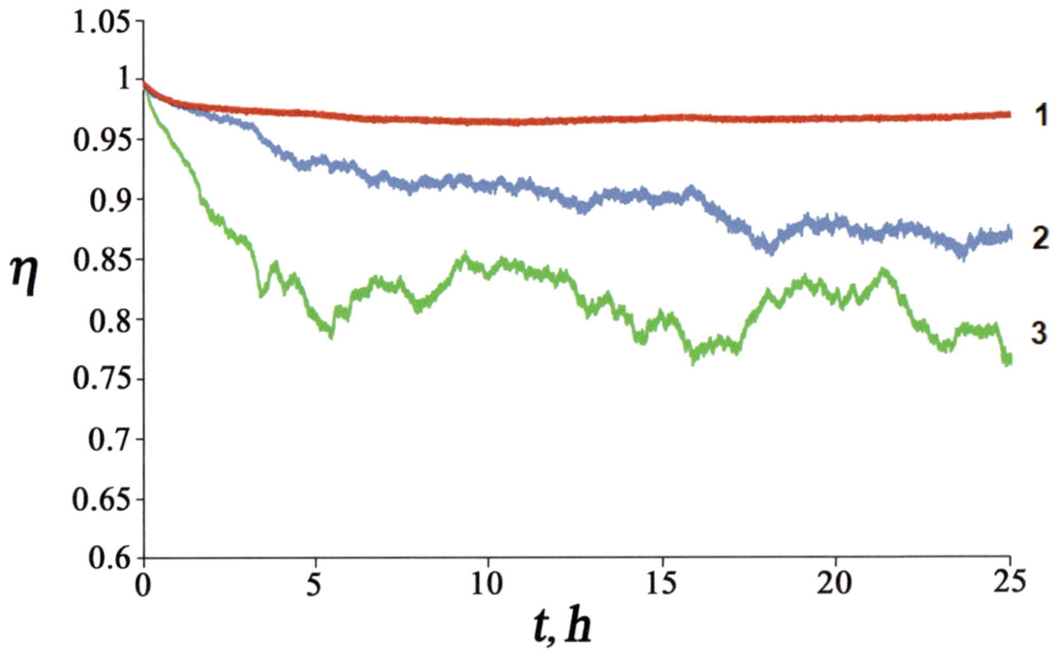

Figure 23. Typical dependences $\eta(t)$ for Vero line: (1) nutrient solution, (2) cells without virus, (3) cells with virus.

http://dx.doi.org/10.5772/66712

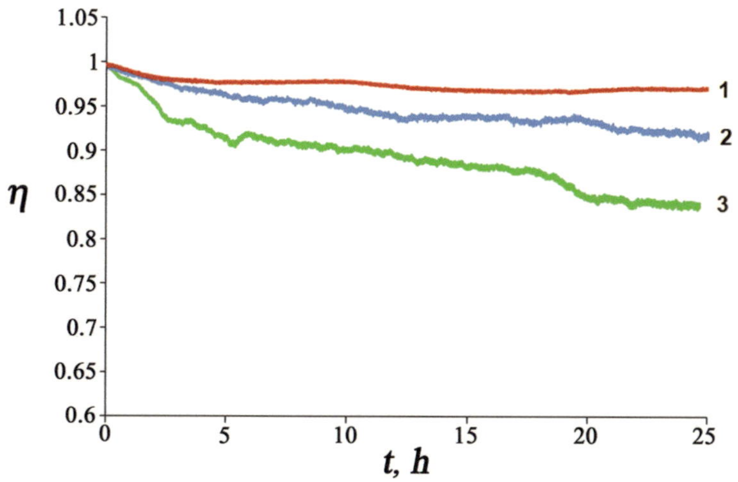

Figure 24. Typical dependences $\eta(t)$ for HLE-3 sells: (1) nutrient solution, (2) cells without virus, (3) cells with virus.

Every picture presents three graphs corresponding to the nutrient solution, virus-free and virus-infected cells. It is seen from the pictures that dependences (1), (2), and (3) differ considerably in the numerical sense. Dependences (2) and (3) are nonstationary processes, but their forms are similar: first value η decreases rapidly, then its decrease slows down. Respectively, σ_u found by Eq. (30) supposing that $\langle x_1 \rangle = \langle x_2 \rangle$ first rapidly grows, and then its growth slows down. We evaluated the multiple correlation coefficient of three masses $\eta(t)$ corresponding to one cell type, and also to different virus-free and virus-infected cell types. For the three masses, the coefficient was in the range from 0.82 to 0.96. This character of curves $\eta(t)$ was probably related with the fact that in the solution, the amount of nutrients gradually decreases and the concentration of harmful cell activity products increases.

Analysis of dependences $\eta(t)$ enabled us to conclude that the presence of the virus can be reliably detected by the curve difference for virus-free cells $\eta_1(t)$ and for virus-infected cells $\eta_2(t)$ 10 min from the experiment start. Twofold excess of the noise amplitude by difference $\eta_1 - \eta_2$ at fixed t was considered the reliability criterion.

7.4. Conclusions

The conducted experiments showed that recording of dependences $\tilde{I}(t)$ and $\eta(t)$ in the image plane of a cell monolayer on a transparent substrate permits reliable recording of difference in the virus-free and virus-infected cell activity. The necessary conditions are the following:

- Rigidity of the optical system providing the value of speckle image fragment correlation coefficient in the absence of the object at 0.99 level for 2 min.
- Maintenance of invariable cuvette temperature selected in the 30–37°C range with ±0.1°C precision.
- Averaging time of radiation intensity equal to 9 s.

8. On some application perspectives of the technique

On the basis of the material discussed above, we can conclude that a dynamic speckle interferometry technique that allows studies of processes in thin transparent biological media has been theoretically substantiated and experimentally tested. Application of the technique is based on the formulas relating the parameters characterizing the target of research and the dynamics of speckle fields. The parameters characterizing the object are mean value $\langle x \rangle$, dispersion σ_u, and relaxation time τ_0 of the optical path difference Δu of sounding wave pairs as well as temporal energetic spectrum of a random process $\Delta u = \Delta u(t)$. The minimum sizes of the averaging regions of the above-named values are transverse and longitudinal resolution of the lens generating the object image. The parameters characterizing speckle dynamics are time-average radiation intensity $\tilde{I}$ at a point in the image plane, relaxation time τ_κ of value $\tilde{I}$, constant level η^* of temporal autocorrelation function $\eta = \eta(t)$ of process $\tilde{I} = \tilde{I}(t)$, and also the temporal energetic spectrum of this process.

Using samples in the form of a cell monolayer cultured on a transparent substrate, or one precipitated on a transparent substrate after defrosting, we demonstrated that value σ_u can be used as a parameter that quantitatively characterizes the activity of live cell. A technique for calibration of a relevant device and a technique for determination of value σ_u were developed. So applications related to assessment of cell activity can be the nearest perspective for the application of the technique. In particular, the technique for determining σ_u can be applied to study of the effect of toxic substances on live cells and determine their science-based maximum allowable doses. Such a technique can be applied to search the optimum drugs preventing penetration of viruses and bacteria into cells.

Determination of value σ_u is based on creating special conditions excluding the influence of parameters $\langle x \rangle$ and τ_0 on speckle dynamics. In the general case, the necessity may arise for simultaneous determination of all the three parameters characterizing the processes occurring in different parts of the same cell. Further studies are needed to solve this problem.

9. Conclusion

On the basis of the model accounting interference of multiple waves with random phases a relation between phase dynamics of the waves sounding a thin transparent object and the speckle dynamics in the object image plane was detected theoretically. General-case formulas were obtained to determine the dependence of time-average intensity $\tilde{I}$ and temporal autocorrelation function $\eta = \eta(t)$ of this intensity at some point in the image plane with mean value $\langle x \rangle$, mean square deviation σ_u, and correlation time τ_0 of optical path difference Δu of wave pairs in the neighborhood of the conjugate point of the object plane. The diameter of this neighborhood equals the linear resolution of the lens that generates the object image. Relation between the temporal spectral function of a random process $\Delta u(t)$ and a similar function of the process $\tilde{I}(t)$ was substantiated.

http://dx.doi.org/10.5772/66712

An optical device relevant to the model used in the theory was developed.

Very good coincidence between the theory and the experiment has been demonstrated by batched random variation of path difference Δu. The procedure of calibrating the optical device for determination of σ_u was developed; its errors and the sensitivity limit of the technique were assessed.

Biological objects in the form of a live cell monolayer on a transparent substrate in a thin cuvette with nutrient solution were used to substantiate application of value σ_u as a cell activity parameter.

It was shown that the technique allows detection of herpes virus in cells as early as 10 min from the experiment start.

Rapid assessment of cell reaction to toxic substances therapeutic management of antibacterial and antiviral drugs can be the nearest perspective for application of the technique. Development of a technique for simultaneous determination of values $\langle x \rangle$, σ_u, and τ_0 in different parts of an individual cell can become a line of further research.

Acknowledgements

The authors thank the management of Yekaterinburg Research Institute of Viral Infections for aid in writing this manuscript. The authors are also grateful to Yu. A. Mikhailova for technical assistance in its preparation.

Author details

Alexander P. Vladimirov* and Alexey A. Bakharev

*Address all correspondence to: vap52@bk.ru

Yekaterinburg Research Institute of Viral Infections, Yekaterinburg, Russia

References

[1] Goldfischer L I. Autocorrelation function and power spectral density of laser-produced speckle patterns. Journal of the Optical Society of America. 1965;**55**(3):247–253.

[2] Enloe L H. Noise like structures in the image of diffusely reflecting objects in coherent illumination. The Bell System Technical Journal. 1967;(7):1479–1489.

[3] Goodman J W. Statistical properties of laser speckle patterns. Laser Speckle and Related Phenomena. 1975;(9):9–75.

[4] Goodman J W. Statistical Optics. New-York: Wiley Classics Library Edition published. 2000; 1985. 567p.

[5] Anisimov I V, Kozel S M, Lokshin G R. Space-time statistical properties of coherent radiation scattered by a moving diffuse reflector. Optics and spectroscopy. 1969;27(3):483–491.

[6] Yoshimura T. Statistical properties of dynamic speckles. Journal of the Optical Society of America. 1986;**3**(7):1032–1054.

[7] Yamaguchi I. Speckle displacement and decorrelation in the diffraction and image fields for small object deformation. Optica Acta. 1981;**28**(10):1359–1376. DOI: 10.1080/713820454

[8] Aleksandrov E B, Bonch-Bruevich A M. Investigation of surface strains by the hologram technique. Soviet Physics-Technical Papers. 1967;12:258–265.

[9] Leendertz J A. Interferometric displacement measurement on scattering surfaces utilizing speckle effect. Journal of Physics E: Scientific Instruments. 1970;**3**(3):214–218.

[10] Vladimirov A P, Mikushin V I. Interferometric determination of vector components of relative displacements: theory and experiment. In: Tuchin V V, Ryabukho V P, Zimnyakov D A, editors. Saratov Fall Meeting '98: Light Scattering Technologies For Mechanics, Biomedicine, and Material Science; 06-09 October 1998; Saratov Bellingham (USA): SPIE; 1999. pp. 38–43. DOI: 10.1117/12.341416

[11] Rabal H J, Braga R A, editors. Dynamic Laser Speckle and Applications. New York: CRC Press; 2008. 282p.

[12] Briers J D, Fercher A F. Retinal blood-flow visualization by means of laser speckle photography. Investigative Ophthalmology and Visual Science. 1982;**22**(2):255–259.

[13] Briers J D. Laser speckle contrast imaging for measuring blood flow. Optica Applicata. 2007;**37**(1-2):139–152.

[14] Ansari M Z, Humeau-Heurtier A, Offenhauser N, Drier J P, Nirala A K. Visualization of perfusion changes with laser speckle contrast imaging using the method of motion history image. Microvascular Research. 2016;**107**:106–109. DOI: 10.1016/j.mvr.2016.06.003

[15] Maxwell J C. Treatise on electricity and magnetism. Moscow: Nauka; 1989. 416p.

[16] Landau L D, Lifshitz E M. The Classical Theory of Fields. 1st ed. USA: Addison-Wesley; 1951. 425p.

[17] Landsberg G S. Optica. 6th ed. Moscow: Fizmatlit; 2003. 848p.

[18] Ginzburg V M, Stepanov V M, editors. Holography. The methods and apparatus. Moscow: Soviet radio; 1974. 376p.

[19] Born M, Wolf E, editors. Electromagnetic Theory of Propagation, Interference and Diffraction of Light. 7th ed. Cambrige: Cambrige University Press; 1999. 952p.

[20] Ryabukho V P. Diffraction of interference fields on random phase objects. In: Tuchin V V, editor. Coherent-Domain Optical Methods: Biomedical Diagnostics, Environmental and

http://dx.doi.org/10.5772/66712

Material Science. 1st ed. Boston, Dordrecht, London: Kluwer Academic Publishers; 2004. pp. 235–318.

[21] Bakut P A, Mandrosov V I, Matveyev I N, Ustinov N D. Theory of coherent images. Moscow: Radio and communication; 1987. 264p.

[22] Vladimirov A P. Three-dimensional diffuser and three-dimensional speckles. Technical physics. 1998;43(12):1454–1458.

[23] Vladimirov A P. Dynamic speckle interferometry of bodies under deformation. Yekaterinburg: UrO RAN Publ.; 2004. 241p.

[24] Hauf V, Grigul U. Optical methods in heat transfer. Moscow: Mir; 1973. 240p.

[25] Vasilyev L A. The shadow methods. Moscow: Nauka; 1968. 400p.

[26] Fomin N A. Speckle Photography for Fluid Mechanics Measurements. 1st ed. Berlin: Springer-Verlag; 1998. 219p.

[27] Jakeman E. Photon correlation. In: Cummins H Z, Pike E R, editors. Photon Correlation and Light Beating. 1st ed. New-York: Springer Science+Business Media; 1974. pp. 75–149.

[28] Vladimirov A P. Speckle dynamics in the image plane of a plastically deformable object [dissertation]. Sverdlovsk: Institute of metal physics; 1989. 106p.

[29] Vladimirov A P, Druzhinin A V, Malygin A S, Mikitas K N. Theory and calibration of speckle dynamics of phase object. Proceedings of SPIE. 2012;**8337**(83370C): 1–15.

[30] Vladimirov A P. Dynamic speckle interferometry of microscopic processes in thin biological objects. Radiophysics and Quantum Electronics. 2015;**57**(8):564–576.

[31] The State System of Measurements. Indirect Measurements. Determination of Measurement Results and Estimation of their Errors MI 2083–90. Moscow. 1991.

[32] Bakharev A A, Burygina N A, Vissarionov VA, Glinskih N P, Karpova E I, Malyushenko O I. Patent RU 2148644: Method of control of the cytotoxicity of the gel using a strain of transplantable human leukocytes L41 CD/84. 1998.

[33] Malygin A S, Bebenina N V, Vladimirov A P, Mikitas K N, Baharev A A. A speckle - interferometric device for studying the cell biological activity. Instruments and experimental techniques. 2012;55(3):415–418.

[34] Barinskij I F, Shubladze A K, Kasparov A A, Gribenyuk S V. Herpes (etiology, diagnosis, treatment). Moscow: Medicina; 1986. 351p.

Chapter 7

Applications of Fiber-Optic Interferometry Technology in Sensor Fields

Lutang Wang and Nian Fang

Additional information is available at the end of the chapter

http://dx.doi.org/10.5772/66276

Abstract

Optical interferometry as a precision metrology has been widely employed in many aspects for accurate measurements of various physical quantities. As an important branch of measurement technology, now the fiber-optic interferometry technology, based on fiber-optic and laser technologies, has been developed and widely applied in sensor fields for detections of various unknown or uncontrolled physical parameters. In this chapter, basic concepts of fiber-optic interferometry are presented and clarified. Also, three novel fiber-optic sensors, based on the optical interferometry technology and mainly used in the power industry for equipment failure monitoring, are demonstrated.

Keywords: optical interferometry, laser, fiber-optic sensor, interference fringe, fiber Bragg grating-based interferometer

1. Introduction

Optical interferometry, as a precision metrology widely used in the optical system for extremely accurate measurements of a variety of physical quantities in laboratories as well as in industry fields [1], has been developed for over a hundred years. The optical interferometry technology is based on the interference of light beams, two beams or multiple beams, launched from same light source, laser source, or other monochromatic source, propagating through space or dielectric mediums such as glass waveguides with different optical paths, and arriving simultaneously at a point in space or on the surface of an object. As a result, the light intensity will vary periodically with the optical path difference, that is the optical phase difference, between the beams [2]. This is the well-known phenomenon of light interference. The periodic change in the intensity of interference light is usually referred to as the interference pattern or the interference fringe [2]. A very small change in the optical path difference in the wavelength scale of light source can induce an obvious and measurable change in the intensity of the interference light. Therefore, by measuring the changes of interference light intensity, one can get the information about the changes of optical paths in an optical measurement system. Based on this mechanism, the optical interferometer is built as an instrument and is widely used for accurate measurements of many physical quantities, such as the distance, displacement, and velocity as well as for tests of optical systems.

With quick developments of laser and fiber optic technologies, the optical interferometry technology also had great progress and evolved from the classical bulk optics to the fiber optics [3]. Based on fiber-optic technologies, applications of the optical interferometers have been expanded to such areas as underwater acoustic detections, voltage and current measurements inside electric power systems [3–7], and biomedical pressure monitoring in living bodies [8]. As one of the most important applications, the optical interferometer is used as the optical interferometer sensor for detections of unknown and uncontrolled physical parameters [3]. The fiber-optic-based interferometer sensor uses optical fibers as light carriers and obtains the detection information from fiber-connected transducers or directly from fibers themselves [3, 8]. Compared with the classical, bulk-optic interferometer sensors, the fiber-optic interferometer sensors can achieve remote sensing and have a number of attractive features, such as excellent sensitivity and large dynamic range, small size with rugged packages, potential for low cost, and high reliability. In general, optical fibers and fiber-optic transducers/sensors are made with totally dielectric materials that are chemically inert and completely immune to electromagnetic interference (EMI), and can also withstand relatively high temperatures [4, 5, 8]. These unique properties make them very favorable to be used in harsh environments [9, 10], such as inside an electric power system in which the strong EMIs often make conventional electronic sensors work unstable and result in the increase of the fault rate.

The main aim of this chapter is to give a brief introduction of optical interferometers based on fiber-optic technology and their practical applications in the electric power industry for monitoring of the power system's running states, as well as for measuring of some crucial physical parameters. In Section 2, we will roughly classify the most commonly used fiber interferometers, according to their architectures, operation principles, and application areas. As a key fiber-optic component, the fiber Bragg grating plays a very important role in the constitution of an in-line fiber-optic interferometer [9, 11, 12], so the principles of the fiber Bragg grating as well as twin-grating-based fiber interferometer also will be introduced briefly. In Section 3, three prototypes of fiber interferometer sensors, developed in our laboratory recently, and intentionally used in the electric power industry for partial discharge (PD) sensing and the measurements of power-frequency electric field strength, are presented. Some preliminary experimental results for demonstrating the performances of these sensors also are presented. In the final section, a conclusion for summarizing our work is given to close this chapter.

http://dx.doi.org/10.5772/66276

2. General principles

As an analog of bulk-optic interferometer, the fiber-optic interferometer follows many elementary physical principles and concepts similarly adopted in the bulk-optic interferometer. Although the optical fibers provide many unique features which make the performances of the interferometer system be improved greatly, the optical fiber properties such as the birefringence, dispersion, and temperature dependence as well as nonlinear effects still influence the ultimate performances of a fiber interferometer system [4, 13]. Therefore, when we design a novel fiber interferometer system, discuss its performances, and explore new applications, a variety of fiber properties have to be taken into account. The fiber interferometers now are most commonly employed for industrial measurements and sensor applications [5, 6, 12, 14]. According to their architectures, these fiber interferometers may be simply classified into four dominant types, as schematically illustrated in **Figure 1**. They are the fiber Mach-Zehnder, Michelson, Sagnac, and Fabry-Perot interferometers.

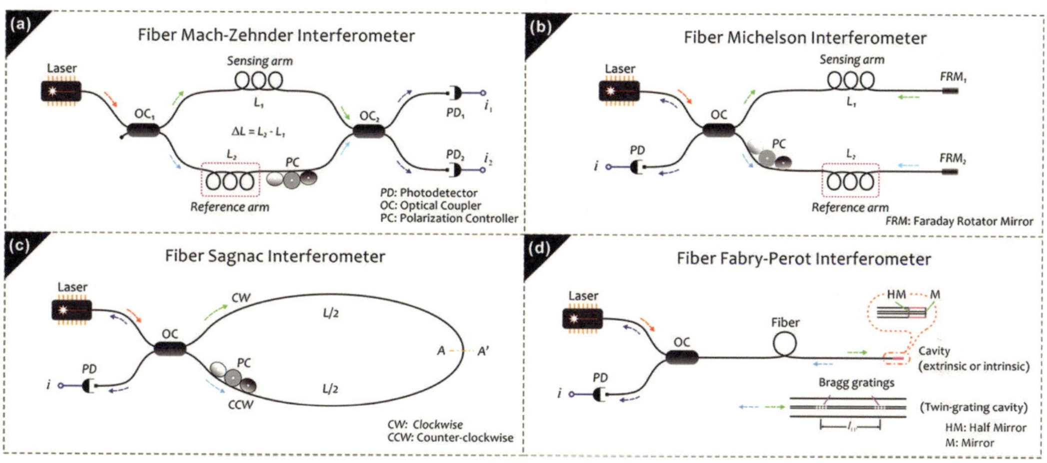

Figure 1. Configurations of main fiber interferometers. (a)–(d) are fiber Mach-Zehnder, Michelson, Sagnac, and Fabry-Perot interferometers, respectively.

2.1. Fiber Mach-Zehnder interferometer

A common two-beam fiber-optic interferometer is the fiber Mach-Zehnder interferometer [2, 4]. As shown in **Figure 1(a)**, the optical paths consist of two fiber arms, one to be assigned as *signal* arm with a length l_s and the other one as *reference* arm with a length l_r. The light beam from a light source is amplitude-divided by a fiber coupler OC_1 into two beams propagating respectively in *signal* and *reference* arms. In general, the *signal* arm is exposed to the external environment to sense disturbances, while the *reference* arm is kept in a relatively constant environment. The phase of the *signal* beam is modified by the external disturbances as a physical measurand when the beam passes through the *signal* arm. It produces a phase difference between two beams, *signal* beam and *reference* beam, which recombine at a second fiber coupler OC_2. Two groups of beams output from the two ports of OC_2 are then detected by two photodetectors, PD_1 and PD_2, and converted into a pair of fringe signals in antiphase. If total optical losses in the interferometer are negligible, the two fringe signal intensities, P_1 and P_2, can be expressed as [4]:

$$P_1 \propto I_1 + I_2 + \xi \cdot \gamma \cdot 2\sqrt{I_1 I_2}\cos(\Delta\varphi) \tag{1}$$

$$P_2 \propto I_1 + I_2 - \xi \cdot \gamma \cdot 2\sqrt{I_1 I_2}\cos(\Delta\varphi) \tag{2}$$

where $\Delta\varphi = \varphi_s - \varphi_r$ is the phase difference, $\varphi_s = 2\pi n l_s/\lambda$ and $\varphi_r = 2\pi n l_r/\lambda$ are the phases of the *signal* beam and *reference* beam, respectively, and n is the refractive index of the fiber core; I_1 and I_2 are the intensities of *signal* beam and *reference* beam arose at output ports of OC_2, respectively; ξ is a ratio, $0 \le \xi \le 1$, reflecting the matching degree of two beams in their polarization states, which is adjustable by an in-line polarization controller PC, as shown in **Figure 1(a)**; γ is defined as the fringe visibility [2], $0 \le \gamma \le 1$, in relation to I_1 and I_2

$$\gamma = 2\sqrt{I_1 I_2} / (I_1 + I_2) \tag{3}$$

Also, γ is a function of the product of the spectral line width $\Delta\nu$ of the light source and the optical path difference $\Delta l = n(l_s - l_r)$, given by

$$\gamma(\Delta\nu \cdot \Delta l) \le 1 \tag{4}$$

For $\Delta\nu \cdot \Delta l = 0$, $\gamma \equiv 1$. In general, for getting a strong contrast on the interference fringe, $\gamma(\Delta\nu \cdot \Delta l) \sim 1$ is required. For a known $\Delta\nu$, when $\gamma = e^{-1}$, Δl corresponds with the coherence length l_c of the light source [4], having $l_c = \Delta l$.

2.2. Fiber Michelson interferometer

The fiber Michelson interferometer also is a two-beam optical interferometer [4], so its output can be similarly described by Eq. (1) or (2). In the system structure, the fiber Michelson interferometer is very similar to the fiber Mach-Zehnder interferometer. As shown in **Figure 1(b)**, the *signal* and *reference* arms are terminated by two mirrors or Faraday rotator mirrors, so that the *signal* beam and *reference* beam, both are reflected by corresponding mirrors back to the coupler OC where they are recombined to generate the interference signal. It is noted that the phase difference $\Delta\varphi = 2(\varphi_s - \varphi_r)$ in the fiber Michelson interferometer is double of that in the fiber Mach-Zehnder interferometer, which, therefore, will effectively double the sensitivity of interferometer.

2.3. Fiber Sagnac interferometer

The fiber Sagnac interferometer is a special and important two-beam, common-path interferometer system [2, 4], in which, as shown in **Figure 1(c)**, two beams from the coupler OC traverse the same fiber loop in opposite directions, usually referred to as the clockwise (CW) and the counter-clockwise (CCW) directions, respectively. The interference fringe is generated when CW and CCW beams, after traveling the whole fiber loop, recombine at OC. Since the optical paths traversed by two beams are very nearly equal, at first sight, the optical phase difference between two beams would always be zero, which represents a static state. However, when the external measurand, such as the acoustic wave, disturbs the fiber close to one end of the loop, an instant phase shift arises. This property makes the Sagnac interferometer be particularly well suited for sensing rapidly varying environmental perturbations. In the early years, the fiber Sagnac interferometer had been developed principally for the purpose of measuring the angle velocity [15], as a fiber gyroscope [4, 16]. Now it has become an important sensor used in the electric power industry for sensing of various physical parameters, such as currents, voltages, electric and magnetic fields, and vibrations as well as acoustic emissions from the partial discharges occurring inside the high-voltage power equipment [17–19].

Similarly, as a two-beam interferometer, the two outputs of fiber Sagnac interferometer also can be expressed by Eqs. (1) and (2). It should be noted that since the optical path difference in the Sagnac interferometer is almost zero, any type of the light source with low or high coherence can be employed.

2.4. Fiber Fabry-Perot interferometer

As shown in **Figure 1(d)**, the fiber Fabry-Perot interferometer actually is a multiple-beam interferometer system and consists of an interferometric cavity formed by two parallel reflectors (half mirrors) with reflectance R on either side of an optically transparent medium with a length l_{FP} [2, 4]. The available cavity structures, usually determined by the applications or design requirements, may be the extrinsic or intrinsic type [12, 20–24]. As illustrated in **Figure 1(d)**, in the extrinsic-type structure, the two reflectors are separated by an air gap or by some solid or liquid material other than the fiber, so that the measurand affects only the optical length of cavity other than the fiber. Here the fiber is used only as a light channel to transport the light beams to and from the interferometer. In contrast, in the intrinsic version, the cavity usually is constructed within the fiber as an integral part of a continuous fiber with two internal reflectors formed by flat-cut fiber ends or by fiber Bragg gratings [12, 22]. Compared with the fiber Mach-Zehnder and Michelson interferometers, the fiber Fabry-Perot interferometer is quite compact in size and therefore available as a point sensor for some applications, such as the smart structure sensing applications [23, 24].

The interference fringe, as a result of multiple reflections of the beam in the cavity, is much narrower than the normal two-beam fringes and becomes much sharper with the increase of the reflectance R. The finesse F, a parameter frequently used to characterize the overall performance of a Fabry-Perot interferometer, in relation to the sharpness of fringe, is defined as the ratio of the separation of adjacent fringes to their width FWHM (full width at half maximum), given by [4]:

$$F = \pi\sqrt{R} / (1 - R) \tag{5}$$

It is clear that the finesse F is completely decided by the reflectance R. Thus, for a lossless cavity, for example, we have $F = 29.8$ for $R = 0.9$ and $F = 312.6$ for $R = 0.99$. Most fiber Fabry-Perot interferometers for sensor applications, however, have low finesses which allow interferometers to operate in a linear region over a larger detection range to the measurand. If $R \ll 1$, the reflectance $R_{\mathrm{F-P}}$ and transmittance $T_{\mathrm{F-P}}$ of fiber Fabry-Perot interferometer can expressed, respectively, as [4]:

$$R_{\mathrm{F-P}} \cong 2R(1 + \cos\phi) \tag{6}$$

$$T_{\mathrm{F-P}} \cong 1 - 2R(1 + \cos\phi) \tag{7}$$

where ϕ is the round-trip propagation phase difference in the interferometer, given by

$$\phi = 4\pi n l_{\mathrm{FP}} / \lambda \tag{8}$$

As sensor applications, the fiber Fabry-Perot interferometer is extremely sensitive to external perturbations that affect the cavity length l_{FP} as well as the refractive index n.

2.5. Fiber Bragg grating-based Fabry-Perot interferometer

As the wavelength selective mirrors, the fiber Bragg gratings also can be used as the intrinsic reflectors in the fiber to construct various types of fiber interferometers, such as the Michelson or Fabry-Perot interferometers.

In the physical structure, as shown in **Figure 2**, a uniform fiber Bragg grating contains a varied refractive index region with a spatial period, distributing along the fiber core within a selected length. A broadband beam propagating in the fiber will interact with each grating plane where only a part of the beam with a specific wavelength matched with the Bragg wavelength of grating will be reflected and propagates in the opposite direction, and the rest of the beam passes through this grating without obvious optical losses [9, 25]. According the mode coupling theory, the Bragg wavelength λ_B may be expressed as [11]:

$$\lambda_{\mathrm{B}} = 2n_{\mathrm{eff}}\Lambda \tag{9}$$

where n_{eff} denotes the effective refractive index of the fiber core; Λ is the grating period. From Eq. (9), it is evident that λ_{B} only depends on n_{eff} and Λ. So λ_{B} is very sensitive to ambient temperature and strain imposed on the fiber, which modify n_{eff} and Λ.

http://dx.doi.org/10.5772/66276

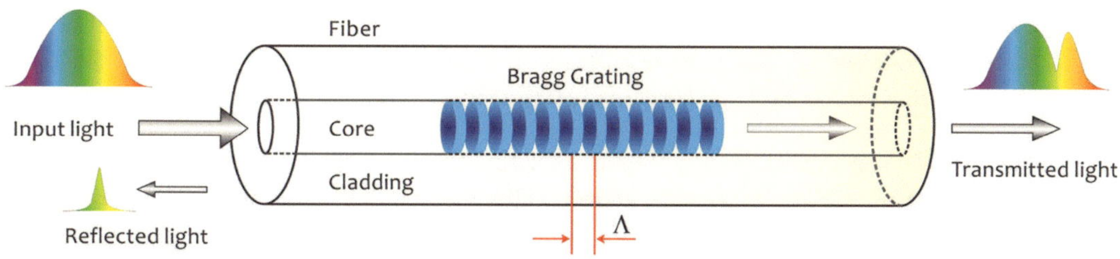

Figure 2. Structure and principle of a fiber Bragg grating.

When two fiber Bragg gratings with identical Bragg wavelengths are imprinted in the same fiber at different positions, as shown in **Figure 1(d)**, a twin-grating-based fiber Fabry-Perot interferometer with an intrinsic cavity is constructed. If a laser beam with wavelength λ is launched into the fiber, when $\lambda = \lambda_B$, two beams partially reflected from both gratings will have a round-trip phase difference $\phi(\lambda)$ as defined in Eq. (8), where l_{FP} is an internal optical path length between two gratings. By assigning the reflectivity and transmission coefficients of each grating as $r_i(\lambda)$ and $t_i(\lambda)$, $i = 1, 2$, respectively, the resultant reflectivity coefficient $r_{FP}(\lambda)$ of twin-grating fiber Fabry-Perot interferometer can be proximately expressed as [26]:

$$r_{FP}(\lambda) \approx r_1(\lambda) + t_1^2(\lambda) r_2(\lambda) e^{-i\phi(\lambda)} \tag{10}$$

When two gratings are of low reflectivity, we have $r(\lambda) = r_1(\lambda) = r_2(\lambda)$, and $t_1(\lambda) = t_2(\lambda) = 1$. So Eq. (10) can be rewritten as

$$r_{FP}(\lambda) \approx r(\lambda)\left[1 + e^{-i\phi(\lambda)}\right] = 2r(\lambda)e^{-i\phi(\lambda)/2}\cos\frac{\phi(\lambda)}{2} \tag{11}$$

The reflectance or power reflectivity $R_{FP}(\lambda) = |r_{FP}(\lambda)|^2$ of twin-grating fiber Fabry-Perot interferometer is obtained, given by

$$R_{FP}(\lambda) = 2R_B(\lambda)R_m(\lambda) \tag{12}$$

where $R_B(\lambda) = |r(\lambda)|^2$ is the reflectance of single grating, which determines the envelop of the reflection spectrum; $R_m(\lambda) = 1 + \cos(4\pi n l_{FP}/\lambda)$ is the transfer function of the interferometer, which forms the periodically alternating spectral peaks in the reflection spectrum of the interferometer. For an input laser beam with optical power P_0 and wavelength λ_0, the reflected light power from the interferometer will be $P_R = P_0 R_{FP}(\lambda_0)$.

When the interferometer is placed in a varying environment, the reflection spectrum will shift as a whole, as a function of measurand without noticeable changes in its envelop [26]. For a Bragg wavelength shift $\Delta\lambda_B$, which is induced by the measurand, such as temperature or strain, a new reflectance R'_{FP} is obtained as $R'_{FP}(\lambda) = R_{FP}(\lambda - \Delta\lambda_B)$. When the measurand fluctuates, the reflected light power $P_{R'}$ in turn, is altered. This feature has been well utilized in sensor applications for monitoring of the changes in the measurand through the detection of the intensity changes of the interference signal [12].

Figure 3 displays a reflection spectrum of a twin-grating fiber Fabry-Perot interferometer with a 10-mm long intrinsic cavity, measured at 24°C. The Bragg wavelength of each grating is 1542.392 nm and the power reflectivity is about 15%.

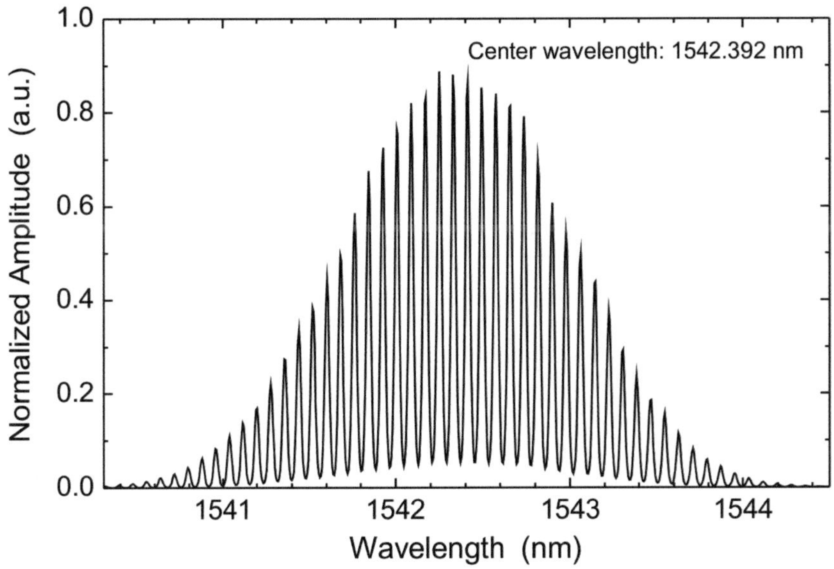

Figure 3. Reflection spectrum of twin-grating fiber Fabry-Perot interferometer.

3. Fiber interferometers for sensor applications

In this section, we will present three selected prototypes of fiber interferometer sensors developed in our laboratory recently, which are intentionally employed in the electric power industry for partial discharge sensing as well as for power-frequency electric field measurements. Some preliminary experimental results for demonstrating the performances of these sensors and experimental systems also are presented. The prototypes to be presented below will include: a fiber Sagnac interferometer-based acoustic sensor, a fiber Fabry-Perot interferometer-based acoustic sensor, and a twin-grating fiber Fabry-Perot interferometer-based power-frequency electric field sensor.

http://dx.doi.org/10.5772/66276

3.1. A fiber Sagnac interferometer-based acoustic sensor

In the previous work, we had proposed a novel fiber Sagnac interferometer-based acoustic sensor for sensing of high-frequency and weak ultrasonic waves produced by the partial discharges [18].

3.1.1. System configuration and operation principles

The scheme of the proposed sensor is presented in **Figure 4(a)**, which is based on a fiber Sagnac interferometer consisting of a 3-dB, 2 × 2 fiber coupler (OC) and two no-isolator, CW DFB lasers (LD_1 and LD_2) with two nearly identical lasing wavelengths at 1528.60 nm and 1529.38 nm, respectively. Two AC current signals i_1 and i_2 are obtained, respectively, from the corresponding photodiodes packaged in laser modules. The final detection signal i_3 is constructed through a cross-correlation operation with i_1 and i_2. A fiber polarization controller PC in the fiber loop is utilized to adjust the total birefringence of the fiber. The fiber loop made of a 1-km long single-mode fiber forms a sensor coil with a diameter of 60 mm. The outer layers of this coil, close to one end of the OC, are utilized as the sensing region, so that the relatively high sensitivity can be achieved.

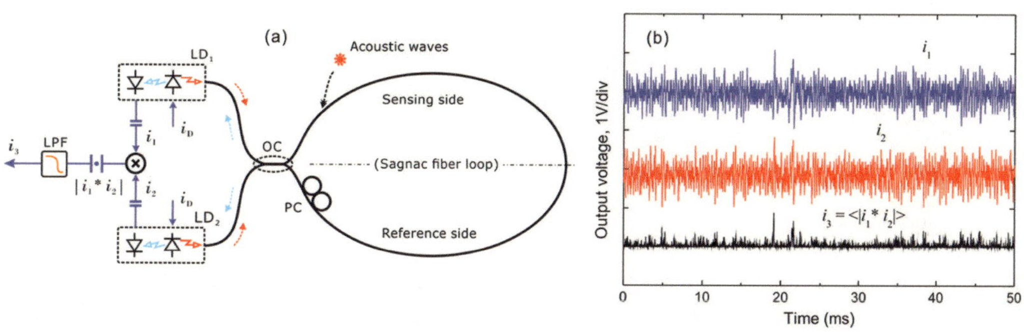

Figure 4. (a) Scheme of proposed fiber acoustic sensor and (b) a group of output waveforms obtained in a static state.

The feature of this sensor is to employ two laser diodes without optical isolators as two individual light sources to illuminate the fiber Sagnac interferometer from two sides, and at the same time, also as two in-line optical amplifiers to enhance the intensities of signal beams from the Sagnac interferometer. So the sensitivity of this sensor, compared with other sensors with the normal configuration, has a significant enhancement.

In principle, this sensor is a balanced fiber Sagnac interferometer, which, in the structural analysis, can be regarded as two individual Sagnac interferometers with a mirror-image relation. According to the analysis methods proposed in Ref. [19], for a balanced fiber Sagnac interferometer, two fringe signals i_1 and i_2 including laser intensity noises can be expressed as

$$i_1 \propto -I_0 \sin\phi_b \sin\Delta\varphi + n_{LD_1} \tag{13}$$

$$i_2 \propto +I_0 \sin\phi_b \sin\Delta\varphi + n_{LD_2} \tag{14}$$

where I_0 is the average intensity of laser beams; ϕ_b is the nonreciprocal phase difference between two beams propagating in the fiber coil, which is a constant and arises from the fiber birefringence; $\Delta\varphi$ is another phase difference between two beams, caused by external disturbances; n_{LD_1} and n_{LD_2} are the intensity noises in two laser diodes, which all can be regarded as a zero-mean, additive white Gaussian noise process.

After taking a cross-correlation operation by simply multiplying i_1 and i_2, the final detection signal i_3 is obtained, expressed as

$$i_3 = \langle |i_1 \cdot i_2| \rangle \propto I_0^2 \sin\phi_b{}^2 \sin\Delta\varphi^2 \tag{15}$$

where $|\cdot|$ represents an absolute value operation; $\langle \cdot \rangle$ denotes a time average realized in electric circuit by a low-pass filter (LPF), and all intensity noise items included in i_1 and i_2 will be removed after this time-average operation. The item $\sin\phi_b$ above is defined as the fringe visibility or the scale factor [19], which can be maximized by adjusting the PC to achieve $\phi_b = \pi/2$.

3.1.2. Experimental results

We carried out several experiments based on this proposed sensor system. **Figure 4(b)** shows a group of signal waveforms obtained in a static state. The amplitude fluctuations in i_1 and i_2 traces reflect the intensity noises in laser beams. However, in i_3 trace, these intensity noises have been obviously suppressed. As a direct benefit, the signal-to-noise ratio (SNR) of this sensor system is significantly improved.

We attempted to detect the acoustic emissions generated by high-voltage discharges between two electrodes. An experimental setup is illustrated in **Figure 5**, in which a pair of pin-type electrodes, connected with an AC high-voltage generator, was placed in an oil tank which was filled with transformer oil. The electrode gap was set at 8 mm. The fiber coil was attached to the shell of the oil tank, facing the electrodes. **Figure 6(a)** shows two groups of acoustic wave signals recorded at two time periods separated by about 10 seconds, when the imposed AC voltage was increased close to 5 kV and the discharges just started. It is obvious that, as shown in two signal traces, in this stage, the discharges in oil present a periodical change in intensity with a 50-Hz repetition frequency and the average intensity increases with time. **Figure 6(b)** displays a signal trace recorded a minute later. In this stage, the oil had become hot and dirty due to the discharge arcs burning the oil. Clearly, in this result, the signal amplitudes increase greatly and the repetition frequency of discharges then changes to 100 Hz, which indicates that the oil will be in a complete breakdown state and an oil-burning will come soon.

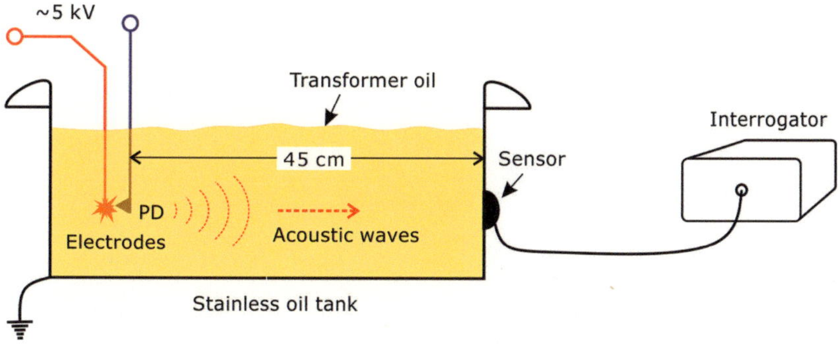

Figure 5. Schematic of an experimental setup for sensing of partial discharges in oil tank.

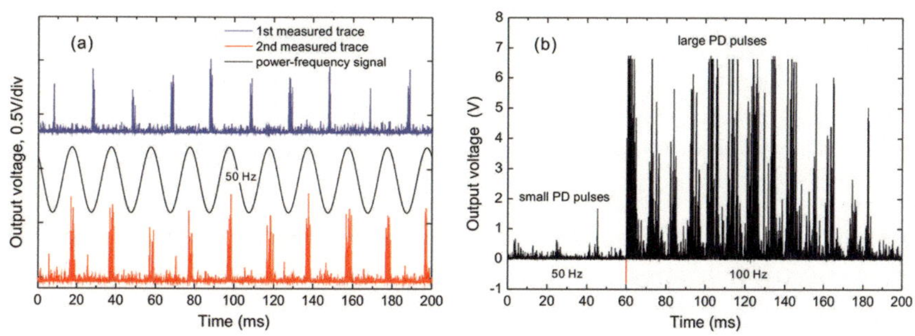

Figure 6. Measured partial discharges occurred in transformer oil. (a) A group of signal waveforms measured when partial discharges just started, and (b) signal waveforms measured a minute later.

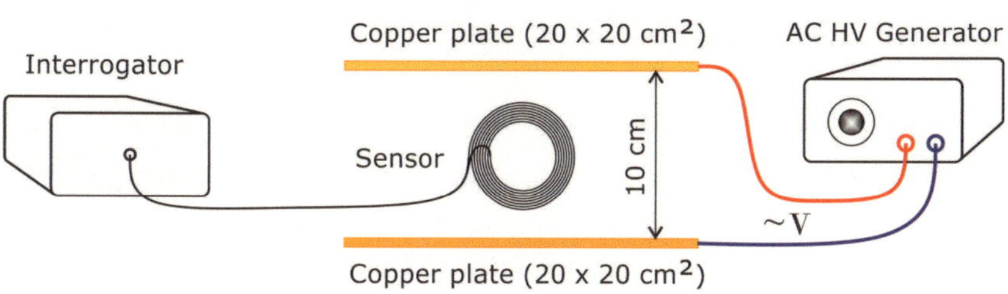

Figure 7. Schematic of an experimental setup for electric field strength measurements.

We also tried to employ this sensor system to measure the strength of power-frequency electric field by detecting of the intensities of ultrasonic waves propagated in air, which were produced between two electrodes during ionization of air by the AC high-voltage electric field. The experimental setup is illustrated in **Figure 7**, in which the fiber coil was placed between two parallel, copper-plate electrodes separated by 10 cm.

In the experiment, the AC voltage imposed on the electrodes was increased gradually and the detection signals were recorded at several specified voltages. **Figure 8(a)** shows three groups of pulse train waveforms, recorded at 1800 V, 2800 V, and 3800 V, respectively. By observing these data, it is obvious that the number of pulses or pulse density within a cycle of AC voltage increases with the imposed AC voltage.

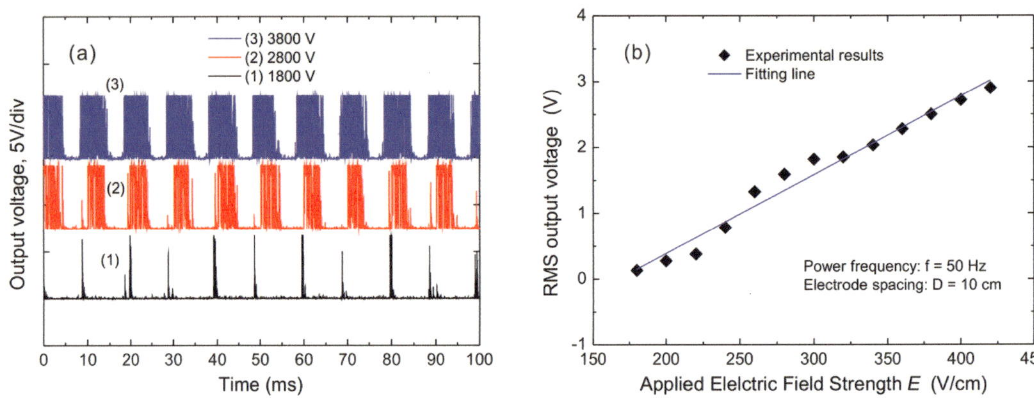

Figure 8. (a) Measured ultrasonic waves generated by air ionization, and (b) relationship between RMS output voltage and applied electric field strength.

Figure 8(b) shows a set of RMS (root-mean-square) voltages measured in a cycle of AC voltage under different electric field strengths from 175 to 475 V/cm. It is obvious that the RMS voltage increases proportionally with the applied electric field strength.

3.2. A fiber Fabry-Perot interferometer-based acoustic sensor

Another novel acoustic sensor based on the fiber Fabry-Perot interferometer is demonstrated. This sensor is specifically designed to be workable in the transformer oil to monitor the partial discharges occurred inside the power transformer.

3.2.1. System configuration and operation principles

The scheme of the proposed sensor system is presented in **Figure 9**, in which a CW DFB laser diode without the optical isolator was employed as the light source, as well as an in-line optical amplifier. The detection signals are obtained from a photodetector packed in laser module. A ferrule-type sensor head is constituted by inserting a single-mode fiber with a flat-cut end into a ceramic ferrule and leaving a 1-mm long space for forming an extrinsic cavity, as shown in **Figure 9**. When the sensor head is immersed into the transformer oil, the oil will permeate into the ceramic ferrule and stop before the fiber end, which forms an extrinsic cavity with two low-reflectance reflectors, the fiber end and the oil surface. The refractive index of transformer oil is estimated in the range 1.40–1.48, close to that of the fiber core, varying with the oil type and quality as well as with the oil temperature. According to the analyses in Section 2.4, the output current signal I_{out} can be expressed as

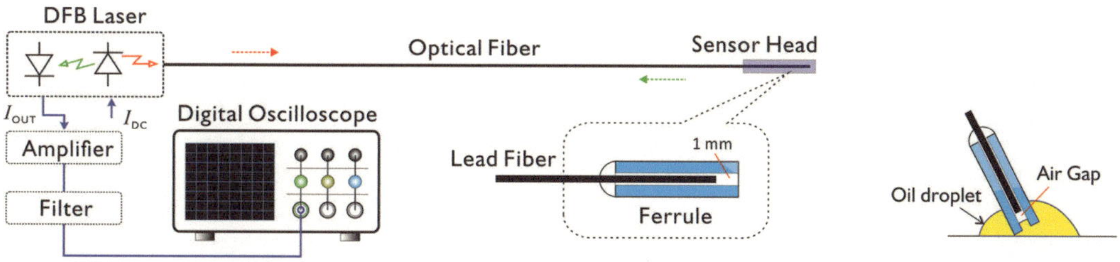

Figure 9. Schematic of a fiber Fabry-Perot interferometer-based acoustic sensor and structure of sensor head.

$$I_{out} \propto 2GP_0R(1+\cos\phi) \tag{16}$$

where G is the gain of laser diode when it is taken as an in-line optical amplifier; P_0 is the optical power launched into the sensor; R is an average reflectance of two reflectors; $\phi = 4\pi l_{FP}/\lambda$ is the round-trip phase difference; and l_{FP} denotes the cavity length. When the partial discharges arise inside the power transformer, ultrasonic waves induced by discharges will propagate in the transformer oil in all directions. The energy of these induced ultrasonic waves mainly distributes in a frequency range of 50–150 kHz [14]. When some of ultrasonic waves arrive at the sensor, the acoustic pressures will modulate the phase difference ϕ by changing the oil level inside the ferrule, in turn, the cavity length l_{FP}, which, as a result, causes the signal I_{out} to change its amplitude. So that by detecting I_{out}, the partial discharges occurred in the power transformer can be monitored.

3.2.2. Experimental results

Figure 10 shows a set of the spectra of the sensor, measured when the sensor was placed in the air as well as was immersed into the transformer oil at different depths (>10 cm). From these results, clearly, the spectrum of the sensor at each depth is of a periodical change characteristic, and the spectral interval between adjacent dips becomes wider as the increase of immersion depth. It indicates the fact that the oil entering into the ceramic ferrule and the fiber end virtually had formed an extrinsic cavity of the fiber Fabry-Perot interferometer.

We experimentally investigated the sensor sensitivity with a setup shown in **Figure 11**, in which the sensor head was placed on a Bakelite plate table and inserted into a silicon oil droplet, here as acoustic wave couplant and also as optical reflection medium. In the experiment, a tiny, 2-cm long metal pin freely falling down from 2-cm height hit the table to generate the weak Lamb waves which propagated in the Bakelite plate and finally arrived at the sensor head in a 45-cm distance. A series of wave signals detected by our sensor is clearly shown in **Figure 12**, which demonstrates that the proposed sensor has a very high sensitivity in detection of weak Lamb waves propagating in the Bakelite plate.

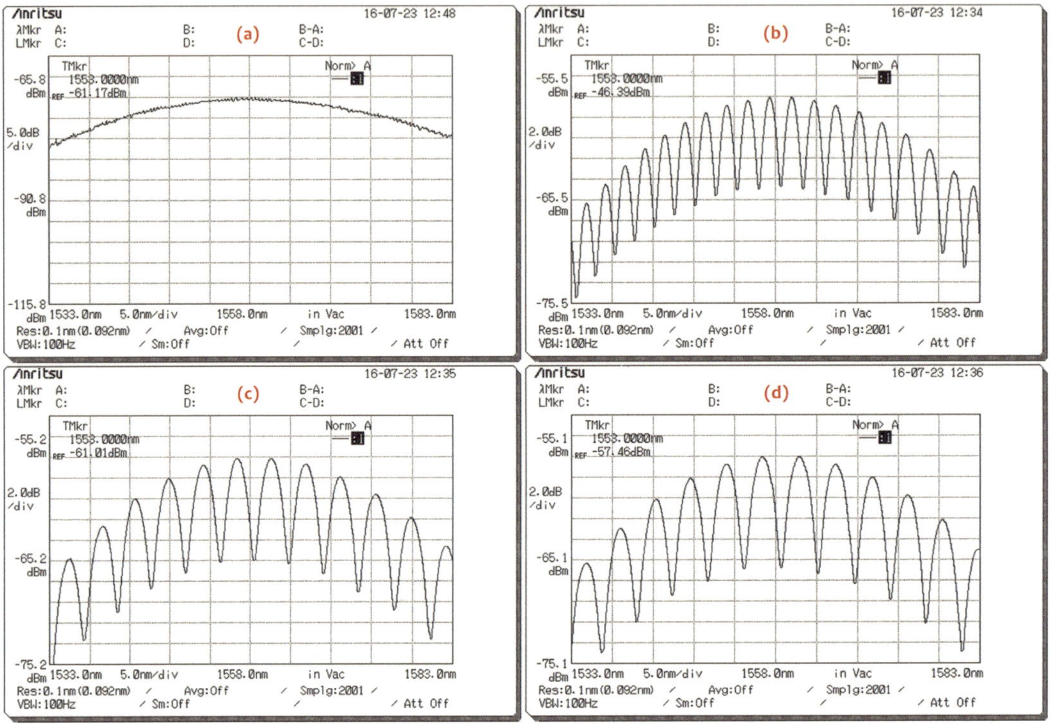

Figure 10. Sensor output spectra measured in air (a) and in transformer oil (b–d) at different depths.

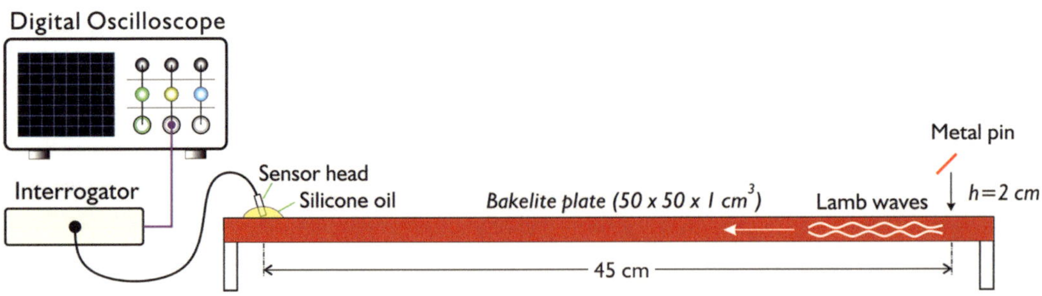

Figure 11. Schematic of an experimental setup for evaluating of sensor sensitivity.

An experiment was carried out for testing the sensor performances in detections of the partial discharges occurred in the transformer oil. **Figure 13** shows a set of photos on this experimental setup. One shown in **Figure 13(a)** is a photo of the hand-made sensor prototype. As shown in **Figure 13(b)**, the sensor was inserted into an 850-mm long PVC pipe filled with transformer

http://dx.doi.org/10.5772/66276

oil and a pair of pin-type electrodes was inserted into the other end of the pipe. In the experiment, the AC voltage imposed on the electrodes was increased gradually until the discharges arose. During this process, the sensor outputs were monitored and recorded.

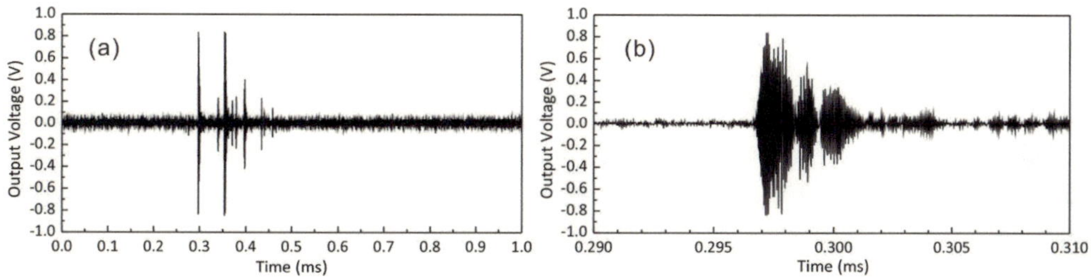

Figure 12. (a) Measured Lamb wave signals and (b) waveforms around 0.3 ms.

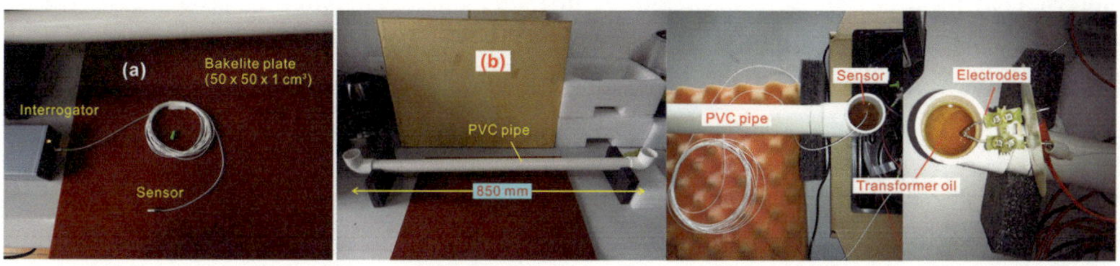

Figure 13. A group of photos on (a) sensor prototype and (b) experimental setup and relevant parts.

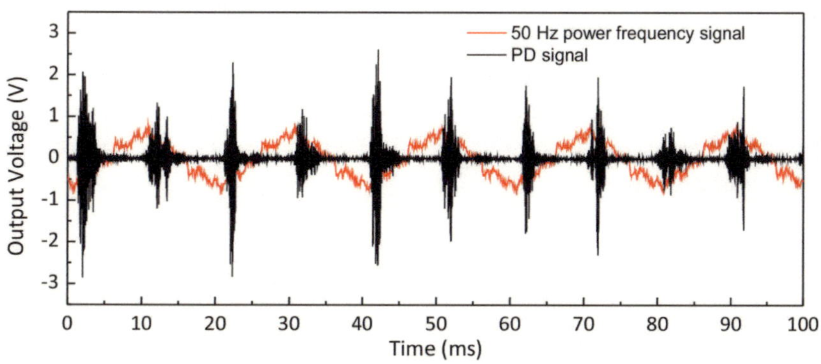

Figure 14. Measured partial discharge signals in transformer oil.

Figure 14 shows a set of measured signal waveforms. The red-line trace is a power-frequency signal measured as a phase reference here, and the black-line trace is the acoustic signals measured when the discharges arose in the transformer oil. From these data, clearly, the proposed sensor can work in the transformer oil to directly detect the internal ultrasonic waves induced by discharges.

Compared with other sensors proposed in [20, 21] with similar structures, this sensor does not need the diaphragm as a reflector, which resultantly make it simpler in fabrication and more sensitive to ultrasonic waves propagating in the transformer oil.

3.3. A twin-grating fiber Fabry-Perot interferometer-based power-frequency electric field sensor

We will demonstrate an all-fiber power-frequency electric field sensor, based on a twin-grating fiber Fabry-Perot interferometer. This sensor technology intentionally is applied in the power industry for measuring of the power-frequency electric field strength near the high-voltage equipment.

Compared with other types of optical electric field sensing technologies based on several physical effects, such as Pockels, Kerr, piezoelectric, or electrostrictive effect [27–30], our electric filed sensing technology is based on the electrical polarization properties [28, 31] of the dielectric coating material with a variety of attractive features, such as small size, low cost, electrodeless, single-lead, high sensitivity, flexibility in a variety of environments and capability of remote sensing.

3.3.1. System configuration and operation principles

As schematically illustrated in **Figure 15(a)**, this fiber electric field sensor actually is a twin-grating fiber Fabry-Perot interferometer-based vibration sensor [32]. In this sensor structure, a polyimide-coated, single-mode fiber including two identical gratings is clamped at a position close to the first grating by a fiber holder to form a fiber vibration sensor with the cantilever beam structure [33, 34]. The length of cantilever beam is chosen so that the cantilever beam has a natural frequency close to the power frequency of 50 Hz or 60 Hz. The polyimide coating adopted in this structure is first to protect the fiber gratings as well as to enhance the mechanical strength of fiber cantilever beam, and second to get an excitation power for vibration (see **Figure 15(b)**) from the applied alternating electric filed by means of the electrical polarization properties of polyimide film.

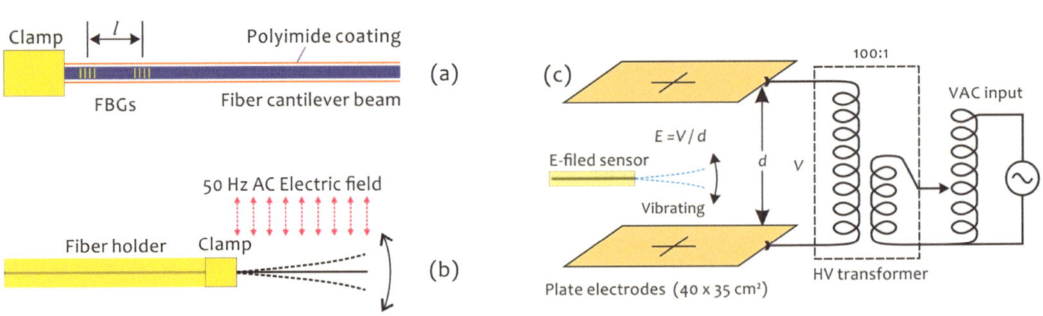

Figure 15. Schematics of electric field sensor structure (a), principle (b), and test setup (c).

The polyimide-coated fiber actually holds a trace of charged ions stored among polyimide crystalline layers, which make this fiber appear to have weak electrical polarization properties [31, 35]. As we assume, the charged ions may originate from the electrification occurred in the coating process, when the fiber was repeatedly painted with a brush filling with polyimide resin, or come from the electrostatic charging [36–38] under a high potential field environment. According to our observations and tests, the polyimide favors to trap the negative ions, so it is the polar dielectric [31], showing the electronegative characteristic. Since the polyimide is a dielectric with a relative permittivity $\varepsilon_r = 3.4$ at a frequency of 1 kHz at room temperature, and also as one kind of high-quality electrical insulation material, widely used in the power industry, some of the trapped net negative ions probably can be stored permanently or semi-permanently in polyimide crystalline layers after the polyimide resin is heat-cured.

When a fiber with polyimide coating in a loose state is placed into a power-frequency electric field, it starts to jitter under the action of alternating electric field force. However, this jittering is very small in amplitude and irregular, which can not bring us any useful information about surrounding electric field strength. A cantilever beam structure, as employed in many types of vibration sensors [32], can be utilized, in this case, to regularize the fiber's jittering and convert it into a regular mechanical vibration. Furthermore, the vibrating amplitude can be mechanically amplified over hundred times by adjusting the natural frequency of the cantilever beam and bringing it close to the power frequency.

Since the vibrating amplitude of the cantilever beam is proportional to the induced electric field force, in turn, the electric field strength, it is possible to evaluate the electric field strength by means of interferometric methods to measure the vibrating amplitude of cantilever beam. This can be realized with our sensor system to be introduced next by detecting the fringe signal intensity proportional to the vibrating amplitude of cantilever beam.

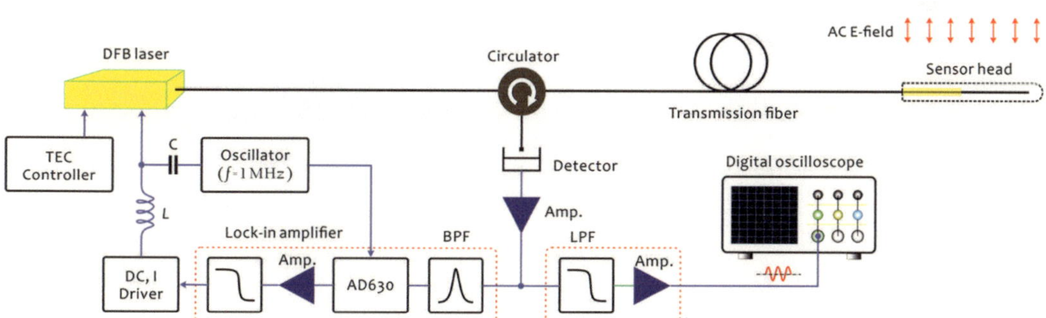

Figure 16. Schematic of electric field sensor system configuration.

The configuration of a sensor system is schematically shown in **Figure 16**, which consists of a CW DFB laser diode, an optical circulator, and a photodetector. The reflection spectrum of sensor is shown in **Figure 3**. The sensor operation is maintained at a quadrature point of the fringe signal by adjusting the laser wavelength with a feedback arrangement to get a quasi-linear dependence of reflected light power on the phase shift produced by vibration. The detected electrical signals after passing a low-pass filter are sent into a digital oscilloscope for waveform display and signal processing. An experimental setup for evaluating sensor performances is shown in **Figure 15(c)**. In the experiments, the sensor was placed between two parallel electrodes which generated a uniform electric field in this space.

In operating principle, when the sensor vibrates under the action of alternating electric field force, the Bragg wavelength of twin-grating-based fiber Fabry-Perot interferometer will be modulated by varying axial strains in the fiber, created by periodic bending of fibers, proportional to the vibrating amplitude, which in turn changes the reflected light power and generates a sinusoid fringe signal with an average amplitude proportional to the electric field strength.

3.3.2. Experimental results

A set of photos showing sensor prototype, experimental setup and sensor vibrating under the action of alternating electric field force is shown in **Figure 17**. The sensor shown in **Figure 17(a)** has a 40-mm long cantilever beam with a natural frequency at 50 Hz. The beam length depends on the power frequency and the hardness of coating material, varying from 39 to 40 mm. The frequency responses of the sensor can be investigated with a frequency-swept shaker. A photo in **Figure 17(a)** shows the sensor vibrating under a 50-Hz, 5-kV electric field environment **Figure 17(b)** is a photo of the experimental setup.

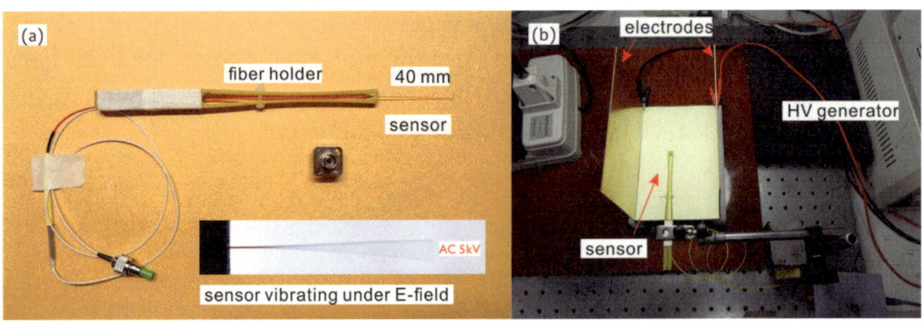

Figure 17. A set of photos on sensor (a) and experimental setup (b).

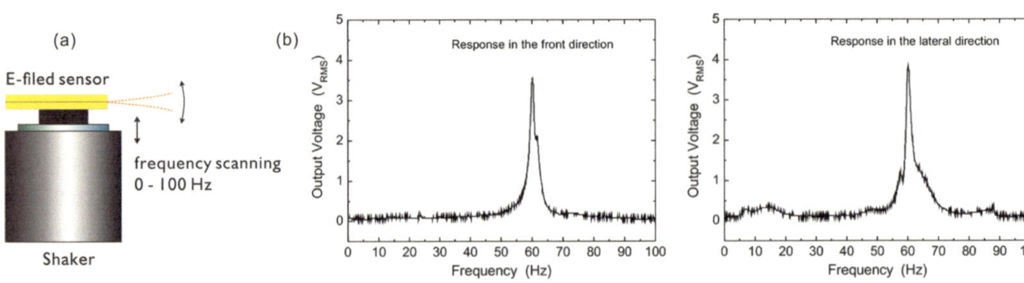

Figure 18. Test for investigating frequency response of sensor with a shaker (a) and results (b).

http://dx.doi.org/10.5772/66276

Figure 18(b) shows two test results on the frequency response distributions of the sensor in two orthogonal directions, the front direction and lateral direction, obtained with a frequency-swept shaker as schematically shown in **Figure 18(a)**. This experiment was to investigate the natural frequency of the sensor, which was designed initially at 60 Hz. The beam length in this case was 39.2 mm. From the test results, clearly, this sensor only has a natural frequency at 60 Hz in both orthogonal directions in the frequency range of 0–100 Hz.

With the experimental setup shown in **Figure 15(c)** as well as in **Figure 17(b)**, we carried out several experiments for investigating sensor performances as well as for verifying the feasibility of the proposed sensor with respect to electric field sensing.

Figure 19(a) shows a measured electric field signal (inset graph) and its frequency distribution from 0 to 100 Hz. In this measurement, the sensor was placed between two electrodes separated by 10 cm, which were connected to 60-Hz, 3000-V voltage. **Figure 19(b)** shows a group of measured RMS voltages of signal amplitude under different AC voltages, varying from 0 to 5 kV. The test arrangement is schematically delineated in **Figure 19(b)**. From these results above, clearly, the 60-Hz sinusoid wave, as a typical vibration signal, really reflects a power-frequency electric field force acting on the sensor, and the RMS voltage of signal amplitude is proportional to the applied electric field strength in a relation $E = \mathrm{V}/\mathrm{D}$.

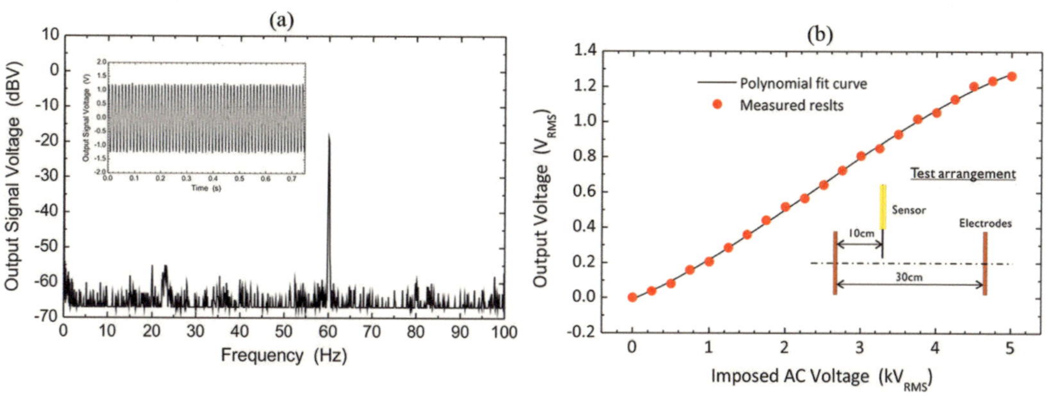

Figure 19. (a) Measured electric field signal waveform at 60-Hz power frequency (inset graph) and its FFT, and (b) relationship between measured RMS voltage and imposed AC voltage.

4. Conclusion

In this chapter, basic concepts of optic interferometry have been presented and clarified. Four main types of fiber interferometers with the help of simple mathematical descriptions also have been briefly introduced. By means of the wavelength of the light source as the length unit as well as the period of light wave as the time unit, optical interferometry is a precision metrology. Combined with the merits of optical interferometry and optical fiber, fiber-optic interferometers have many unique features capable of satisfying a variety of needs in industrial measurement fields, and can provide a much higher sensitivity than their bulk-optic versions, since the optical fiber can be rolled to form a coil which extremely extends the optical path in a limited space. The main applications of fiber-optic interferometers now are in sensor fields as fiber-optic sensors to detect various physical parameters with high accuracy and sensitivity.

Many properties of optical fiber, such as the random birefringence distributions along the fiber, temperature dependency, dispersion, and nonlinearity, however, sometimes become main factors to influence the ultimate performances of a fiber interferometer sensor. Therefore, when we design a fiber interferometer sensor or explore its new applications, these factors should be taken into account. Due to a limit of the number of pages, these properties of the optical fiber as well as their impact on sensor performances, however, have not been mentioned yet in this chapter.

Three novel fiber-optic sensor models, based on optical interferometry technologies and developed in our laboratory recently, have been proposed and demonstrated. Experimental results mainly used for describing basic sensing concepts and sensor structures, as well as for demonstrating sensor performances also were presented, although these are not all of our research achievements. These fiber interferometer sensors are intentionally used, in the future, in the electric power industry as one of most important application fields of fiber-optic interferometer sensing technologies, for various physical parameter measurements, which may be difficult for electric/electronic sensors.

Many research activities are in progress to characterize these proposed fiber sensors in respect of the sensitivity, directionality, frequency responses, and stability. Also how to pack these fiber sensors to make them useful in practical applications is still a challenging work.

Acknowledgements

This work is supported by the National Natural Science Foundations of China (Grant No. 61377082).

Author details

Lutang Wang* and Nian Fang

*Address all correspondence to: ltwang@mail.shu.edu.cn

Key Laboratory of Specialty Fiber Optics and Optical Access Networks, School of Communication and Information Engineering, Shanghai University, Shanghai, China

References

[1] Langenbeck, P., editor. Interferometry for precision measurement. Bellingham, Washington USA: SPIE; 2014. 244 p. DOI: 98227–0010.

[2] Hariharan, P., editor. Basics of interferometry. 2nd ed. Boston: Academic Press; 2007. 213 p.

[3] Culshaw, B. Fiber optics in sensing and measurement. IEEE Journal of Selected Topics in Quantum Electronics 2000; 6: 1014–1021.

[4] Udd, E., editor. Fiber optic sensors: An introduction for engineers and scientists. Hoboken, New Jersey, USA: John Wiley & Sons, Inc.; 2006. 467 p.

[5] Berthold III, J.W. Industrial applications of fiber optic sensors. In Fiber optic sensors: An introduction for engineers and scientists; Udd, E.; New Jersey USA: John Wiley & Sons, Inc.; 1991, 2006; pp. 409–437.

[6] Udd, E. Fiber-optic acoustic sensor based on the Sagnac interferometer. In Single Mode Optical Fibers 90, Proc. of SPIE 1983; 0425: 90–95.

[7] Rajkumar, N.; Jagadeesh Kumar, V.; Sankaran, P. Fiber sensor for simultaneous measurement of current and voltage in a high-voltage system. Applied Optics. 2005; 32(7): 1125–1128.

[8] Lopez-Higuera, J.M., editor. Handbook of optical fiber sensing technology. Chichester, England: John Wiley & Sons Ltd.; 2002. 789 p.

[9] Mihailov, S.J. Fiber Bragg grating sensors for harsh environments. Sensors. 2012; 12: 1898–1918. DOI: 10.3390/s120201898

[10] Blackburn, T.R.; Phung, B.T.; James, R.E. Optical fibre sensor for partial discharge detection and location in high-voltage power transformer. In Sixth International Conference on Dielectric Materials, Measurements and Applications 1992; 33–36.

[11] Kersey, A.D.; Davis, A.M.; Patrick, J.H.; LeBlanc, M.; Koo, K.P.; Askins, C.G.; et al. Fiber grating sensors. Journal of Lightwave Technology. 1997; 15(8): 1442–1463.

[12] He, X.; Taylor, H.F. Intrinsic fiber Fabry-Perot temperature sensor with fiber Bragg grating mirrors. Optics Letters. 2002; 27(16): 1388–1390.

[13] Rashleigh, S.C. Origins and control of polarization effects in single-mode fibers. Journal of Lightwave Technology. 1983; LT-1: 312–331.

[14] Posada, J.E.; Rubio-Serrano, J.; Garcia-Souto, J. A. All-fiber interferometric sensor of 150 kHz acoustic emission for the detection of partial discharges within power transformers. In 21st International Conference on Optical Fiber Sensors, Proc. of SPIE 2011; 7753: 77531S:1–77531S:4.

[15] Harvey, D.; McBride, R.; Barton, J.S.; Jones, J.D.C. A velocimeter based on the fiber optic Sagnac interferometer. Measurement Science and Technology. 1992; 3: 1077–1083.

[16] Burns, W.K.; Chen, C.L; Moeller, R.P. Fiber-optic gyroscopes with broad-band sources. Journal of Lightwave Technology. 1983; LT-1: 98–105.

[17] Carolan, T.A.; Reuben, R.L.; Barton, J.S.; Jones, J.D.C. Fiber-optic Sagnac interferometer for noncontact structural monitoring in power plant applications. Applied Optics. 1997; 36: 1504–385.

[18] Wang, L.T.; Fang, N.; Huang, Z.M. A novel Sagnac-based fiber-optic acoustic sensor using two laser diodes with external optical injections. In: Kennis Chen, editor. Testing and Measurement: Techniques and Applications; 16–17 Jan. 2015; Phuket Island, Thailand. London: CRC Press/Balkema; 2015. pp. 145–150.

[19] Wang, L.T.; Fang, N.; Wu, C.; Qin, H.; Huang, Z. A fiber optic PD sensor using a balanced Sagnac interferometer and an EDFA-based DOP tunable fiber ring laser. Sensors. 2014; 14(5): 8398–8422. DOI: 10.3390/s140508398

[20] Chin, K.K.; Sun, Y.; Feng, G.; Georgiou, G.E.; Guo, K.; Niver, E. et al. Fabry Perot diaphragm fiber-optic sensor. Applied Optics. 2007; 46: 7614–7619.

[21] Cibula, E.; Donlagic, D. Miniature fiber-optic pressure sensor with a polymer diaphragm. Applied Optics. 2005; 44(10): 2736–2744.

[22] Satoshi, A.; Tanaka, S.; Takahashi, N. High-sensitivity vibration sensing using in-fiber Fabry-Perot interferometer with fiber-Bragg-grating reflectors. In 20th International Conference on Optical Fibre Sensors, Proc. of SPIE 2009; 7503: 75033L:1–75033L:4.

[23] Xie, Z. Two applications of the Fabry-Perot interferometric sensor. Ph.D. Thesis, Texas A&M University: College Station, TX, USA, August 2006.

[24] Dong, B.; Han, M.; Wang, A. Two-wavelength quadrature multipoint detection of partial discharge in power transformers using fiber Fabry-Perot acoustic sensors. In Fiber Optic Sensors and Applications IX, Proc. of SPIE 2012; 8370: 83700K:1–83700K:11.

[25] Majumder, M.; Gangopadhyay, T.K.; Chakraborty, A.K.; Dasgupta, K.; Bhattacharya, D.K. Fiber Bragg gratings in structural health monitoring–present status and applications. Sensors and Actuators A. 2008; 147:150–164.

[26] Miridonov, S.V.; Shlyagin, M.G.; Tentori, D. Twin-grating fiber optic sensor demodulation. Optics Communications. 2001; 191(2001): 253–262.

[27] Bohnert, K.M.; Nehring, J. Fiber-optic sensing of electric field components. Applied Optics. 1988; 27(23): 4814–4818.

[28] Zahn, M. Optical, electrical and electromechanical measurement methodologies of field, charge and polarization in dielectrics. IEEE Transactions on Dielectrics and Electrical Insulation. 1998; 5(5): 627–650.

[29] Koo, K.P.; Sigel, G.H. An electric field sensor utilizing a piezoelectric Polyvinylidene Fluoride (PVF2) film in a single-mode fiber interferometer. IEEE Journal of Quantum Electronics. 1982; QE-18(4): 670–675.

[30] Tomic, M.C.; Elaza, J.M. Voltage measurement based on the electrostrictive effect with simultaneous temperature measurement using a 3 × 3 fiber-optic coupler and low coherence interferometric interrogation. Sensors and Actuators A. 2004; 115: 462–469.

[31] Raju, G.G., editor. Dielectrics in electric fields. New York, NY USA: Marcel Dekker, Inc.; 2003. 570 p.

[32] Zhu, Y.; Shum, P.; Lu, C.; Lacquet, B.M.; Swart, P.L.; Spammer, S. J. Temperature-insensitive fiber Bragg grating accelerometer. IEEE Photonics Technology Letters. 2003; 15(10): 1437–1439.

[33] Todd, M.D.; Johnson, G.A.; Althouse, B.A.; Vohra, S. T. Flexural beam-based fiber Bragg grating accelerometers. IEEE Photonics Technology Letters. 1998; 10(11): 1605–1607.

[34] Leissa, A.W.; Sonalla, M.I. Vibrations of cantilever beams with various initial conditions. Journal of Sound and Vibration. 1991; 150(1): 83–99.

[35] Gerhard, R. A matter of attraction: electric charges localised on dielectric polymers enable electromechanical transduction. In: 2014 Annual Report Conference on Electrical Insulation and Dielectric Phenomena; IEEE; 2014. pp. 1–10.

[36] Watson, P.K. The transport and trapping of electrons in polymers. IEEE Transactions on Dielectrics and Electrical Insulation. 1995; 2(5): 915–924.

[37] Albrecht, V.; Janke, A.; Nemeth, E.; Spange, S.; Schubert, G.; Simon, F. Some aspects of the polymers' electrostatic charging effects. Journal of Electrostatics. 2009; 67: 7–11.

[38] Bartnikas, R. Performance characteristics of dielectrics in the presence of space charge. IEEE Transactions on Dielectrics and Electrical Insulation. 1997; 4(5): 544–557.

Chapter 8

Interferometry Applications in All-Optical Communications Networks

Rogerio Pais Dionisio

Additional information is available at the end of the chapter

http://dx.doi.org/10.5772/66133

Abstract

Throughout the years, the expanded search and flow of information led to an expansion of traffic intensity in today's optical communication systems. Coherent communications, using the amplitude and phase of the optical wave, resurface as one of the transmission methods to increase the effective bandwidth of optical channels. In this framework, this chapter presents a study on all-optical format conversion of modulated signals, using exclusively interferometric techniques through wavelength conversion, based on Mach-Zehnder interferometers with semiconductor optical amplifiers (MZI-SOA). This technique, when applied in interconnection nodes between optical networks with different bit rates and modulation formats, allows a better efficiency and scalability of the network. The chapter presents an experimental characterization of the static and dynamic properties of the MZI-SOA and explores all-optical techniques for the conversion from amplitude modulation to phase modulation. Finally, it briefly presents the potential of MZI-SOAs for the conversion of amplitude signals to more advanced modulation formats, such as quadrature phase shift keying (QPSK) and quadrature amplitude modulation (QAM) signals.

Keywords: all-optical networks, signal processing, Mach-Zehnder interferometer

1. Introduction

From the exponential popularity of Internet and its derived new services, recently deployed communication networks should be capable to provide more information throughput than ever. On a recent study [1], CISCO has estimated that global Internet traffic has increased by a factor of height over the past five years and will increase nearly by a factor of four over the next five years, with an annual global Internet protocol (IP) traffic exceeding the tens of zettabyte mark by the end of 2017.

High definition (HD) streaming video and peer-to-peer applications use the majority of bandwidth in most broadband networks today. When we add the bandwidth requirements of online social networks, Internet browsing or online gaming, broadband service providers have to supply an increasingly larger amount of bandwidth. Besides human-made Internet traffic, machine-to-machine (M2M) communications [2], fostered by smart city applications [3] and the opening of new radio channel opportunities [4] are spreading globally. The Internet of things (IOT) [5] will contribute even more to increase the quantity of data exchanged in communication networks, and especially on optical communication networks, used both by wireless and cable communications.

Photonics technologies have largely contribute to the considerable development of telecommunication networks, since they appearance 30 years ago, and one can predict that they will serve as ground for most of the network revolutions still to come. A recent study [6] of the drivers of photonics suggests that its future development will be made along four main paths: to make optical networks faster, more transparent, more dynamic and greener. With current technology, it may be difficult to follow simultaneously all four path, which is a major constrain to implement a novel generation of optical networks. However, these difficulties may be relieved by recent developments in the field of wavelength conversion, all-optical signal processing and flexible techniques to generate enhanced modulation formats in optical signals.

In this context, this chapter presents several techniques on format conversion of modulated signals, using Mach-Zehnder interferometers with semiconductor optical amplifiers (MZI-SOA). The MZI-SOA show very attractive properties, and therefore, the goal is to investigate their potential as an optical node for the dynamic conversion and generation of optical signals. As such, two techniques to implement all-optical modulation format conversion are explored. These techniques, when applied in interconnection nodes between different optical networks with variable bit rates and modulation systems, allow a better efficiency and scalability of the network.

2. MZI-SOA features

This section presents an experimental methodology to study the static and dynamic properties of the MZI-SOA. The device under test is a commercial hybrid-integrated device, manufactured by the centre for integrated photonics (CIP) [7], consisting of a passive, planar silica balanced MZI with nonlinear SOAs and phase shifters (PSs) assembled in each interferometer arm. Sections 2.1 and 2.2 describe the characterization of active and passive components of the MZI-SOA, in static operational conditions. The methodology uses simple optical power measurements to extract operative parameters. Then, section 2.3 investigates the dynamic properties of the MZI-SOA as an optical gate. Overall conclusions about this topic are presented at the end.

http://dx.doi.org/10.5772/66133

2.1. Static properties

MZI-SOAs are devices with a small footprint, but a huge potential for application in many optical domains. They can be used in optical gates [8], digital phase and amplitude modulation [9], wavelength conversion and switching [10], signal regeneration and all-optical processing [11], among others. In all these applications, the operational parameters of MZI-SOA active elements (SOA and PSs) must be previously calibrated, to set an optimal setting point in terms of power/phase variations and mean output power, according to the available inputs and the expected application. The majority of the biasing point procedures require extensive and complex initial setup stage, which changes from device to device, due to their intrinsic production yields.

Figure 1 show a photograph of the MZI-SOA under test in this experimental study. The package contains two MZI-SOA structures on a single chip [12]. The MZI-SOA is hybridly incorporated with a silica substrate (motherboard) that contains the optical waveguides, such as Y junctions and couplers; the SOAs are built-in on an independent silicon board, which is passively assembled in the motherboard; the active elements, i.e. the PSs and the SOAs, are assembled in the daughterboard; a Vgroove to simplify fibre pigtailing of the motherboard [13]. MZI-SOA with hybrid integration result in more flexibility but increased yields, where each active and passive element has its own asymmetries and tolerances (e.g. asymmetric splitting ratio of the couplers). These issues have various implications in some MZI-SOA functionalities, for example, on the maximization of extinction ratio (ER) between the output ports. For experimental testing, the chip was installed on a prototype box that included SOA current and phase shifter voltage electronics together with temperature control. A temperature control system takes measurements from a thermistor and actuates on a Peltier cell to keep the chip temperature constant at the desired value. All measurements were made at a temperature of 25°C.

Figure 1. Twin MZI-SOA device used for experimental measurements. A strip of eight fibres enters the MZI-SOAs on the right (four fibres per MZI-SOA), and a strip of four fibre exits on the left (two fibres per MZI-SOA).

Due to the interferometric configuration of the MZI-SOA, the power distribution along the interferometer arms will affect the interference strength at the output and its eventual optimization. Thus, it is essential to characterize all passive parts. To execute the experimental analysis on the device asymmetry, we use the setup depicted in **Figure 2**. Each arm of the MZI incorporates one SOA and one PS. For each input-output path, two sets of switches are used, as seen in **Figure 2**. This configuration allows measurements of the gain of one of the arms when the other arm is blocked by a switched off SOA, and the SOA gain dependence on the biasing current.

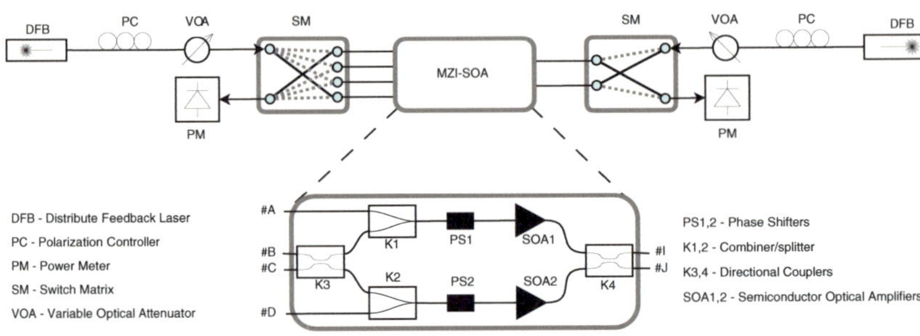

Figure 2. Experimental setup used to characterize the MZI-SOA with power measurements.

A distributed feed-back (DFB) laser with wavelength 1546.12 nm is used as the input signal and will be kept fixed in all upcoming experimental tests (The wavelength was chosen according to the MZI-SOA manufacturer specifications). All power measurements were obtained through a power meter (PM).

2.1.1. Internal couplers

In order to carry out the analysis of all MZI-SOA internal couplers, and their asymmetry properties, each input-output path is analysed independently, biasing one SOA at time accordingly to the path. For all coupling factor measurements, continuous wave (CW) laser input power is constant at 3 dBm. When simply biased, the SOAs current value is set to 200 mA. PS1 and PS2 are always switched off, with voltage source = 0 V, since a PS has little impact on couplers characterization, and only one SOA is biased at each measurement.

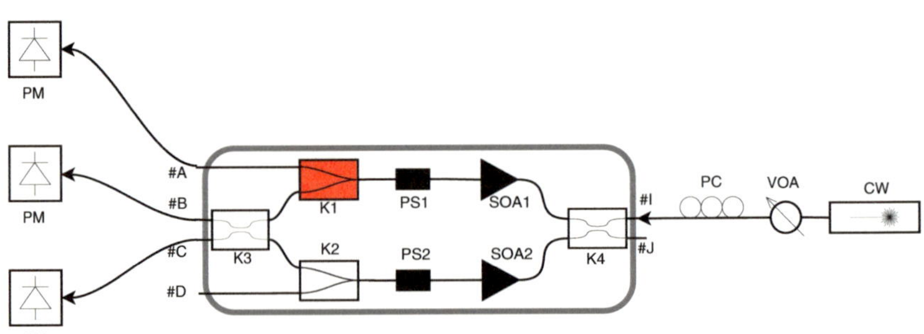

Figure 3. Setup for the characterization of K1 coupling factor.

http://dx.doi.org/10.5772/66133

Coupling factor of coupler K1 is characterized using the setup depicted in **Figure 3**. SOA1 is biased at 200 mA and SOA2 is switched off, i.e. with 0 mA current.

Port #I is used as the input, but port #J could also be used as well. Power measurements must be carried out at ports #A, #B and #C, with a PM. Coupling factor α_1 is obtained as

$$\alpha_1 = \frac{P_{\#A}}{\left(P_{\#A} + P_{\#B} + P_{\#C}\right)} \tag{1}$$

using a linear scale to express power values.

With SOA1 unbiased (0 mA) and SOA2 biased at 200 mA, K2 coupling factor is measured when light is injected through MZI-SOA port #J and optical power measurements are made at ports #B, #C and #D, as depicted in **Figure 4**.

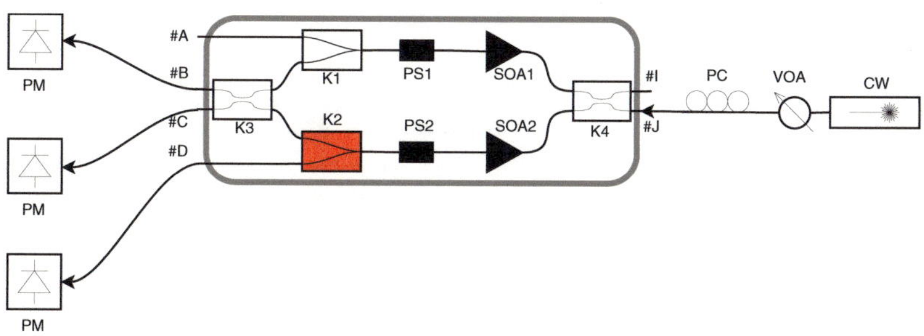

Figure 4. Setup for the characterization of K2 coupling factor.

Coupling factor α_2 is computed as,

$$\alpha_2 = \frac{P_{\#D}}{\left(P_{\#B} + P_{\#C} + P_{\#D}\right)}. \tag{2}$$

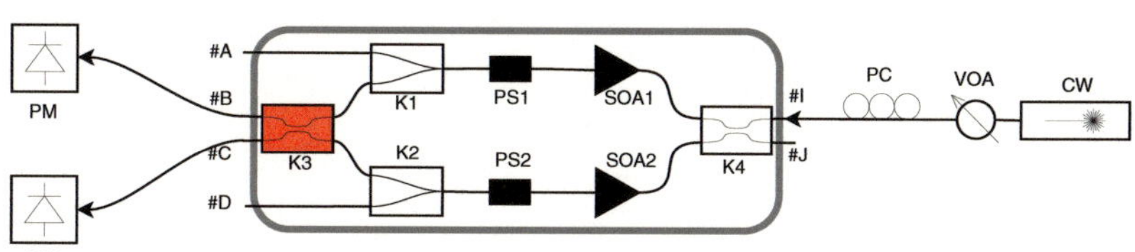

Figure 5. Setup for the characterization of K3 coupling factor.

The coupling factor of coupler K3 is computed from two power measurements, at port #B and port #C, as shown in **Figure 5**. Either SOA1 or SOA2 could be biased at 200 mA, but one SOA must be switched off.

Input power can be injected from port #I or port #J. Coupling factor α_3 is obtained through the following expression,

$$\alpha_3 = \frac{P_{\#B}}{\left(P_{\#B} + P_{\#C}\right)}. \tag{3}$$

At last, K4 coupling factor calculation follows the setup depicted in **Figure 6**. When the input optical power is injected on Port #D, then we should bias SOA2, and SOA1 should remain unbiased. If the selected input is port #A, then SOA1 should be biased and SOA2 switched off. If the input is either port #B or #C, then anyone of the two SOAs can be used.

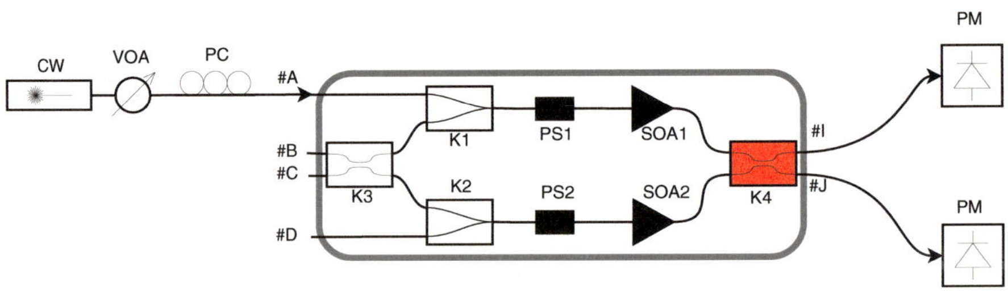

Figure 6. Setup for the characterization of K4 coupling factor.

Power measurement is taken at ports #I and #J. Coupling factor α_4 is given by,

$$\alpha_4 = \frac{p_{\#I}}{\left(p_{\#I} + p_{\#J}\right)}. \tag{4}$$

From the measurement results presented in **Table 1**, we conclude that couplers are not symmetrical, as ideally would be expected on this device. Furthermore, splitting factors are all different. This asymmetry induces an uneven power distribution along the interferometer arms that in any operational point will cause side effects, such as different saturation levels of the SOAs from each arm.

Coupler	CW port	PM port	I_{SOA1}	I_{SOA2}	Formulation	Value
K1	#I	#A, #B, #C	200 mA	0 mA	Eq. (1)	52%
K2	#J	#B, #C, #D	0 mA	200 mA	Eq. (2)	42%
K3	#I	#B, #C	200 mA	0 mA	Eq. (3)	52%
K4	#A	#I, #J	200 mA	0 mA	Eq. (4)	60%

Table 1. Procedure for measurements and calculations.

http://dx.doi.org/10.5772/66133

From here, we can observe the coupling factor of all passive couplers inside the interferometric structure and compute the overall asymmetry on the power distribution along the interferometer arms.

2.1.2. SOA

The next stage of the interferometer characterization is the analysis of SOA output power dependence on input power and SOA bias current. These measurements were realized, considering the two arms individually, first with a light path from input #B to output #J and SOA1 unbiased; then, with a light path from input #A to output #I and SOA2 unbiased, as depicted, respectively, in **Figures 7** and **8**. It is important to emphasize that this analysis refers to the whole light path where the optical wave propagates, so it includes all power losses and any other type of asymmetry of the interferometer chip.

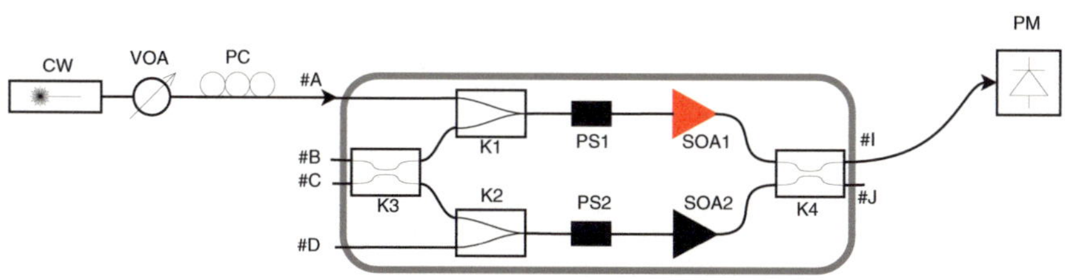

Figure 7. Setup for the characterization of SOA1.

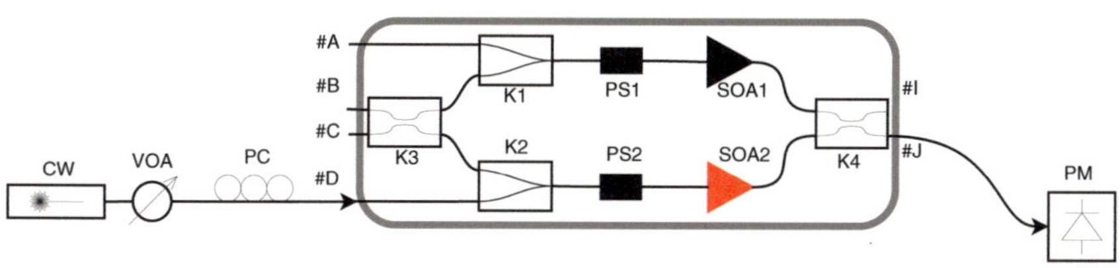

Figure 8. Setup for the characterization of SOA2.

For the evaluation of the SOA dependence from bias current, the injected power of CW power is set previously at a low value and the bias current is swept from 150 to 400 mA. For each current step, the optical power value on the output port is first measured and then saved. Then, input power is gradually increased, respecting the maximum permissible power given by the device specifications, and the previous process and measurements are repeated. **Figure 9** shows the results of the dependence of the SOA output power as a function of the SOA bias current.

To characterize the influence of the input power on the SOA gain, the power of a CW laser is gradually increased from 1 to 10 mW, while the SOA bias current is kept constant. The output power is then recorded for each measurement. From **Figure 10**, we observe the SOA output power dependence on the input power, for both SOAs and several bias currents. From the resulting curves, we obtain the necessary setting points (input power and SOA bias) when the SOA gain begin to saturate.

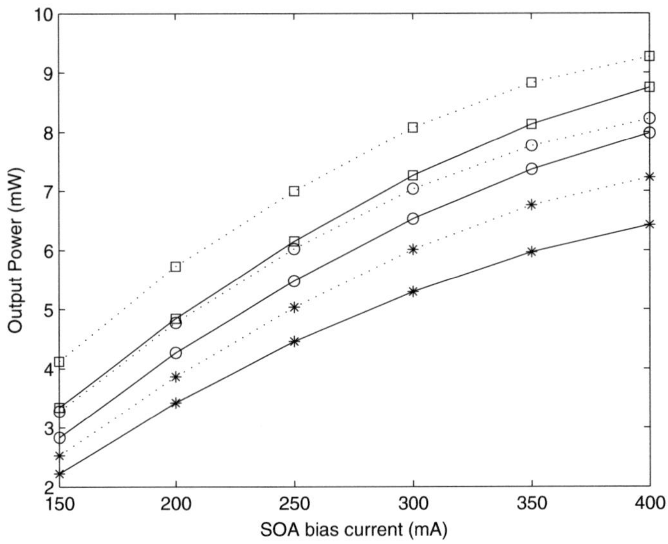

Figure 9. MZI-SOA output power at port #I (full line) and #J (dashed line), as a function of SOA1 and SOA2 bias current, respectively. Input power equal to –8, –2 and 10 dBm (asterisks, circles and squares, respectively). The lines are guides for the eyes.

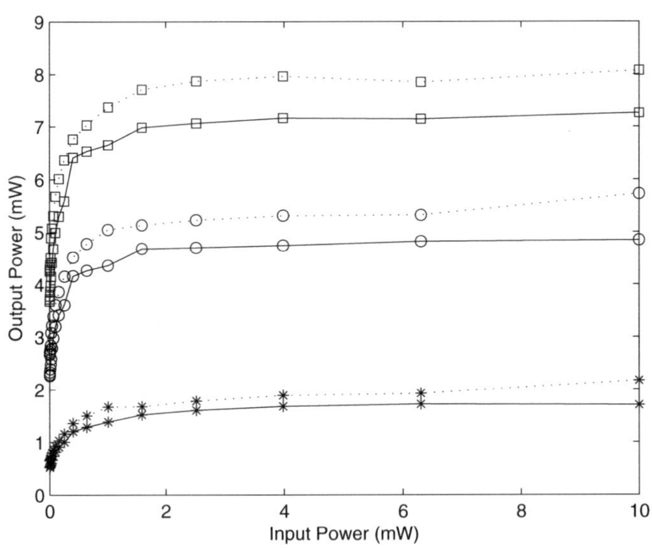

Figure 10. MZI-SOA output power at port #J (dashed line) and port #I (full line), vs. input CW power at ports #D and #A, respectively. SOA bias currents are adjusted to 100, 200 and 300 mA (asterisks, circles and squares, respectively). The lines are guides for the eyes.

http://dx.doi.org/10.5772/66133

2.1.3. Extinction ratio characterization

The power at each output ports of the MZI-SOAs under test (ports #I and #J) is a result of an interference process taking place in coupler K4. The intensity and phase of the optical signal of coupler K4 inputs will set the conditions for the measured outputs. In particular, when used as an intensity modulator, one key factor to take into account is the ER between the output signals. By definition, the ER is the ratio of the optical power levels measured at output ports #J and #I,

$$ER = 10\log_{10}\left(\frac{P_{\#I}}{P_{\#J}}\right). \tag{5}$$

The ER value is given in dB and power $P_{\#I}$ and $P_{\#J}$ are given on a linear scale.

As an example, the power distribution on the MZI-SOA internal waveguides can be varied even more, through SOAs current variation, leading to the subsequent change of the interference settings on coupler K4. To observe the effect of bias currents on the ER, we employ the setup from **Figure 11**. Port #B receives a CW signal, and both PSs of the MZI-SOA remains unbiased (voltage set to 0 V).

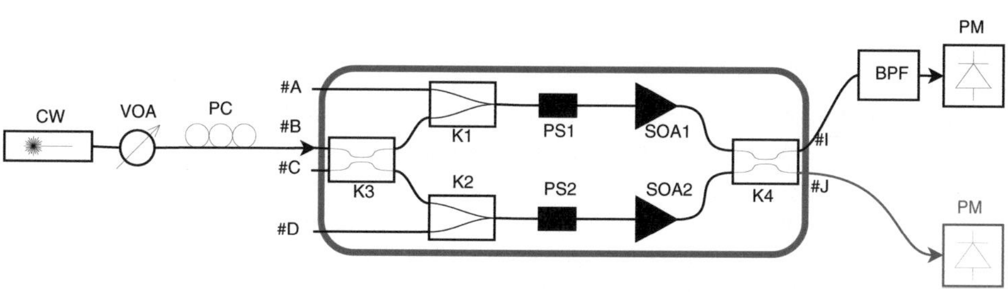

Figure 11. Setup for the characterization of ER dependence on bias current.

In **Figure 12**, we observe the dynamic of both the constructive and destructive interference, respectively, at the output port #J and #I, by gradually increasing SOA1 bias current, with SOA2 current constant at a reference value (200 mA).

This process is repeated with SOA1 bias current constant at 200 mA, while SOA2 current is swept, using a CW laser beam connected at input port #B. With this methodology and after measuring the power at the output port #I and #J, we observe in **Figure 11** that there is a misalignment between #J and #I maximum and minimum power levels. This is caused by the gain variation in the SOA with the varying bias current, together with the asymmetrical coupling factors of the couplers and the phase shift of the electromagnetic field. The best operational point, without phase shifters corrections, is established searching for SOA1 bias current for the maximum ER.

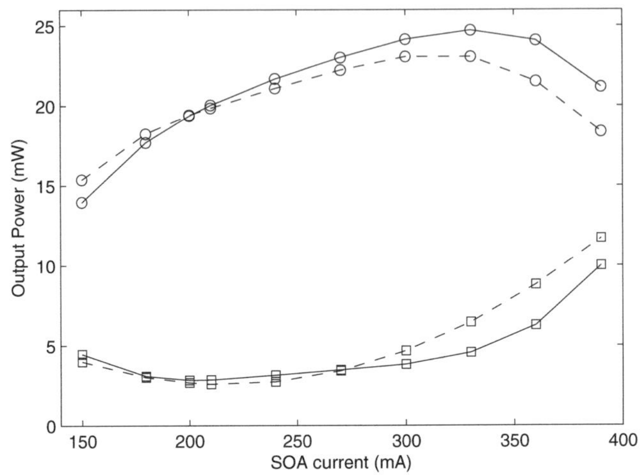

Figure 12. MZI-SOA output power at port #I (squares) and port #J (circles), as a function of SOA1 (dashed line) and SOA2 (full line) bias current. Input power is injected at port #B. The lines are guides for the eyes.

2.2. Dynamic characterization

After the static characterization, this section investigates the dynamic properties of the MZI-SOA. This study is focused on the properties of the MZI-SOA as an optical gate.

2.2.1. Experimental setup

Figure 13 shows the proposed setup to characterize the dynamic properties of the MZI-SOA. The setup follows the wavelength conversion design presented in Ref. [14], in a co-propagation design. This setup implements an all-optical exclusive OR (XOR) logical gate using a single MZI-SOA chip. From the MZI arms, both optical signals are launched into the SOAs, where their carrier densities and thereby the refractive index are modulated. This result in a phase modulation of the CW probe signal propagating through the SOAs by cross-phase modulation (XPM), according to the intensity variations of the input control signals. By carefully setting the input optical powers and controlling the SOA bias current, the control signal from the two SOAs interferes either destructively or constructively at the output of the MZI in order to provide the logical XOR operation of the two data sequences on the optical probe signal. From the XOR truth table, when both data signals injected at ports #D and #A are time synchronized, no pulses are observed on the probe signal at the MZI-SOA output (port #I). On the other hand, as the data signals give up time overlapping, some pulses with increasing intensity will appear on the probe signal, at the same MZI-SOA output.

The experimental setup consists of an external cavity laser peaking at 1549.32 nm (λ_1), followed by a polarization controller (PC) and an external Mach-Zehnder modulator (MZM). The non-return-to-zero (NRZ) data signal generated by a serial bit error rate tester (ref. Agilent N4901B) is then optically amplified by an erbium-doped fiber amplifier (ref. IPG-EAD-500-C3-W) and divided into two identical signals using a 50:50 coupler (COUPLER1) with symmetrical outputs. Both optical signals coming from COUPLER1 are synchronized using optical delay lines. Polarization controllers are inserted on the light path to ports #D and #A of the MZI-SOA (ref. CIP 40G-2R2-ORP), to optimize the destructive interference at port #I. The probe signal, a CW light beam with 0 dBm and lasing at 1546.12 nm (λ_2), is launched into port #B of the MZI-SOA in a co-propagating direction with the data control signals. Different data patterns may be obtained by delaying signals at port #A and port #D. Finally, the probe signal is recovered at port #J, using a filter with a 40 GHz bandwidth (X-tract Net Test). The setup uses two instruments to analyse and measure the optical output signal: an optical complex spectrum analyser (OCSA) (APEX AP 2441A) to gather power and phase information of the output signal for time domain characterization and a sampling oscilloscope (Agilent 86100A) connected through a photodiode (HP-11982A).

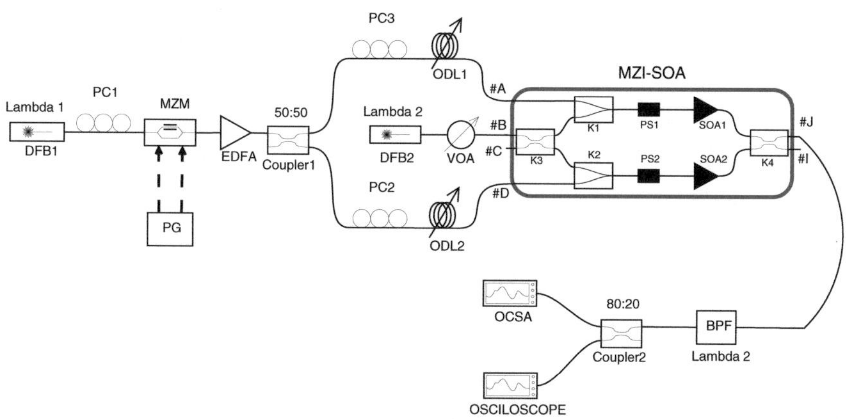

Figure 13. All-optical XOR gate setup, based on a MZI-SOA, in a co-propagation scheme.

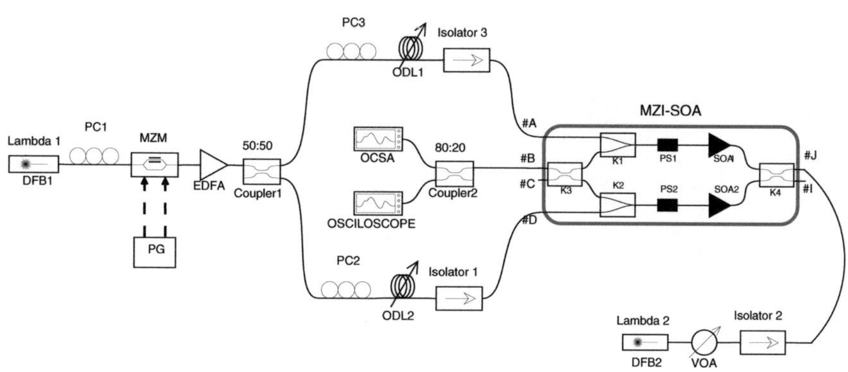

Figure 14. Wavelength and format conversion setup in counter propagation scheme.

The setup for the counter-propagation scheme uses the same probe signal, but now the signal is injected into port #I. As shown in **Figure 14**, the output signal is recovered at the constructive interference output (port #B). An isolator is placed at ports #I, #D and #A to protect all laser sources from back propagation signals.

2.2.2. *Experimental results and discussion*

Figure 15 shows the data input signals (λ_1) injected into the arms #A and #D of the MZI-SOA, each with 2 dBm mean power, and the corresponding XOR gate output (λ_2), at 10 Gbps at port #J, in a co-propagating scheme. In this experimental scenario, the results are in conformity with the truth table of an XOR gate: the output presents a logical zero (0) if both the operands have identical value and a logical one (1) otherwise.

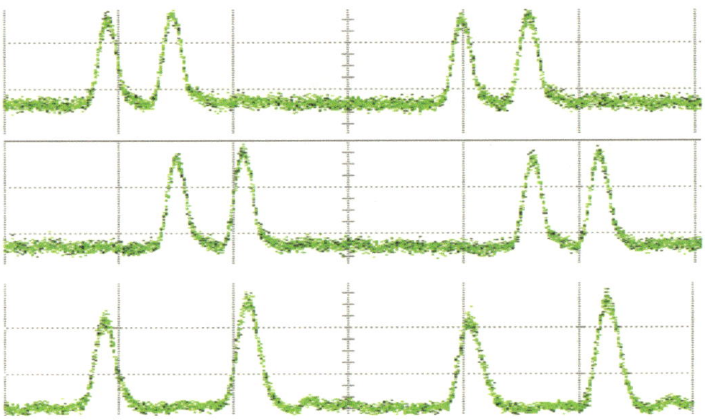

Figure 15. Optical sequences at MZI-SOA input ports #D and #A (first two signals from top) and resulting XOR output at port #J (bottom sequence). Horizontal scale: 500 ps/div. Vertical scale is arbitrary.

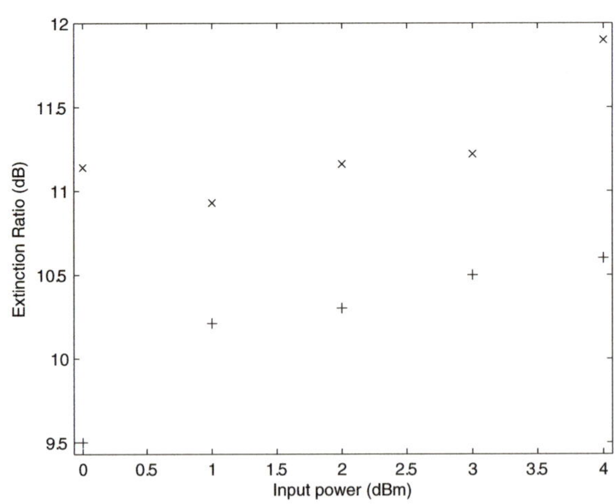

Figure 16. Experimental measurements of the ER of the output signal, as a function of the input power for co-propagation (+ sign) and counter-propagation (× sign) schemes.

Figure 16 shows the dependence of the ER at the output signal as a function of the power of the NRZ input signals, varying from 0 to 4 dBm. We set the CW probe signal power at 0 dBm. For both co- and counter-propagation proposals, we observe that changes on the input power do not affect the overall performance of the all-optical logical XOR gate, since the power variation of the two control signals involved in the assessment process is equivalent. But when compared with the co-propagation scheme, the counter-propagation scheme present better results, increasing the ER from 0.72 to 1.64 dB and improving the performance of the MZI-SOA as an optical gate. These results are in line with other experimental studies [15].

2.3. Summary

Having in mind next generation optical networks, which are meant to be as flexible and transparent as possible, this section has characterized the static properties and the operating conditions of the MZI-SOA working as an optical gate, which results will be useful for the following section, dealing with phase modulation and other advanced modulation format conversion techniques.

3. All-optical format conversion techniques

In the last two decades, the information volume flowing on communication networks increased exponentially, fostering the search for fast optical switching, gating and transmission techniques, along with equipment with integration facilities and low power consumption. Among those techniques, phase modulation of optical signals is an option that allows greater transmission distances in both analogue and digital transmission systems. Other method for more advanced modulation format conversion based on interferometric techniques will be also briefly described.

3.1. Format conversion from amplitude to phase modulation

Phase modulation generates signals with logical value 0 or 1, by varying the phase of light, while allowing it to be in the ON position. As opposed to intensity modulation, phase modulation has superior bandwidth efficiency and is not easily affected by signal distortions caused by relay nodes and transmission fibres. Several techniques have already been proposed to implement optical phase modulators, based on frequency shifters [16], lithium niobate ($LiNbO_3$) waveguide [17], gain transparent SOA [18] or using highly nonlinear fiber (HNLF) as the optical medium to phase modulate a CW laser [19].

Following the experimental findings with the XOR gate, this section characterizes the phase modulation properties of a MZI-SOA, using both interferometric arms, in co-propagation schemes. The setup is the same as the XOR gate of **Figure 13**, with MZI-SOA operational parameters (SOA bias current, input optical power) tuned to create a destructive output at port #I. When both control signals are synchronized, the optical CW coming out of the output of the two SOAs have opposite phases and interfere destructively when combined at COUPLER4. According to the XOR truth table in **Figure 17**, the resulting optical signal at port #I has no observable amplitude variation [20]. Though, the phase ϕ of the probe signal λ_2 will vary in accordance to the input pattern, as depicted in **Figure 17** [21].

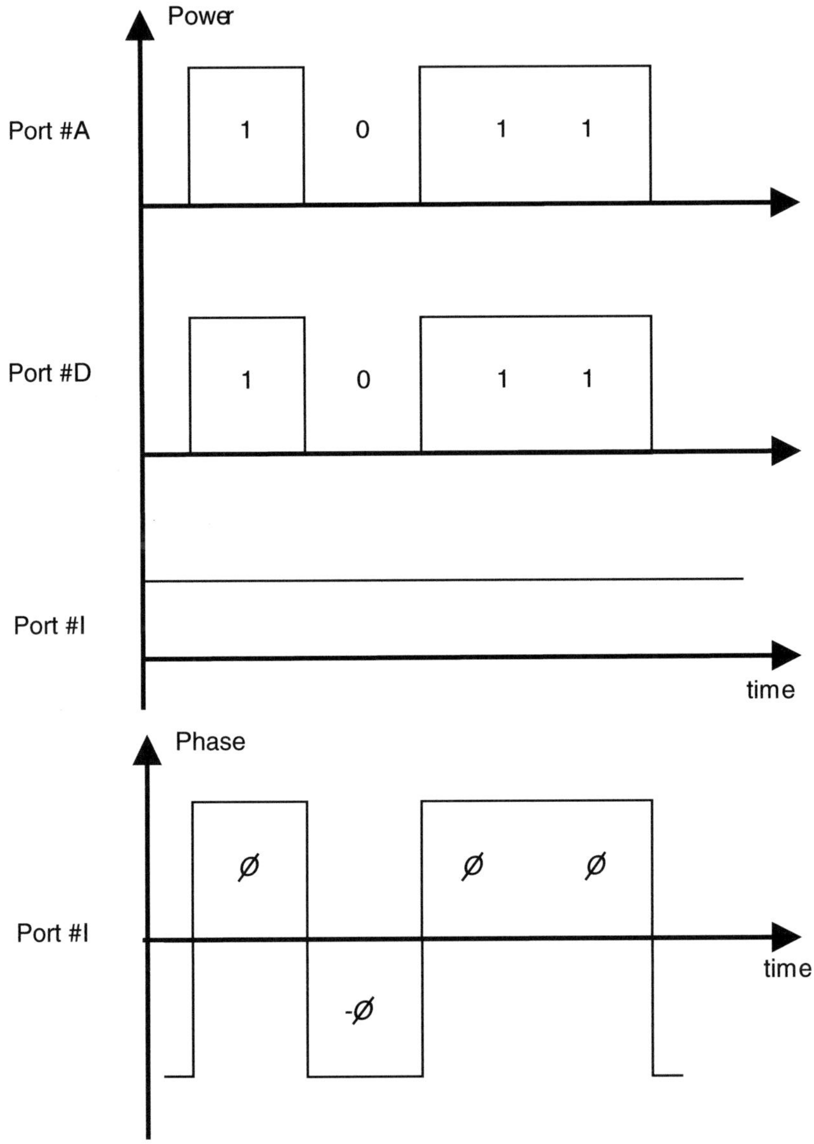

Figure 17. Principle of operation diagram of the conversion technique.

3.1.1. Experimental results and discussion

To investigate the phase modulation performance of an MZI-SOA for different bit rates, the operational parameters and the input power were optimized according to the phase eye

http://dx.doi.org/10.5772/66133

diagram opening, also called phase span in the subsequent paragraphs. BER measurements were not performed due to setup limitations imposed by the coherent receiver. The OCSA limits the size of the data sequence length to 4 bits at 2.5 Gbps or 16 bits at 10 Gbps [22], thus BER measurements are not feasible.

In order to confirm the feasibility of the interferometer as a phase modulator, experiments were carried out at bitrates of 2.5 and 10 Gbps. Data signals are launched into ports #D and #A with 2.5 dBm mean power and an average ER of 11.3 dB. The bias currents of both SOAs (I_{SOA}) were increased at the same time, from 150 to 300 mA at 2.5 Gbps and from 150 to 400 mA at 10 Gbps bit rate. For each bias value, the mean power of the control signal (P_{CW}) was swept from −6 to 2 dBm. The voltage applied to the PS was set to maximize the destructive interference at output port #I.

Figure 18 (left) presents a 4 bit pattern signal injected at ports #D and #A. **Figure 18** (right) shows the output signal at port #I, when control signal mean power is −4 dBm and SOA bias current is 250 mA. The phase variation associated to different logic levels is well pronounced but reversed when compared with the control signal intensity. Phase span and the mean power on the output signal are also proportional to the power of the CW probe signal and the bias current of the SOASs, as long as the SOAs are not in the saturated regime [9].

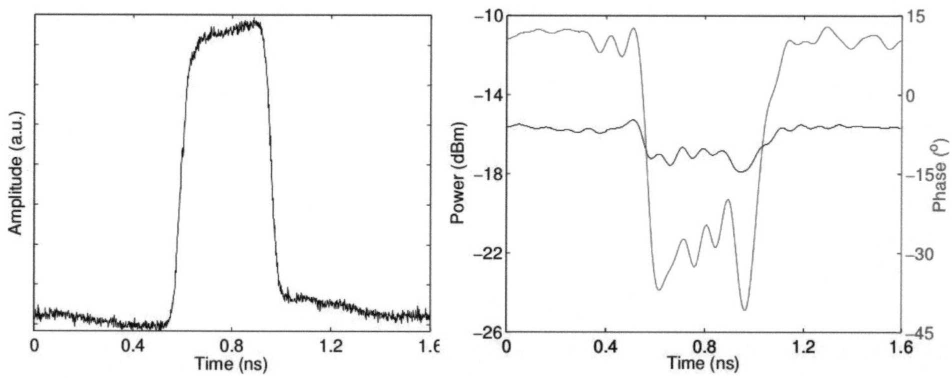

Figure 18. Left—Input sequence "0100"; Right—Phase and power output, with SOA input current I_{SOA} equal to 250 mA and input laser power P_{CW} equal to −4 dBm.

The proposed optical phase modulator was also characterized at 10 Gbps, and the tests were performed using data sequences 16 bits long [22]. **Figure 19** (left) shows the bit pattern launched at the input ports (#D and #A) of the interferometer. With SOA biased at 150 mA and control signal power launched at 0 dBm, the resulting output signal is depicted in **Figure 19** (right). Output power fluctuations are primarily due to noise. Similar to the previous experiments at 2.5 Gbps, phase shifts are inverted when compared with the logic levels of the data (control) signals. However, due to the carrier recovery time and the dynamics of the SOA, output phase levels are less pronounced at 10 Gbps when fast transitions occurs at the MZI-SOA input signals. Moreover, if the power of the probe signal is increased above 0 dBm, the

SOAs saturates and the conversion process is less efficient, which reduces the output mean power and phase span [9].

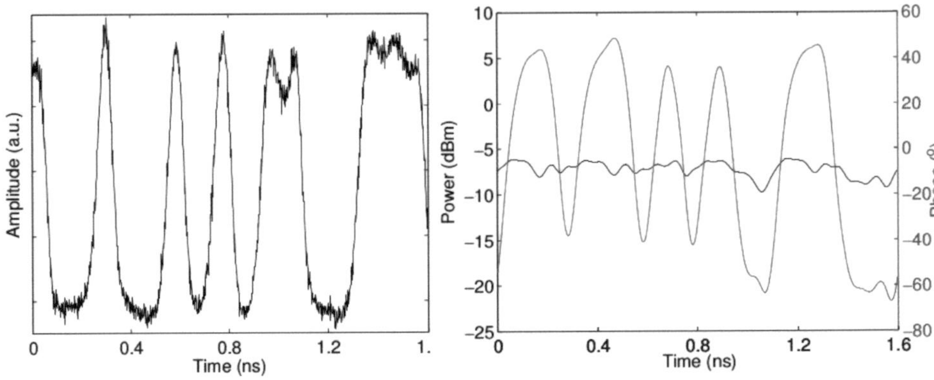

Figure 19. Left—Input sequence "1110010010101100"; Right—Phase and power output, with SOA input current I_{SOA} equal to 150 mA and input laser power P_{CW} equal to 0 dBm.

3.1.2. Summary

We have presented a method to perform optical phase modulation, from an all-optical XOR gate configuration. We measure the influence of input CW power and SOAs bias current on the signal phase at the MZI-SOA output port. We verify that an increase in the SOA bias current produces higher values of the mean power and the phase span of the output signal, but SOAs gain saturation has an inverse result on the same output signal. Overall, the experimental results show the viability of the MZI-SOA as device capable of all-optical modulation and format conversion.

3.2. Other advanced format conversions techniques

Higher order all-optical modulation formats can be generated using the phase modulator setup of **Figures 15** and **16** as a building block. For example, all-optical OOK to quadrature phase shift keying (QPSK) converter can be constructed with nested amplitude to binary phase shift keying (BPSK) converter pairs [23]. When the overall phase difference between the two nested BPSK pairs is set to $\pi/2$ rad, a QPSK signal is generated. Kang et al. [24] demonstrate the viability of this conversion scheme to generate a QPSK signal with 173 Gbps.

Furthermore, one may build a modulator for generating even higher order modulation formats, including quadrature amplitude modulation (QAM), following a similar method by which the amplitude to QPSK format converter is constructed [25]. Other methods of all-optical format conversion techniques using MZI-SOA are subject of further research in Ref. [26].

http://dx.doi.org/10.5772/66133

4. Conclusions

A MZI-SOA is a compact semiconductor device capable of performing many different all-optical modulations format conversion and generation functions. Its use has a building block for future all-optical networks can avoid the electronic bottleneck affecting opaque optical network nodes. It is estimated that 80% of the traffic flowing into a node is a pass-through traffic, with a destination located in another node of the network [27]. It is then particularly efficient to maintain that traffic flow in the optical domain, without optoelectronic conversion or packet processing. Transparent optical systems have the advantage of contributing to more energy-efficient networking without decreasing flexibility and agility. This clear and challenging objective is mandatory to cope with the traffic increase, while maintaining the cost and energy of the transported bit at an acceptable level.

Author details

Rogerio Pais Dionisio

Address all correspondence to: rdionisio@ipcb.pt

Escola Superior de Tecnologia, Instituto Politecnico de Castelo Branco, Castelo Branco, Portugal

References

[1] CISCO. The ZettabyteEra [Internet]. May 2012. Available from: http://www.cisco.com

[2] G. Wu, S. Talwar, K. Johnsson, N. Himayat and K. Johnson. M2M: from mobile to embedded internet. IEEE Communications Magazine. 2011;49(4):36–43.

[3] IEEE Communications Magazine. 2011;49(4):60–65.

[4] H. Karimi, M. Fenton, G. Lapierre, and E. Fournier. European Harmonized Technical Conditions and Band Plans for Broadband Wireless Access in the 790–862 MHz Digital Dividend Spectrum. In: IEEE, editor. Symposium on New Frontiers in Dynamic Spectrum; April 2010; Singapore. IEEE. 2010. pp. 1–9.

[5] Internet of Things [Internet]. July 2012. Available from: http://www.internet-of-things.eu/

[6] Celtic-Plus. Purple Book—Celtic Plus Programme of Possible and Recomended Research Items [Internet]. Available from: http://www.celtic-initiative.org/PurpleBook+/Purplebook.asp [Accessed: February 2012]

[7] CIP. CIP Technologies [Internet]. July 2012. Available from: http://www.ciphotonics.com/

[8] C. Reis, R. P. Dionisio, B. Neto, A. Teixeira and P. Andre. All-optical XOR based on integrated MZI-SOA with co- and counter-propagation scheme. In: IEEE, editor. ICTON Mediterranean Winter Conference; 10-12 December 2009; Angers, France. Warsaw, Poland: National Institute of Telecommunications. 2009. pp. 1–4.

[9] R. Dionisio, C. Reis, P. Andre, R. Nogueira and A. Teixeira. Experimental study of a phase modulator using an active interferometric device. In: IEEE, editor. MELECON 2010—15th IEEE Mediterranean Electrotechnical Conference; 26–28 April 2010; La Valleta, Malta; 2010. pp. 1142–1146.

[10] N. Yan, J. del Val Puente, T. G. Silveira, A. Teixeira, A. P. S. Ferreira, E. Tangdiongga, P. Monteiro and A. M. J. Koonen. Simulation and experimental characterization of SOA-MZI-Based multiwavelength conversion. Journal of Light wave Technology. 2009;27(2): 117–127.

[11] A theoretical and experimental study on modulation-format-independent wavelength conversion. Journal of Light wave Technology. 2010;28(4):587–595

[12] G. Maxwell, A. Poustie, C. Ford, M. Harlow, P. Townley, M. Nield, I. Lealman, S. Oliver, L. Rivers and R. Waller. Hybrid integration of monolithic semiconductor optical amplifier arrays using passive assembly. In: IEEE, editor. 55th Electronic Components & Technology Conference; 30 May–2 June 2006. Lake Buena Vista, Fl, USA. 2005. pp. 1349–1352.

[13] G. Maxwell, B. Manning, M. Nield, M. Hariow, C. Ford, M. Clements, S. Lucas, P. Townley, R. McDougall, S. Oliver, R. Cecil, L. Johnston, A. Poustie, R. Webb, I. Lealman, L. Rivers, J. King, S. Perrin, R. Moore, I. Reid and D. Scrase. Very low coupling loss, hybrid-integrated all-optical regenerator with passive assembly. In: IEEE, editor. European Conference on Optical Communication—ECOC 2002; 8–12 September 2002; Copenhagen, Denmark. 2002. pp. 1–2.

[14] T. Fjelde, D. Wolfson, A. Kloch, B. Dagens, A. Coquelin, I. Guillemot, F. Gaborit, F. Poingt and M. Renaud. Demonstration of 20 Gbit/s all-optical logic XOR in integrated SOA-based interferometric wavelength converter. Electronics Letters. 2000;36(22): 1863–1864.

[15] M. Hattori, K. Nishimura, R. Inohara and M. Usami. Bidirectional data injection operation of hybrid integrated SOA-MZI all-optical wavelength converter. IEEE Journal of Light wave Technology. 2007;25(2):512–519.

[16] B. Qi, L. L. Huang, H. K. Lo, and L. Qian. Polarization insensitive phase modulator for quantum cryptosystems. Optics Express. 2006;14(10):4264–4269.

[17] C. Langrock, E. Diamanti, R. V. Roussev, Y. Yamamoto, M. M. Fejer, and H. Takesue. Highly efficient single-photon detection at communication wavelengths by use of up

conversion in reverse-proton-exchanged periodically poled LiNbO3 waveguides. Optics Letter. 2005;30(13):1725–1727.

[18] W. Hong, D. Huang, X. Zhang, and G. Zhu. Simulation and analysis of gain-transparent SOA used as optical phase-modulator in DPSK applications. SPIE. 2007;6782:67.

[19] V. Marembert, C. Schubert, C. Weinert, H. G. Weber, K. Schulze, F. Futami and S. Watanabe. Investigations of fiber Kerr switch: nonlinear phase shift measurements and optical time division demultiplexing of 320 Gbit/s DPSK signals. In: Conference on Lasers and Electro-Optics CLEO 2005; 24–26 May 2005; Baltimore, USA. 2005. pp. 1432–1434.

[20] R. Vilar, J. M. Martinez, F. Ramos and J. Marti. All-optical DGD monitor for packet-switched networks based on an integrated active Mach-Zehnder interferometer operating as logic XOR gate. Optics Communications. 2008;281(21):5330–5334.

[21] S. C. Cao and J. C. Cartledge. Measurement-based method for characterizing the intensity and phase modulation properties of SOA-MZI wavelength converters. IEEE Photonics Technology Letters. 2002;14(11):1578–1580.

[22] APEX, editor. APEX Optical Spectrum Analyzer AP2041B. November 2009.

[23] K. Mishina, S. M. Nissanka, A. Maruta, S. Mitani, K. Ishida, K. Shimizu, T. Hatta and K. I. Kitayama. All-optical modulation format conversion from NRZ-OOK to RZ-QPSK using parallel SOA-MZI OOK/BPSK converters. Optics Express. 2007;15(12):7774–7785.

[24] I. Kang, M. Rasras, L. Buhl, M. Dinu, S. Cabot, M. Cappuzzo, L. Gomez, Y. Chen, S. Patel, N. Dutta, A. Piccirilli, J. Jaques, and C. Generation of 173-Gbits/s single-polarization QPSK signals by all-optical format conversion using a photonic integrated. In: ECOC 09—European Conference on Optical; September; p. 1.

[25] R. Dionisio, R. Nogueira and A. Teixeira. Advanced optical modulation and format conversion. In: SPIE, editor. AOP 2013—Conference on Application of Optics and Photonics; 26 May 2014; Aveiro, Portugal. doi:10.1117/12

[26] R. Dionisio, R. Nogueira and A. Teixeira. Parametric impairments analysis of all-optical format conversion techniques with MZI-SOA. In: SPIE, editor. Conference on Application of Optics and Photonics; 26 May 2014; Aveiro, Portugal. 2014. doi:10.1117/12

[27] Y. Zhang, P. Chowdhury, M. Tornatore, and B. Mukherjee. Energy efficiency in telecom optical networks. IEEE Communications Surveys Tutorials. 2010;12(4):441–458.

Chapter 9

Point Diffraction Interferometry

Daodang Wang and Rongguang Liang

Additional information is available at the end of the chapter

http://dx.doi.org/10.5772/65907

Abstract

The point diffraction interferometer (PDI) employs a point-diffraction spherical wavefront as ideal measurement reference, and it overcomes the accuracy limitation of reference optics in traditional interferometers. To overcome the limitation of measurement range either with pinhole (low light transmission) or with single-mode fiber (low NA), a single-mode fiber with narrowed exit aperture has been proposed to obtain the point-diffraction wavefront with both high NA and high power. It is a key issue to analyze the point-diffraction wavefront error in PDI, which determines the achievable accuracy of the system. The FDTD method based on the vector diffraction theory provides a powerful tool for the design and optimization of the PDI system. In addition, a high-precision method based on shearing interferometry can be applied to measure point-diffraction wavefront with high NA, in which a double-step calibration including three-dimensional coordinate reconstruction and symmetric lateral displacement compensation is used to calibrate the geometric aberration. The PDI is expected to be a powerful tool for high-precision optical testing. With the PDI method, a high accuracy with RMS value better than subnanometer can be obtained in the optical surface testing and submicron in the absolute three-dimensional coordinate measurement, demonstrating the feasibility and wide application foreground of PDI.

Keywords: point diffraction, optical testing, pinhole, optical fiber, wavefront sphericity

1. Introduction

The development of optical design and fabrication such as projection optics for extreme ultraviolet lithography (EUVL) and laser fusion, etc. places ultrahigh requirement on the optical testing precision and accuracy. In the EUVL operating at a wavelength of 13.5 nm [1–4], the projection optics is composed of 4–6 aspheric mirrors and each mirror needs to be finished with the figure error less than 0.2–0.3 nm (RMS), requiring a higher testing accuracy (e.g., 0.1 nm RMS). The optical interferometers, including the Twyman-Green [5] and Fizeau interferometers [6], have been widely used in the testing lenses and mirrors for figure metrology, in which standard lenses are applied to produce the necessary reference wavefront. Due to the fabricating error of the reference optics, the achievable measurement accuracy of the traditional interferometers is generally limited within $\lambda/50$ (λ is the operating wavelength). As a novel optical technique, the point-diffraction interferometer (PDI) [7–12] has been developed to overcome the accuracy limitation in traditional interferometers, and it can reach the measurement accuracy in the order of subnanometer. The PDI method employs point-diffraction spherical wavefront as ideal measurement reference, and it can achieve the accuracy better than $10^{-3}\lambda$. The PDI does not require precise and costly standard parts, and it provides a feasible way to overcome the accuracy limitation of the reference optics in traditional interferometers. The process of point diffraction determines the reachable accuracy of PDI, thus, a good reproducibility of the measurement accuracy can be achieved with PDI. The ideal spherical wave can be generated from the point-diffraction source such as circle pinhole [13–15] and optical fiber [16–18]. Nikon realized that the high-precision testing of spherical and aspherical surfaces with the pinhole PDI [14, 15], and the Lawrence Livermore National Laboratory measured surface figure with the single-mode-fiber PDI [16]. The diffracted wave in both PDI systems serves as testing wave as well as reference wave. To extend the aperture of the optics, a fiber PDI with two optical fibers serving as point sources was proposed for large optics measurement [18], in which the diffracted wave from one fiber acts as testing wave and that from the other fiber as reference wave. Considering the fact that the diffraction light power from pinhole is poor (transmittance <0.1‰) and the numerical aperture (NA) of diffracted wavefront with single-mode fiber is quite low (<0.20), both approaches limit the measurement range of the existing PDIs. A submicron-aperture (SMA) fiber with cone-shaped exit end has been proposed to obtain both the high diffraction light power and high-NA spherical wavefront [19–21], and it is considered as a feasible way to extend the measurement range of the system. Due to the fact that the achievable testing accuracy of PDI is mainly determined by the sphericity of diffracted wavefront, the analysis of the diffracted wavefront error has become a fundamental way to evaluate the performance of PDI. In our research, both numerical analysis and experimental measurement have been performed to reconstruct the point-diffraction wavefront aberration. Besides, various setups of PDI have been developed to realize the precise surface testing with adjustable contrast [12, 22] and three-dimensional (3D) coordinate measurement [20, 23, 24]. The new PDI for three-dimensional (3D) coordinate measurement avoids the measurement uncertainty introduced by the imperfect target in multilateration and allows the target (made of two point-diffraction sources) to take free movement within a volumetric space over NA of point-diffraction wave. Section 2 presents a brief review of the PDI for testing of wavefront and the optical surface. Section 3 presents the basic theory of the SMA fiber PDI for 3D coordinate measurement. Sections 4 and 5 explain the numerical method and the experimental measurement method for the evaluation of diffracted wavefront sphericity, as well as the corresponding analyzing results.

http://dx.doi.org/10.5772/65907

2. PDI for wavefront and surface testing

2.1. Wavefront testing with PDI

The PDI was first proposed by Smartt and Strong in 1972 [7]. The early version of the PDI was an interferometer using a PDI plate with partial transmission. **Figure 1** shows the principle of the PDI with partially transparent PDI plate. The PDI plate consists of an absorbing metal coating mask on a clear substrate and a tiny pinhole placed near focus to divide the wave into two parts, namely the testing wave and reference wave. The pinhole picks off part of the incident focused light wave and generates the diffracted spherical wavefront as reference wavefront. The test wave passes through the PDI plate and then interferes with diffracted spherical wave. The PDI can be used to measure the wavefront error from imaging optics and flow field, etc. In the design of the PDI shown in **Figure 1**, it is a key issue to determine the pinhole size and plate transmittance [9]. Due to the reduction in light intensity incident on the tiny pinhole, the controllable reduction in the test wave intensity by mask attenuation is required to get maximum fringe contrast.

Another PDI plate with double apertures has been applied in the PDI for the evaluation of wavefront error in imaging optics [25], as shown in **Figure 2**. A beam splitter such as the

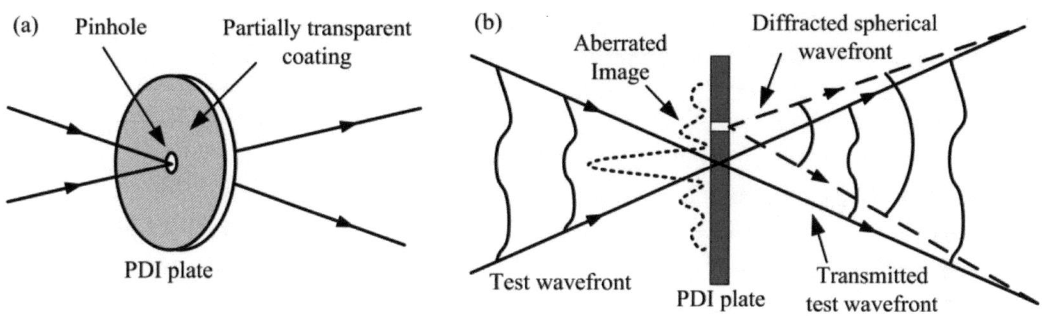

Figure 1. Principle of PDI with PDI plate. (a) PDI plate and (b) PDI operation.

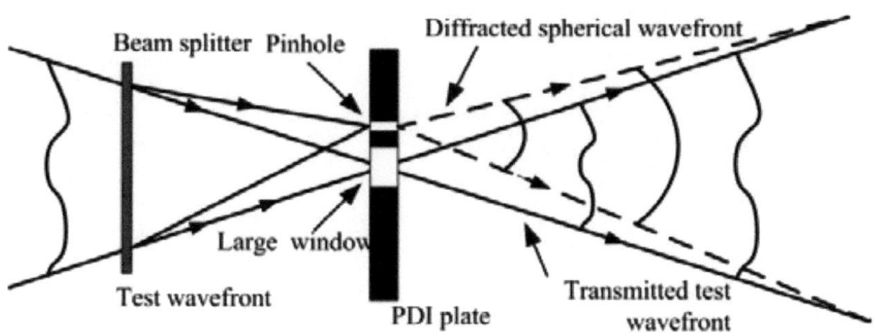

Figure 2. PDI plate with double aperture.

transmission grating is used to divide the test wave into two waves with a small angular separation. The PDI plate consists of one tiny pinhole and one large window on an opaque mask, and both of them are placed at the respect focal points of two beams. The pinhole is applied to generate the spherical reference wavefront by diffraction and the window transmits the test wave. The intensity of the diffracted wave relative to that of the conventional PDI is increased by several orders of magnitude. With the application of double-aperture PDI plate, there is no need to further attenuation in the test wave to match the intensities of interfering waves. Besides, the beam division enables the potential to introduce the various phase shifts between the interfering waves.

2.2. Optical surface testing with PDI

Figure 3 shows the typical configuration of PDI for optical surface testing. Either the pinhole (**Figure 3a**) [13–15] or the single-mode optical fiber (**Figure 3b**) [16–18] can be used as the point-diffraction source to generate the required spherical wavefront. The diffracted wavefront is separated to two parts, i.e., the test and reference wavefronts. The test wave travels toward the spherical surface under test, and the pinhole (exit end of fiber) is positioned at the curvature center of the test surface, so that the reflected wave from the test surface converges at the pinhole mirror (semitransparent metallic film on the output end of fiber) and then is reflected at the pinhole mirror (semitransparent metallic film on the output end of fiber). The test wave combines and interferes with the reference wave after reflection. By translating the test surface with a precise PZT scanner, the test surface error can be measured with the phase-shifting method. The selection of beam polarization state in the system can significantly influence the measurement. The polarization would affect the sphericity of the diffracted wavefront, and the reflection at mirror over a high NA can also introduce polarization-dependent phase shifts. In the PDI system, the polarization state of diffracted spherical wave is generally adjusted to be circularly polarized [26], in which the effect of polarization on the measurement precision is negligible.

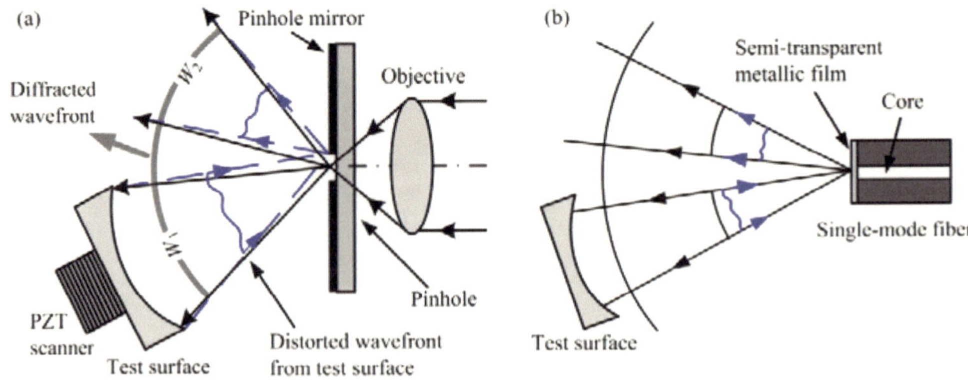

Figure 3. Configuration of PDI for optical surface testing. (a) Pinhole PDI [13], (b) single-mode fiber PDI [16].

http://dx.doi.org/10.5772/65907

In the PDI with pinhole method, the circular pinhole of submicron (or even smaller) diameter is adopted to obtain high measurable NA. **Figure 4(a)** shows a scanning-electron microscope (SEM) picture of the pinhole fabricated by etching the chromium film with the focused ion-beam etching (FIBE) method [22], in which the metallic layer is sputtered onto the silica substrate. A nearly perfect circular pinhole can be obtained with the FIBE method. The high measurable NA can be achieved with the pinhole method; however, the nonadjustable fringe contrast would limit the measurement accuracy in the testing of low-reflectivity surface, due to the poor fringe contrast and difficulty in fringe processing.

A pinhole PDI with adjustable fringe contrast can be adopted for the testing of high-NA spherical surfaces with low reflectivity [22]. The polarizing elements are applied to transform

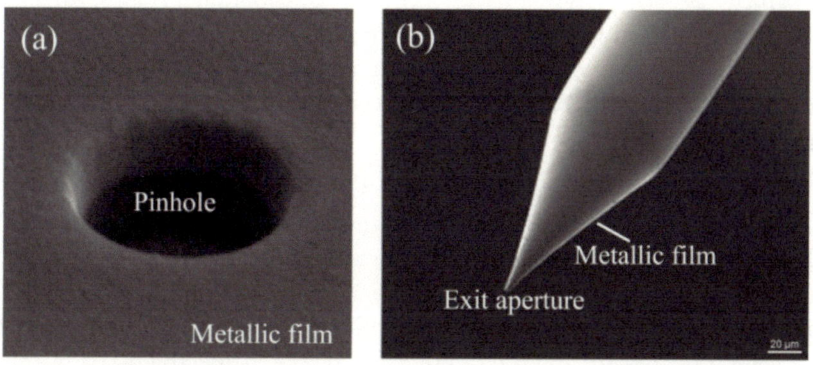

Figure 4. Point diffraction sources in PDI. (a) Pinhole [22], and (b) submicron-aperture fiber [21].

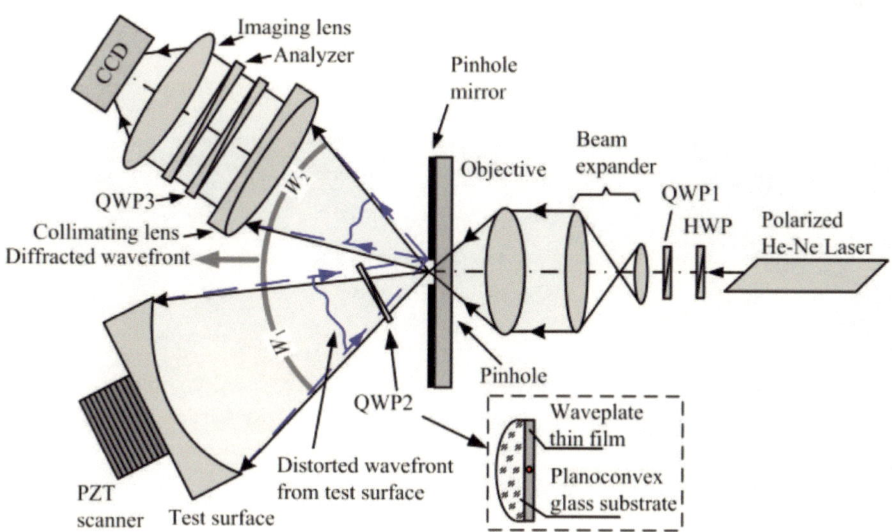

Figure 5. Configuration of pinhole PDI with adjustable fringe contrast [22].

the polarization states and adjust the relative intensities of the interfering beams, by which the adjustable fringe contrast can be achieved. **Figure 5** shows the optical configuration of the pinhole PDI with adjustable fringe contrast. A quarter-wave plate (QWP2) with special structure (consisting of a thin waveplate film and a plano-convex substrate) is placed at the test path, with the fast axis oriented at −45° to horizontal. The diffracted wave is adjusted to be circularly polarized. The test wave W_1 travels toward the test surface and passes through QWP2 twice, respectively, before and after the reflection at the test surface, then the test wave W_1 becomes opposite circularly polarized with respect to the reference wave W_2. The relative intensities of the test and reference beams can be adjusted by rotating the transmission axis of the analyzer, realizing the adjustable fringe contrast. Due to the fact that the diffracted wave is divergent, QWP2 in the test path would introduce different phase retardations in various directions. To minimize the effect of QWP2 on the divergent test wave, a true zero-order waveplate, which has advantages of less sensitive to variation in angle of incidence and wavelength, is employed in the system. In the true zero-order waveplate, a thin waveplate film is cemented on the glass substrate. To minimize the aberrations introduced by the glass substrate in the case of divergent waves, a plano-convex substrate is used in the true zero-order waveplate QWP2.

Figure 6 shows the testing results about a spherical surface with the reflectivity 4%, NA 0.40 and aperture diameter 137.7 mm. **Figure 6(a)** is the surface error measured with the adjustable-contrast pinhole PDI system shown in **Figure 5**, and **Figure 6(b)** is that obtained with the ZYGO interferometer. According to **Figure 6**, a good agreement between the PDI result and that from the ZYGO interferometer is achieved, and the PV and RMS differences between the testing results are about 0.0214λ and 0.0025λ, respectively. Thus, high-accuracy testing of the surface with low reflectivity is realized with the adjustable-contrast PDI system.

The light transmission through the pinhole is quite low (<0.1‰), and it significantly limits the achievable measurement range of the pinhole PDI. In the PDI with single-mode fiber, the

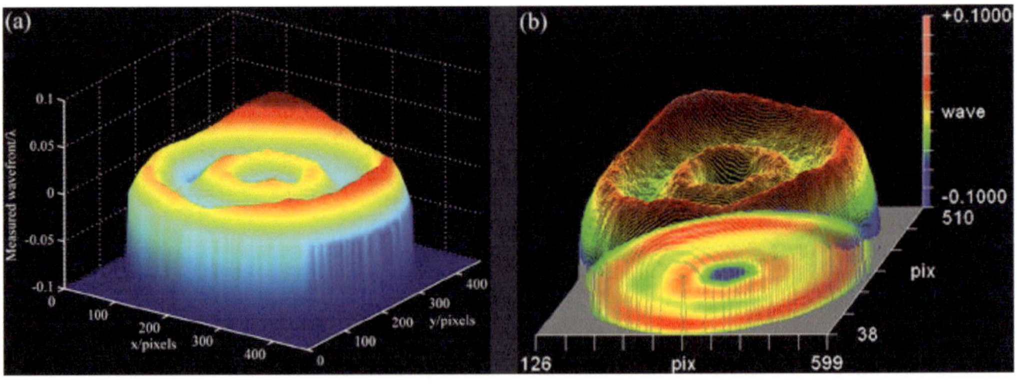

Figure 6. Measured surface error of test spherical surface with (a) the adjustable-contrast pinhole PDI and (b) the ZYGO interferometer [22].

adjustable fringe contrast is easy to realize and high light transmittance (>10%) can be obtained; however, the measurable NA in the fiber method is limited by the NA of the fiber, which is commonly less than 0.2. Besides, the measurable aperture of the test surface is approximately half that of the diffraction wave, both for the pinhole PDI and single-mode fiber PDI shown in **Figure 3**. A PDI system with two optical fibers as point-diffraction sources was developed for full use of the diffracted wavefront [18], in which the diffracted wave from one fiber serves as reference wavefront and that from the other fiber as test wave. Besides, a novel submicron-aperture (SMA) fiber with cone-shaped exit end, as shown in **Figure 4(b)**, has been proposed to obtain both the high diffraction light power and high-NA spherical wavefront [19–21]. The SMA fiber taper surface is coated with metallic film and the exit aperture is formed from the polished tip, it is formed with the same processing technology as manufacturing of fiber-based probes for the scanning near-field optical microscopy. With the SMA fiber, the measurable range can be almost extended to a half space, and the corresponding light transmittance over 50% is obtained [20]. Thus, it is considered as a feasible way to extend the measurement range of the system.

3. PDI for 3D coordinate measurement

In this section, a fiber PDI with submicron aperture [20], which is based on the single-mode fiber with a narrowed exit aperture, for the absolute 3D coordinate measurement within large angle range is described, and the system configuration is shown in **Figure 7**. After passing through the half-wave plate (HWP1), the frequency-stabilized laser beam is separate into p-polarized and s-polarized beams by the polarized beam splitter (PBS), and they are coupled into the single-mode fibers SF1 and SF2, respectively. The exit ends of the fibers SF1 and SF2 are integrated into a target with certain lateral offset, and the output point-diffraction waves W_1 and W_2 interfere on the CCD detector. By translating the PZT scanner, the point-diffraction interference field can be retrieved with phase-shifting method, and the corresponding 3D coordinates of target can be obtained with numerical iterative reconstruction algorithm.

To increase the NA of diffracted wavefront with optical fiber, a single-mode fiber with submicron aperture (as shown in **Figure 4b**) is applied as point-diffraction source. Both the exit ends

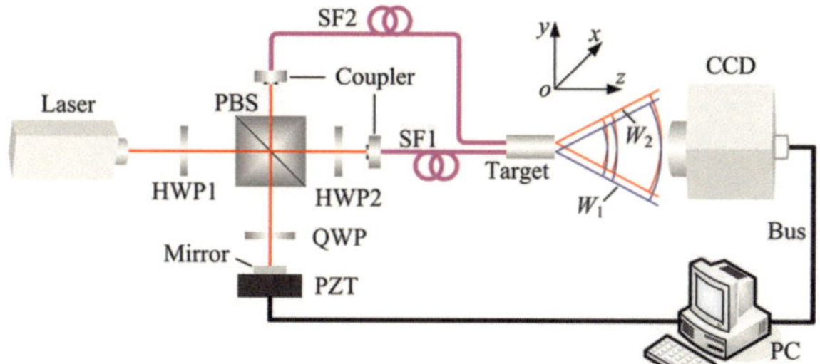

Figure 7. System configuration of the SMA fiber PDI for 3D coordinate measurement [20].

of two single-mode fibers are integrated in a target with certain lateral offset. According to the PDI system shown in **Figure 7**, the absolute 3D coordinates of target can be measured from the phase distribution in interference field corresponding to the optical path difference (OPD), as shown in **Figure 8**.

Denoting the plane of CCD detector as the xy plane and the central position o, with the distances r_1 and r_2 between an arbitrary point $P(x, y, z)$ on the CCD detector and the exit apertures of two fibers on the target, as shown in **Figure 8**, we have the phase difference $\varphi(x, y, z)$,

$$\varphi(x, y, z) = k[r_1(x_1, y_1, z_1; x, y, z) - r_2(x_2, y_2, z_2; x, y, z)], \tag{1}$$

where $k = 2\pi/\lambda$, (x, y, z) is the known 3D coordinate of the point P, (x_1, y_1, z_1) and (x_2, y_2, z_2) are those of two fiber apertures in the target under measurement. According to the one-to-one correspondence relationship of the phase φ at position P (that is the i_{th} pixel on CCD) and 3D coordinates of two fiber apertures in the target, the phase difference can be written as follows:

$$f_i(\mathbf{\Phi}) = k(\varphi^i - \varphi^0) - \widehat{\xi}_i, \tag{2}$$

where the vector $\mathbf{\Phi}$ indicates the coordinates of two fiber apertures under measurement that is $\mathbf{\Phi} = \{(x_1, y_1, z_1); (x_2, y_2, z_2)\}$; $\widehat{\xi}_i$ is the measured phase difference. The phase differences φ^i and φ^0 can be written as follows:

$$\begin{cases} \varphi^i(x^i, y^i, z^i) = k[r_1(x_1, y_1, z_1; x^i, y^i, z^i) - r_2(x_2, y_2, z_2; x^i, y^i, z^i)], \\ \varphi^0(x^0, y^0, z^0) = k[r_1(x_1, y_1, z_1; x^0, y^0, z^0) - r_2(x_2, y_2, z_2; x^0, y^0, z^0)], \end{cases} \tag{3}$$

where (x^i, y^i, z^i) is the 3D coordinate of the i_{th} pixel on CCD. From Eq. (2), various equations can be established corresponding to the pixel positions on the CCD, and we have the matrix equations,

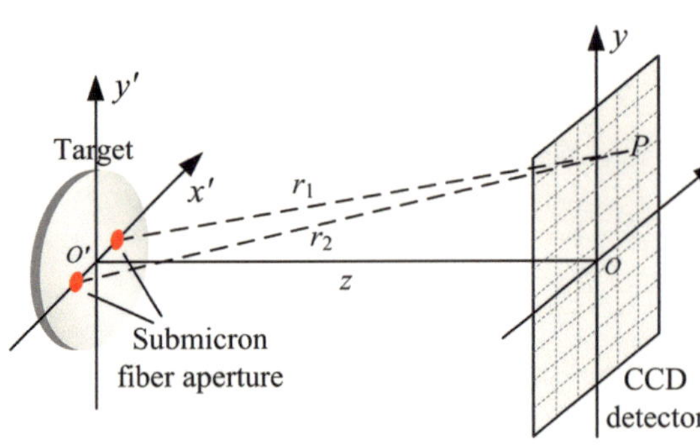

Figure 8. Model for 3D coordinate reconstruction [20].

$$f(\mathbf{\Phi}) = \{f_i(\mathbf{\Phi})\} = \begin{cases} k(\varphi^1 - \varphi^0) - \widehat{\xi}_1 \\ \vdots \\ k(\varphi^i - \varphi^0) - \widehat{\xi}_i \\ \vdots \\ k(\varphi^m - \varphi^0) - \widehat{\xi}_m \end{cases}, \tag{4}$$

where the subscript m is the number of the pixels selected for coordinate reconstruction. For the 6 unknowns in Eq. (2), at least 6 pixels are needed to obtain the 3D coordinates of two fiber apertures. To realize the accurate measurement of 3D coordinates with the PDI system, over 6 pixels could be applied to reconstruct the 3D coordinates, and a quadratic function corresponding to Eq. (4) can be obtained as:

$$F(\mathbf{\Phi}) = \frac{1}{2} f(\mathbf{\Phi})^T f(\mathbf{\Phi}) = \frac{1}{2} \sum_{i=1}^{m} f_i^2(\mathbf{\Phi}). \tag{5}$$

Thus, the space coordinates of two fiber apertures can be determined from the global minimum $\mathbf{\Phi}^*$ of the function F (that is the least-square solution of Eq. (4)), with a single true solution set.

An experimental SMA-fiber PDI system with aperture size about 0.5 μm, has been set up for the measurement of 3D coordinates. For comparison, the target is installed on the probe of a CMM with the positioning accuracy about 1.0 μm, and the 3D coordinates measured with CMM is taken as the nominal value. **Figure 9** shows the measurement results about the 3D coordinate deviations in x and z directions corresponding to the initial target position (−0.375, 15, and 200) mm. According to **Figure 9**, a good agreement between the CMM results and those from the SMA-fiber PDI system is achieved, and the coordinate measurement errors in x and z directions are 0.47 μm and 0.68 μm RMS, respectively. Due to the decreasing in the light intensity and enlargement in fringe spacing, the measurement error also grows with the increase of measurement distance. Several hardware factors can limit the measurement range

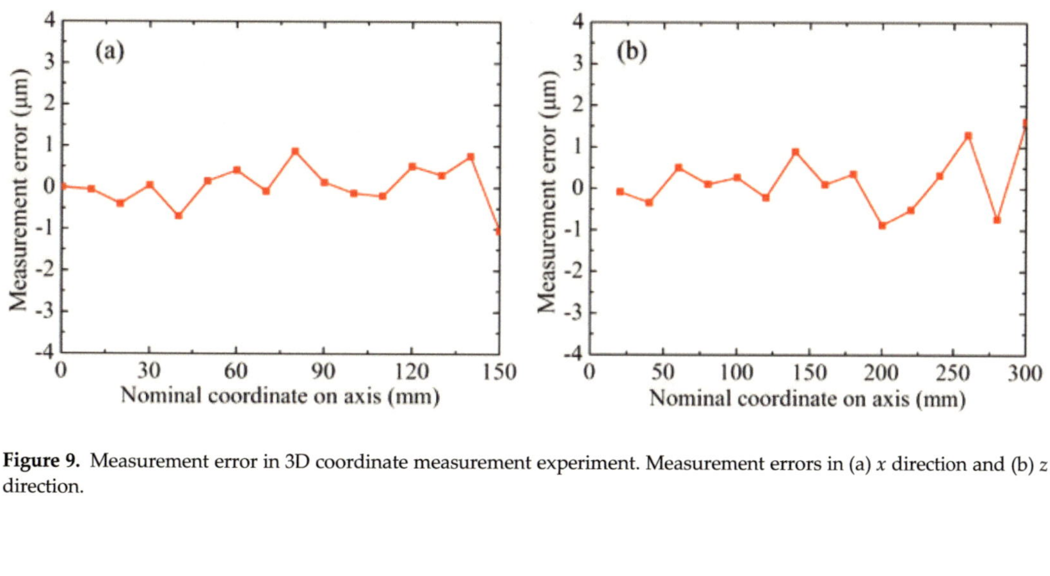

Figure 9. Measurement error in 3D coordinate measurement experiment. Measurement errors in (a) x direction and (b) z direction.

of the PDI system in practical application, including the detector size and light power, though there is no theoretical limit in principle. The PDI system can be further improved by increasing CCD detector size, lateral off set of two SMA fibers and source power. Compared to the traditional single-mode fiber, the SMA fiber provides a feasible way to obtain the ideal spherical reference wavefront, whose aperture angle almost covers half space, and the lateral measurement range is greatly extended.

4. Numerical analysis of point-diffraction wavefront

As one of the most important elements in the PDI, the point source for the diffracted reference wavefront determines the achievable accuracy in the measurement. Purely empirical design of point source parameters is both time consuming and costly. The numerical method based on diffraction theory is a feasible way for the analyzing point-diffraction wavefront. The scalar diffraction theory is valid only when the pinhole size is several times larger than the operating wavelength. For the high-NA spherical wavefront emerging from a tiny aperture with the size comparable with or less than operating wavelength, the vector diffraction theory (that is a nonapproximate method) is required to realize the accurate estimation of diffracted wavefront error. In this section, numerical analysis based on finite difference time domain (FDTD) method (that is a vector diffraction theory) [27] is presented. **Figure 10** shows the flow diagram for the simulation of point-diffraction wavefront based on FDTD method. Due to limitations of computer memory capacity and runtime, FDTD cannot be directly applied to calculate the far-field distribution of pinhole diffraction. In the first step, the near-field distribution of point diffraction is analyzed with the FDTD method, and then the near-to-far field transition based on Huygens' principle is performed to obtain the far-field distribution of point-diffraction wavefront at the position under study. Finally, the sphericity evaluation is carried out to get the departure of point-diffraction wavefront from an ideal sphere.

4.1. Simulation of pinhole diffraction

Figure 11 shows the amplitude and phase distribution in the near-field diffraction, respectively, corresponding to the pinholes of 0.5 μm and 1 μm diameters, respectively. According to **Figure 11**, the phase distribution within Airy disk range is close to an ideal sphere, and the angle of Airy disk increases with the decrease of pinhole diameter.

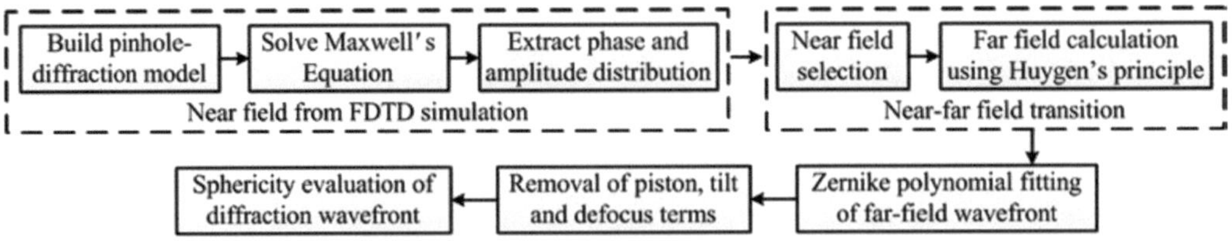

Figure 10. Procedure for simulation of point-diffraction wavefront based on FDTD method.

http://dx.doi.org/10.5772/65907

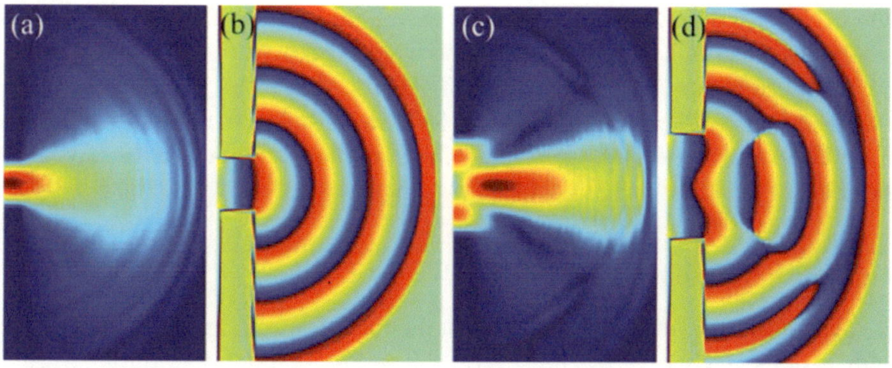

Figure 11. Near-field distribution of pinhole-diffraction wavefront. (a) Amplitude and (b) phase distribution with 0.5 μm-diameter pinhole; (c) amplitude and (d) phase distribution with 1 μm-diameter pinhole [27].

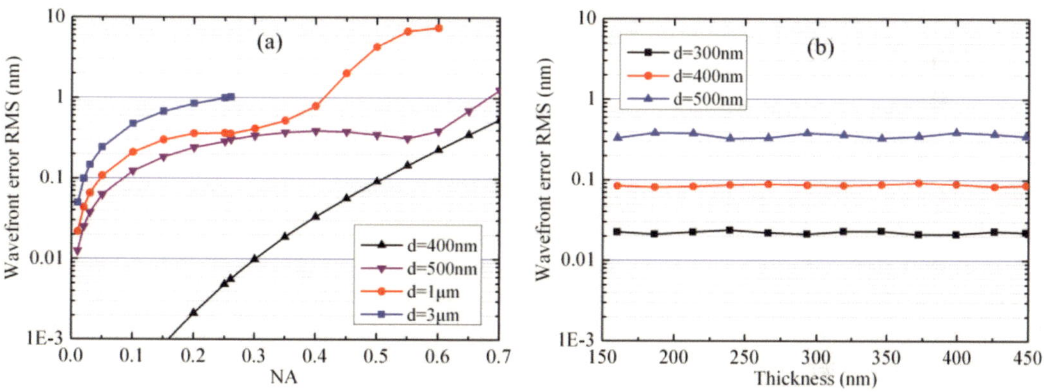

Figure 12. Simulation results for pinhole-diffraction wavefront. (a) Diffracted wavefront error over various NAs for different pinhole diameters, and (b) diffracted wavefront error under various film thicknesses [27].

To analyze the wavefront error where the test surface with 500 mm curvature radius is placed, far field is positioned at 500 mm away from pinhole. The wavefront errors over various NAs of diffracted wavefronts corresponding to different pinhole diameters are shown in **Figure 12(a)**. According to **Figure 12(a)**, the point-diffraction wavefront error grows both with the pinhole size and NA range, and the wavefront error RMS for the 1 μm pinhole diameter over 0.35 NA is smaller than 0.53 nm. Thus, it can be taken as ideal reference wavefront and applied to realize the measurement accuracy reaching to RMS $\lambda/1000$ (λ = 532 nm). Besides, larger diffraction angle could be obtained with smaller pinhole, and it allows much higher measurable NA. The diffraction angle corresponding to 1 μm pinhole diameter is about 75°, and that for 3 μm pinhole diameter is about 30°. Thus, the small pinhole is required in the testing of surfaces with high NA. However, the small pinhole size would result in significant reduction of diffraction light intensity. The pinhole size could be optimized according to the requirement of testing accuracy and achievable diffracted wavefront precision corresponding to various pinhole diameters. The pinhole in PDI plays not only the role as a filter to remove the aberration in diffracted wavefront, but also as waveguide. **Figure 12(b)** shows the numerical results about the effect of film thickness on diffracted wavefront for different pinhole diameters, with the thickness ranging from 150 nm to 450 nm. The mean values of wavefront error RMS corresponding to 300 nm, 400 nm, and 500 nm pinhole diameters are about 0.02 nm, 0.08 nm, and 0.36 nm, respectively, and the corresponding fluctuation ranges are within 10%. Thus, the effect of variation in film thickness ranging from 150 nm to 450 nm is negligible.

4.2. Simulation of SMA fiber diffraction

Figure 13(a) shows the simulation results about the full wavefront aperture angle for the SMA fibers with various exit apertures [20]. According to **Figure 13(a)**, the full aperture angle of diffracted wave obviously increases with the decreasing of fiber aperture, and that corresponding to the 0.5 μm fiber aperture are about 160°, providing the necessary conic boundary to extend the measurement range almost within a half space. **Figure 13(b)** shows the wavefront errors over various NAs of diffracted wavefronts corresponding to different SMA fiber apertures [20]. From **Figure 13(b)**, the increase in NA range and fiber aperture could lead to the growth of diffracted wavefront error. The wavefront error over 0.70 NA range for the 0.5 μm fiber aperture is better than $\lambda/1000$ RMS, and it can be applied as ideal spherical reference wave in the PDI system.

Figure 14 shows the analyzing result about light transmittance in the point diffraction both with the SMA fiber method and pinhole method. According to **Figure 14**, the light transmittance in both the SMA fiber diffraction and pinhole diffraction grows with the increase in aperture size. It can be seen from **Figures 12(a)** and **13(a)** that the similar aperture angles can be obtained with the same diffraction aperture size, however, the light transmittance in the SMA fiber diffraction is far larger than that in pinhole diffraction according to **Figure 14**. The light transmittance corresponding to the 0.5 μm aperture in the SMA fiber method and pinhole method is about 67% and 6%, respectively. Thus, the SMA fiber provides a feasible way to

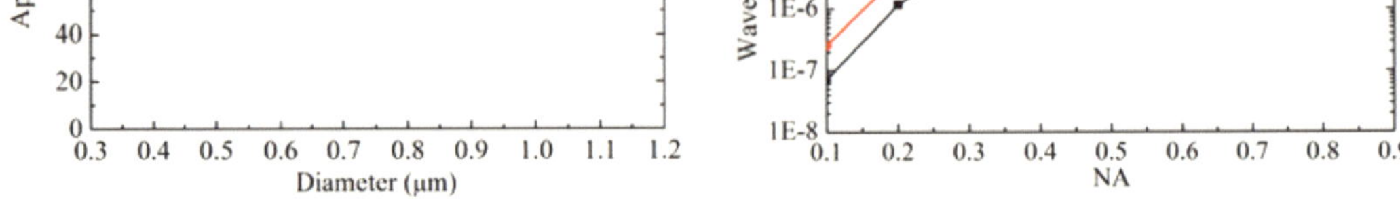

Figure 13. Simulation results for SMA-fiber-diffraction wavefront. (a) Aperture angle of diffracted wave and light transmittance for various fiber apertures, and (b) diffracted wavefront error within various NA ranges for different fiber apertures [20].

http://dx.doi.org/10.5772/65907

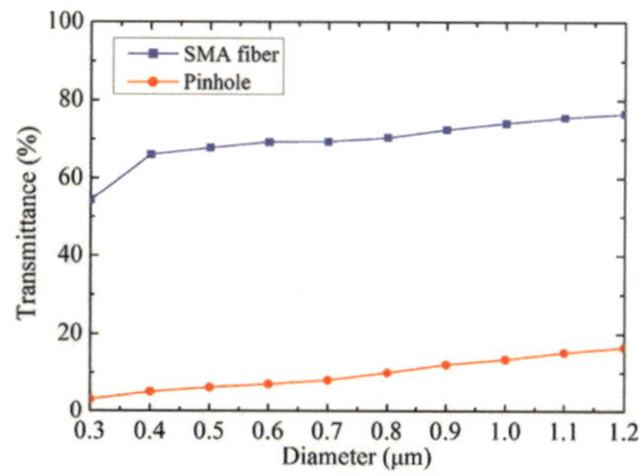

Figure 14. Simulation results for the light transmittance in pinhole diffraction and SMA fiber diffraction.

obtain the high light intensity required in the optical testing, enabling the extension of measurement range with PDI system.

5. Experimental measurement of point-diffraction wavefront

The numerical simulation based on the vector diffraction theory such as FDTD method provides an easy and efficient way to estimate the point-diffraction wavefront error. However, both the computational accuracy and complex practical factors (such as environmental disturbance and performances of various optical parts) can introduce significant deviation from an ideal case. Due to the accuracy limitation of standard optics, the traditional interferometers fail to measure the point-diffraction wavefront error, which is expected to be in the order of subnanometer or even smaller. Various experimental testing methods have been proposed to measure the point-diffraction wavefront error, the majority of which are based on the hybrid method [15] and null test [28]. Typically, the hybrid method requires several measurements with the rotation and displacement of the optics under test, it is sensitive to environmental disturbance and cannot completely separate the systematic error. The null test is a self-reference method, and it is widely applied to reconstruct the point-diffraction wavefront and calibrate the PDI. However, it requires the foreknowledge about the system configuration to remove the high-order aberrations, which are introduced by the point-source separation and cannot be negligible in the case of high-NA wavefront. In this section, a high-precision measurement method, which is based on shearing interferometry, is presented to analyze the point-diffraction wavefront from SMA fiber [21].

Figure 15 shows the schematic diagram of SMA fiber projector used to evaluate the sphericity of point-diffraction wavefront. The SMA fiber can be applied to obtain both high diffraction light power and high-NA spherical wavefront. Four parallel SMA single-mode fibers S_1, S_2, S_3, and S_4 with coplanar exit ends are integrated in a metal tube. To obtain the identical wavefront parameters, the four SMA fibers with same aperture size and cone angle are carefully chosen in the experiment to minimize the measurement error. The coherent beams are coupled into the SMA fibers, and the point-diffraction waves from the fibers interfere on a detection plane, as shown in **Figure 15(b)**. By alternatively switching on the waves from the fiber pairs G_1 (S_1 and S_2) and G_2

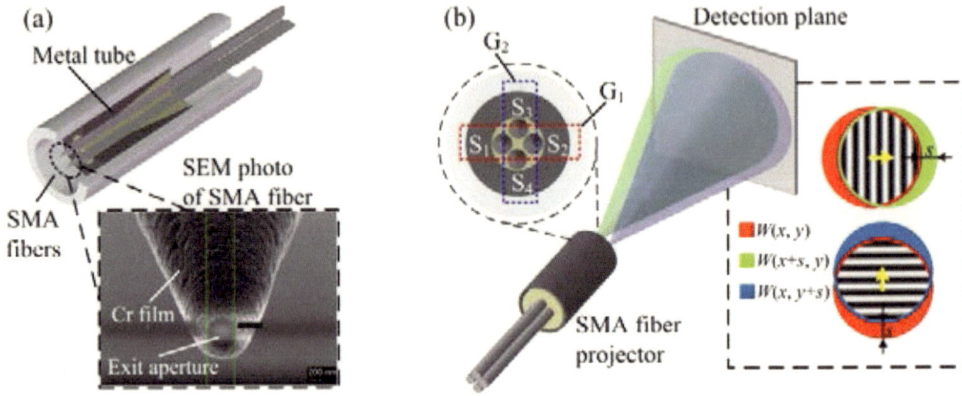

Figure 15. Schematic diagram of SMA fiber point-diffraction wavefront measurement system. (a) Structure of SMA fiber projector and (b) schematic diagram of SMA fiber point-diffraction wavefront measurement [21].

(S_3 and S_4), the shearing interferograms in x and y directions can be obtained and the corresponding shearing wavefronts can be measured with a phase-shifting method, respectively.

5.1. Point-diffraction wavefront retrieval method

With the shearing wavefronts obtained from the projector shown in **Figure 15**, the differential Zernike polynomials fitting method can be applied to retrieve the SMA fiber point-diffraction wavefront. Denoting the point-diffraction wavefront under test as $W(x, y)$, we have the shearing wavefronts $\Delta W_x(x, y)$ and $\Delta W_y(x, y)$ in x and y directions,

$$\begin{cases} \Delta W_x(x,y) = W(x,y) - W(x+s,y) \\ \Delta W_y(x,y) = W(x,y) - W(x,y+s) \end{cases}, \tag{6}$$

where s is the shearing amount as shown in **Figure 15(b)**. The wavefront $W(x, y)$ can be described with Zernike polynomials and can be decomposed into a series (N terms) of orthonormal polynomials $\{Z_i(x, y)\}$ with the corresponding coefficients $\{a_i\}$,

$$W(x,y) = \sum_{i=1}^{N} a_i Z_i(x,y). \tag{7}$$

According to Eqs. (6) and (7), the shearing wavefronts $\Delta W_x(x, y)$ and $\Delta W_y(x, y)$ can be expressed as follows:

$$\begin{cases} \Delta W_x(x,y) = \sum_{i=1}^{N} a_i \Delta Z_{i,x}(x,y) \\ \Delta W_y(x,y) = \sum_{i=1}^{N} a_i \Delta Z_{i,y}(x,y) \end{cases}, \tag{8}$$

where the differential Zernike polynomials $\Delta Z_{i,x}(x, y)$ and $\Delta Z_{i,y}(x, y)$ can be written as follows:

$$\begin{cases} \Delta Z_{i,x}(x,y) = Z_i(x,y) - Z_i(x+s,y) \\ \Delta Z_{i,y}(x,y) = Z_i(x,y) - Z_i(x,y+s) \end{cases}. \tag{9}$$

Denoting the shearing wavefronts, differential Zernike polynomials and the coefficients in Eq. (8) as $\Delta\mathbf{W} = (\Delta W_x, \Delta W_y)^T$, $\Delta\mathbf{Z} = (\Delta Z_{i,x}, \Delta Z_{i,y})^T$ and $\mathbf{a}$, respectively, Eq. (8) can be transformed into a matrix form,

$$\Delta\mathbf{W} = \Delta\mathbf{Z}\mathbf{a}. \tag{10}$$

Thus, the coefficients $\{a_i\}$ of Zernike polynomials in Eq. (7) can be obtained from the least-squares solution of Eq. (10),

$$\mathbf{a} = (\Delta\mathbf{Z}^T\Delta\mathbf{Z})^{-1}\Delta\mathbf{Z}^T\Delta\mathbf{W}. \tag{11}$$

According to Eq. (11), the retrieval of point-diffraction wavefront $W(x,y)$ depends on the measurement precision of shearing wavefronts $\Delta W_x(x,y)$ and $\Delta W_y(x,y)$. Traditionally, the systematic error introduced by lateral displacement between SMA fibers can be calibrated by removing Zernike tilt and power terms. However, the residual high-order aberrations can significantly influence the measurement precision, especially in the case of high NA and large lateral displacement. Thus, a general and rigorous method for geometric error removal is required to realize the high-precision measurement of point-diffraction wavefront.

5.2. High-precision method for systematic error calibration

In the null test of pinhole diffraction wavefront and single-mode fiber diffraction wavefront, the high wavefront NA in a pinhole PDI and large lateral displacement between fibers in fiber PDI could introduce some high-order geometric aberrations, respectively. Different from traditional pinhole PDI and single-mode fiber PDI, the null test of SMA fiber diffraction wavefront involves both high NA and large lateral displacement, placing much higher requirement on the calibration of the systematic error introduced by lateral displacement between SMA fibers. A double-step calibration method based on three-dimensional coordinate reconstruction and symmetric lateral displacement compensation can be applied to completely remove systematic error. It should be noted that the possible longitudinal displacement between SMA fibers may also introduce certain systematic error; however, it can be well minimized with the fine adjusting mechanism. Besides, the error introduced by the longitudinal displacement is low-order aberration, and it can be well calibrated with traditional misalignment calibration method by subtracting the Zernike piston, tilt and power terms.

5.2.1. First-step calibration based on three-dimensional coordinate reconstruction

Without loss of generality, we take the displacement in x direction between SMA fibers as the geometric error calibration model to be analyzed, as shown in **Figure 16**.

Supposing that the distance between exit apertures of fiber pairs G_1 (S_1 and S_2) and an arbitrary point $P(x,y,z)$ on the detection plane are R_1 and R_2, we have the corresponding optical path difference OPD,

$$\mathrm{OPD} = R_1(x_1, y_1, z_1; x, y, z) - R_2(x_2, y_2, z_2; x, y, z), \tag{12}$$

where (x_1, y_1, z_1) and (x_2, y_2, z_2) are the three-dimensional coordinates of S_1 and S_2, (x, y, z) is that of the arbitrary known point P. To simplify the analysis, the origin of the coordinate system is located at S_1, the distance between the SMA fiber projector and CCD detector is D. Thus, the OPD in Eq. (12) can be simplified as follows:

$$\mathrm{OPD} = \sqrt{x^2 + y^2 + D^2} - \sqrt{(x + s)^2 + y^2 + D^2}. \tag{13}$$

According to the one-to-one correspondence of the OPD distribution on the detection plane and the coordinate of fiber apertures, the 3D coordinate measurement method with PDI introduced in Section 3 can be applied to reconstruct the 3D coordinates in Eq. (12) that is the global minimum $\mathbf{\Phi}^*$ of the residual function $F(\mathbf{\Phi})$,

$$F(\mathbf{\Phi}) = \sum_k f_k^2(\mathbf{\Phi}) = \sum_k (\mathrm{OPD}_k - \widehat{\xi}_k)^2, \tag{14}$$

where the vector $\mathbf{\Phi} = \{(x_1, y_1, z_1); (x_2, y_2, z_2)\}$ is the coordinates of SMA fibers under measurement, the subscript k indicates the point number on the detection plane, $\widehat{\xi}_k$ is the measured OPD and OPD_k is the OPD reconstructed from coordinates $\mathbf{\Phi}$ according to Eq. (13). With the reconstructed coordinates $\mathbf{\Phi}^*$, the systematic error can be preliminarily calibrated,

$$\mathrm{OPD}_1 = \mathrm{OPD}(\mathbf{\Phi}) - \mathrm{OPD}(\mathbf{\Phi}^*). \tag{15}$$

However, the reconstruction accuracy of fiber coordinates can only reach the order of submicron in the practical case, resulting in obvious residual error in the calibration result. The root mean square (RMS) value of the residual error in the preliminary calibration is 0.0077λ corresponding to the 0.5 μm coordinate reconstruction error for 0.60 NA fibers and 250 μm lateral displacement between two fibers. To further remove the residual systematic error, a second-step calibration, which is based on symmetric lateral displacement compensation, needs to be carried out.

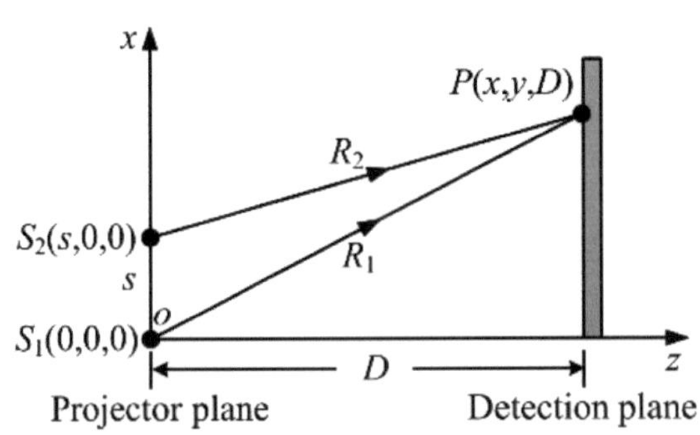

Figure 16. Geometry for systematic error analysis [21].

5.2.2. Second-step calibration based on symmetric lateral displacement compensation

The expression for OPD in Eq. (12) can be simplified as follows:

$$\mathrm{OPD} \cong a_2 Z_2 + a_9 Z_9 + a_{19} Z_{19} + a_{33} Z_{33}, \tag{16}$$

where Z_2 refers to x tilt terms, Z_9, Z_{19}, and Z_{33} are Zernike primary, secondary, and tertiary x coma terms, respectively, a_2, a_9, a_{19}, and a_{33} are the corresponding coefficients, and

$$\begin{cases} a_2 = s[-t-ts^2/(2D^2) + t^3/3 + 3t^3s^2/(4D^2) - 3t^5/16 - 15t^5s^2/(16D^2) + t^7/8] \\ a_9 = s[t^3/6 + 3t^3s^2/(8D^2) - 3t^5/20 - 15t^5s^2/(20D^2) + t^7/8] \\ a_{19} = s[-3t^5/80 - 3t^5s^2/(16D^2) + 3t^7/56] \\ a_{33} = s(t^7/112) \end{cases}. \tag{17}$$

Similarly, the reconstructed optical path difference $\mathrm{OPD}(\mathbf{\Phi}^*)$ (though with certain coordinate reconstruction error) follows the relationship given by Eq. (16). Thus, the residual systematic error OPD_1 after first-step calibration, according to Eqs. (15) and (16), can be written as follows:

$$\mathrm{OPD}_1 \cong \Delta a_2 Z_2 + \Delta a_9 Z_9 + \Delta a_{19} Z_{19} + \Delta a_{33} Z_{33}, \tag{18}$$

where $\{\Delta a_i\}$ (i = 2, 9, 19, 33) are the corresponding Zernike coefficients. With the implementation of the first-step calibration, the coordinate reconstruction error reaches the order of submicron, and the corresponding approximation error in Eq. (18) can be well restricted and is negligible. According to Eq. (16), the major systematic error introduced by the lateral displacement includes tilt and coma terms, they cannot be completely removed with traditional misalignment calibration method by subtracting the Zernike piston, tilt and power terms. From Eq. (16), the residual coma aberration due to lateral displacement depends on the lateral displacement s, NA, and distance D, the corresponding Zernike coefficients are odd functions about s. Thus, the superposition of geometric errors $\mathrm{OPD}^{(s)}$ and $\mathrm{OPD}^{(-s)}$ corresponding to opposite shear directions can be expressed as follows:

$$\mathrm{OPD}^{(s)} + \mathrm{OPD}^{(-s)} \cong 0. \tag{19}$$

According to the analysis above, the geometric error can be further reduced by superposing two measurements with opposite shear directions, corresponding to lateral displacement s and $-s$, respectively. Denoting the preliminarily calibrated wavefront data corresponding to lateral displacement s and $-s$ as $W_{m1}^{(s)}$ and $W_{m1}^{(-s)}$, and the true shearing wavefront under test ΔW, we have

$$\begin{cases} W_{m1}^{(s)} = \Delta W + \mathrm{OPD}_1{}^{(s)} \\ W_{m1}^{(-s)} = \Delta W + \mathrm{OPD}_1{}^{(-s)} \end{cases}. \tag{20}$$

According to Eqs. (19) and (20), the shearing wavefront ΔW can be obtained with

$$\Delta W \cong (W_{m1}^{(s)} + W_{m1}^{(-s)})/2. \tag{21}$$

To get the measurement data with the lateral displacement $-s$, one may rotate the SMA fiber projector by 180°. **Figure 17(a)** and **(b)** shows the residual systematic errors corresponding to various lateral displacement amounts with different NAs with traditional method and the double-step calibration method. The traditional method is not valid when the lateral displacement is over 50 μm because the residual error RMS is larger than $2.0\times10^{-3}\lambda$ with the 0.10 NA fiber. The residual error also increases significantly with the lateral displacement, especially for the case with high NA fiber. In **Figure 17(b)**, the residual error is less than $1.0\times10^{-4}\lambda$ RMS within a 300 μm lateral displacement even for the 0.60 NA fiber, it confirms the feasibility of the double-step calibration in Section 5.2.

To obtain point-diffraction spherical wavefront in the experiment, the projector should be placed at the far-field zone that is the distance $D >> 2(2d)^2/\lambda$ from the fibers [29], where d is the aperture size. With the unwrapped phase distribution, the systematic error calibration is carried out to obtain the shearing wavefront with the high-precision double-step calibration method introduced in Section 5.2. The experimental measurement of SMA fiber point-diffraction wavefront has been carried out with the system diagram shown in **Figure 15**. The unwrapped original wavefront before and after 180° rotation in x and y directions is shown in **Figure 18(a)–(d)**, and the corresponding precalibrated shearing wavefronts after the first-step calibration are shown in **Figure 18(f)–(i)**, respectively. **Figure 18(e)** and **(j)** is the true shearing wavefronts in x and y directions obtained with the double-step calibration method. According to **Figure 18**, obvious residual errors can be seen in the obtained wavefronts after first-step calibration, the precalibrated shearing wavefront error RMS in x and y directions are 0.0254λ and 0.0194λ, respectively. The residual error is further removed after the second-step calibration, and true shearing wavefront error RMS in x and y directions are $1.38\times10^{-4}\lambda$ and $1.36\times10^{-4}\lambda$, respectively. Besides, there is no significant odd error in the finally retrieved wavefronts in **Figure 18(e)** and **(j)**. Thus, the high-accuracy calibration is realized with the proposed method.

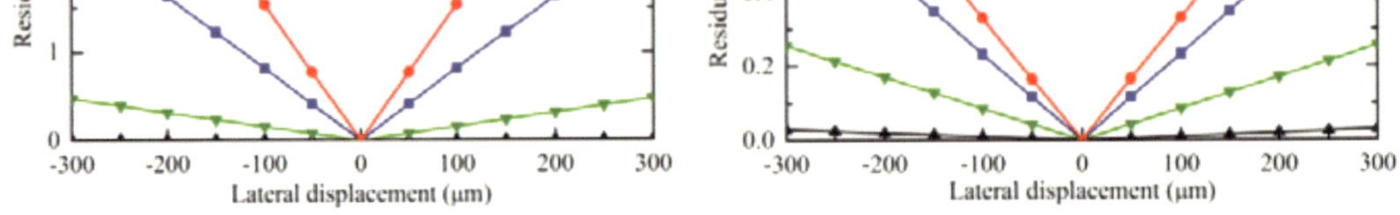

Figure 17. Residual errors in the calibration of systematic error under various lateral displacements and NAs in computer simulation. (a) RMS value with traditional method and (b) RMS value with the proposed double-step calibration method [21].

With the true shearing wavefronts in x and y directions after systematic error calibration, the point-diffraction wavefront can be reconstructed based on the differential Zernike polynomials fitting method introduced in Section 5.1. The measured wavefront error compared with an ideal sphere is shown in **Figure 19(a)**, PV and RMS are $9.20\times10^{-4}\lambda$ and $1.54\times10^{-4}\lambda$, respectively. **Figure 19(b)** and **(c)** shows the measured wavefronts after the 45° rotation and translation along z direction of projector, respectively. It can be seen from **Figure 19** that the measured wavefront shape in the original position agrees well with those obtained with projector rotation and translation, demonstrating the measurement repeatability and accuracy. The protective glass on CCD detector could introduce additional deformation on point-diffraction wavefront, and the measurement accuracy is expected to be further improved by removing the protective glass. Due to the smoothing effect of each pixel, the relatively large pixel acts as a low-pass filter and the measured wavefront should be taken as a macroparameter. By adopting

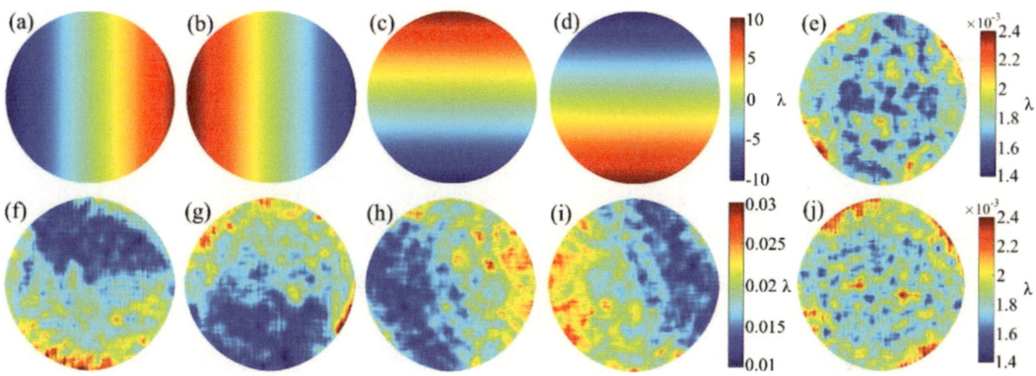

Figure 18. Measurement results of shearing wavefront retrieval in SMA fiber point-diffraction wavefront measurement. Unwrapped original wavefronts, including (a) x direction and (c) y direction before 180° rotation, (b) x direction and (d) y direction after 180-degree rotation; shearing wavefronts after the first-step calibration, including (f) x direction and (h) y direction before 180° rotation, (g) x direction and (i) y direction after 180-degree rotation; finally retrieved shearing wavefronts after the double-step calibration, including (e) x and (j) y directions [21].

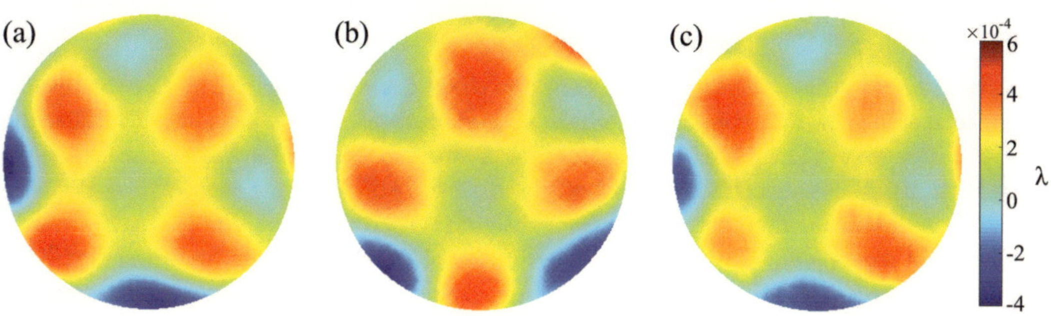

Figure 19. Experimental result of point-diffraction wavefront measurement. (a) Measured point-diffraction wavefront in original position, (b) measured wavefront after 45° projector rotation, and (c) measured wavefront after projector translation [21].

the CCD with smaller pixel size and placing the SMA fiber projector at a position farther from CCD detector, the smoothing effect can be further decreased.

6. Conclusion

In this chapter, the application of PDI system in the wavefront and surface testing is introduced. Various PDI setups, including the fiber PDI and pinhole PDI, have developed for high-accuracy optical testing, and the measurement accuracy up to subnanometer was reported. Different from the pinhole PDI, the adjustable fringe contrast is easily realized with fiber PDI, and it can be applied to test the surface with various reflectivities. To realize the adjustable fringe contrast with pinhole method, the polarizing elements with special structure is needed to transform the polarization states and adjust the relative intensities of the interfering testing and reference beams. High measurable NA can be achieved with the pinhole; however, the light transmission is quite low; high light transmittance can be obtained with single-mode fiber; however, the measurable NA is limited by the NA of the fiber. To obtain both the high diffraction light power and high-NA spherical wavefront, a SMA fiber with cone-shaped exit end has been proposed, and it is considered as a feasible way to extend the measurement range of the system. Based on the SMA fiber, a PDI system for the absolute 3D coordinate measurement was proposed. The system utilizes the SMA fiber to get spherical wave with both high NA and high light intensity, and large measurable range is achieved.

The numerical study of point-diffraction wavefront both with pinhole and SMA fiber is performed, which is based on the FDTD method. The aperture angle, light transmittance, and wavefront error are analyzed. According to simulation results, the ideal spherical wavefront with the accuracy better than subnanometer can be obtained both with pinhole and SMA fiber, and the increase in NA range and diffraction aperture size could lead to the growth of diffracted wavefront error. It can be seen from the comparison of pinhole diffraction and SMA fiber diffraction, the light transmittance in both the methods grows with the increase in aperture size, and the similar aperture angles can be obtained with the same diffraction aperture size. However, the light transmittance in SMA fiber diffraction is far larger than that in pinhole diffraction. Thus, the SMA fiber provides a feasible way to obtain the high light intensity required in the optical testing, enabling the extension of measurement range with PDI system.

The experimental measurement of SMA fiber point-diffraction wavefront is carried out to evaluate its accuracy in practical application, which is based on shearing interferometry. To realize the high-precision calibration of the systematic error introduced by the lateral displacement between SMA fibers, a double-step calibration method based on three-dimensional coordinate reconstruction and symmetric lateral displacement compensation is used to remove the geometric aberration from the shearing wavefront. Besides, the differential Zernike polynomials fitting method is used to reconstruct the point-diffraction wavefront. The spherical wavefront with the accuracy reaching the order of $10^{-4}\lambda$ is obtained in the experiment. The calibration method can be performed without any preknowledge about the measurement system configuration. It realizes the high-accuracy measurement of point-diffraction wavefront and also provides a feasible method for geometric aberration calibration in interferometric system.

Acknowledgements

This work was partially supported by the National Natural Science Foundation of China (11404312), Zhejiang Provincial Natural Science Foundation of China (LY17E050014), and Zhejiang Key Discipline of Instrument Science and Technology (JL150508). Authors are grateful to editors for giving them the chance to contribution to this book.

Author details

Daodang Wang[1]* and Rongguang Liang[2]

*Address all correspondence to: wangdaodang@sina.com

1 College of Metrology and Measurement Engineering, China Jiliang University, Hangzhou, China

2 College of Optical Sciences, University of Arizona, Tucson, AZ, USA

References

[1] Stulen R H, Sweeney D W. Extreme ultraviolet lithography. IEEE Journal of Quantum Electronics. 1999; **35**(5): 694–699. DOI: 10.1109/3.760315

[2] Goldberg K A, Naulleau P P, Denham P E, Rekawa S B, Jackson K, Anderson E H, et al. At-wavelength alignment and testing of the 0.3 NA MET optic. Journal of Vacuum Science & Technology B. 2004; **22**(6): 2956–2961. DOI: 10.1116/1.1815303

[3] Otaki K, Zhu Y, Ishii M, Nakayama S, Murakami K, Gemma T. Rigorous wavefront analysis of the visible-light point diffraction interferometer for EUVL. Proceeding of SPIE. 2004; **5193**: 182–190. DOI: 10.1117/12.507046

[4] Takeuchi S, Kakuchi O, Yamazoe K, Gomei Y, Decker T A, Johnson M A, et al. Visible light point-diffraction interferometer for testing of EUVL optics. Proceeding of SPIE. 2006; **6151**: 61510E. DOI: 10.1117/12.656275

[5] Malacara D. Twyman-Green Interferometer. In: Malacara D, editor. Optical Shop Testing. 3rd ed. New Jersey: Wiley; 2007. pp. 46–96.

[6] Mantravadi M V, Malacara D. Newton, Fizeau, and Haidinger interferometers. In: Malacara D, editor. Optical Shop Testing. 3rd ed. New Jersey: Wiley; 2007. pp. 1–45.

[7] Smartt R N, Steel W H. Theory and application of point-diffraction interferometers. Japanese Journal of Applied Physics. 1975; **14**(S1): 351–356. DOI: 10.7567/JJAPS.14S1.351

[8] Wang G Y, Zheng Y L, Sun A M, Wu S D, Wang Z J. Polarization pinhole interferometer. Optics Letters. 1991; **16**(17): 1352–1354. DOI: 10.1364/OL.16.001352

[9] Gong Q, Geary J M. Modeling point diffraction interferometers. Optical Engineering. 1996; **35**(2): 351–356. DOI: 10.1117/1.600903

[10] Millerd J E, Martinek S J, Brock N J, Hayes J B, Wyant J C. Instantaneous phase-shift, point-diffraction interferometer. Proceeding of SPIE. 2004; **5380**: 422–429. DOI: 10.1117/12.557126

[11] Gong Q, Eichhorn W. Alignment and testing of piston and aberrations of a segmented mirror. Proceeding of SPIE. 2005; **5869**: 586912. DOI: 10.1117/12.613657

[12] Neal R M, Wyant J C. Polarization phase-shifting point-diffraction interferometer. Applied Optics. 2006; **45**(15): 3463–3476. DOI: 10.1364/AO.45.003463

[13] Otaki K, Bonneau F, Ichihara Y. Absolute measurement of spherical surface by point diffraction interferometer. Proceeding of SPIE. 1999; **3740**: 602–605. DOI: 10.1117/12.347755

[14] Otaki K, Ota K, Nishiyama I, Yamamoto T, Fukuda Y, Okazaki S. Development of the point diffraction interferometer for extreme ultraviolet lithography: design, fabrication, and evaluation. Journal of Vacuum Science & Technology B. 2002; **20**(6): 2449–2458. DOI: 10.1116/1.1526605

[15] Otaki K, Yamamoto T, Fukuda Y, Ota K, Nishiyama I, Okazaki S. Accuracy evaluation of the point diffraction interferometer for extreme ultraviolet lithography aspheric mirror. Journal of Vacuum Science & Technology B. 2002; **20**(1): 295–300. DOI: 10.1116/1.1445161

[16] Sommargren G E. Phase shifting diffraction interferometry for measuring extreme ultraviolet optics. OSA Trends in Optics & Photonics. 1996; **V4**: 108–112.

[17] Kihm H, Kim S W. Oblique fiber optic diffraction interferometer for testing. Optical Engineering. 2005; **44**(12): 125601. DOI: 10.1117/1.2148367

[18] Matsuura T, Udaka K, Oshikane Y, Inoue H, Nakano M, Yamauchi K, et al. Spherical concave mirror measurement by phase-shifting point diffraction interferometer with two optical fibers. Nuclear Instruments & Methods in Physics Research A. 2010; **616**(2–3): 233–236. DOI: 10.1016/j.nima.2009.12.031

[19] Chkhalo N I, Klimov A Y, Rogov V V, Salashchenko N N, Toropov M N. A source of a reference spherical wave based on a single mode optical fiber with a narrowed exit aperture. Review of Scientific Instruments. 2008; **79**(3): 033107. DOI: 10.1063/1.2900561

[20] Wang D, Chen X, Xu Y, Wang F, Kong M, Zhao J, et al. High-NA fiber point-diffraction interferometer for three-dimensional coordinate measurement. Optics Express. 2014; **22** (21): 25550–25559. DOI: 10.1364/OE.22.025550

[21] Wang D, Xu Y, Liang R, Kong M, Zhao J, Zhang B, et al. High-precision method for submicron-aperture fiber point-diffraction wavefront measurement. Optics Express. 2016; **24**(7): 7079–7090. DOI: 10.1364/OE.24.007079

[22] Rhee H-G, Kim S-W. Absolute distance measurement by two-point-diffraction interferometry. Applied Optics. 2002; **41**(28): 5921–5928. DOI: 10.1364/AO.41.005921

[23] Rhee H-G, Chu J, Lee Y-W. Absolute three-dimensional coordinate measurement by the two-point diffraction interferometry. Optics Express. 2007; **15**(8): 4435–4444. DOI: 10.1364/OE.15.004435

[24] Wang D, Yang Y, Chen C, Zhuo Y. Point diffraction interferometer with adjustable fringe contrast for testing spherical surfaces. Applied Optics. 2011; **50**(16): 2342–2348. DOI: 10.1364/AO.50.002342

[25] Medecki H, Tejnil E, Goldberg K A, Bokor J. Phase-shifting point diffraction interferometer. Optics Letters. 1996; **21**(19): 1526–1528. DOI: 10.1364/OL.21.001526

[26] Johnson M A, Phillion D W, Sommargren G E, Decker T A, Taylor J S, Gomei Y, et al. Construction and testing of wavefront reference sources for interferometry of ultra-precise imaging systems. Proceeding of SPIE. 2005; **5869**: 58690P. DOI: 10.1117/12.623185

[27] Wang D, Wang F, Zou H, Zhang B. Analysis of diffraction wavefront in visible-light point-diffraction interferometer. Applied Optics. 2013; **52**(31): 7602–7608. DOI: 10.1364/AO.52.007602

[28] Naulleau P, Goldberg K, Lee S, Chang C, Bresloff C, Batson P, et al. Characterization of the accuracy of EUV phase-shifting point diffraction interferometry. Proceeding of SPIE. 1998; **3331**: 114–123. DOI: 10.1117/12.309563

[29] Kong J A. Electromagnetic Wave Theory. 1st ed. New Jersey: Wiley; 1986. 667 p.

Chapter 10

Interferometry Using Generalized Lock-in Amplifier (G-LIA): A Versatile Approach for Phase-Sensitive Sensing and Imaging

Aurélien Bruyant, Julien Vaillant, Tzu-Heng Wu,
Yunlong Zhu, Yi Huang and Abeer Al Mohtar

Additional information is available at the end of the chapter

http://dx.doi.org/10.5772/66657

Abstract

A large number of interferometric setups make use of non-linear phase modulators. In the past, specific extraction methods have been proposed mostly to cover the important case of sinusoidal phase modulation with certain limits in term of signal-to-noise ratio. Recently, a detection method based on "Generalized Lock-in Amplifier" (G-LIA) was proposed to extract optimally amplitude and phase information in two-arm interferometers when nearly arbitrary phase modulations are used such as triangular or sinusoidal phase modulations. This method offers the opportunity to develop highly sensitive interferometers with simple-phase modulators such as piezo-actuated mirrors, piezo stretchers, or power-modulated laser diodes in unbalanced interferometers. Here we present the basics of the approach and we give application examples for monitoring displacement, sensing, and digital holography. The case where an amplitude modulation is also present is also detailed and discussed in the context of unbalanced interferometry and near-field nanoscopy.

Keywords: phase extraction method, unbalanced interferometry, digital holography, near-field optics, cost-effective interferometry

1. Introduction

In order to determine amplitude and phase in a two-arm interferometer, a phase modulator is often required to clearly discriminate phase changes from amplitude changes. Such operation is straightforward when the phase modulation is a linear function of time. In this case, a standard Lock-in Amplifier (LIA) gives the required information with an optimal signal to noise ratio (SNR). Unfortunately, a number of phase modulators interesting in term of cost, achromaticity or integration offers non-linear responses, that are even sometimes coupled with unwanted amplitude modulation. A critical question that arises is "How can we extract phase and amplitude information in an optimal way when non-linear phase modulation is used?"

To solve this issue while keeping the benefits of high SNR, approaches have been proposed based on multiple lock-in detection at selected signal harmonics. These approaches were mainly employed in the case where the phase modulation is a sine function [1–6]. Such phase modulation is, for example, achieved using piezo-actuator, fiber stretchers, and other phase modulators where a sine excitation typically offers the best response. The multiple lock-in approach works fine but it is less direct and does not necessarily provide an optimal SNR or a straightforward implementation. Especially, if an amplitude modulation is present at the same frequency as that of the phase modulation. Alternately, the Generalized Lock-in Amplifier (G-LIA) technique was recently introduced [7] to solve this issue with a procedure similar to a single LIA operation. In this chapter, we first detail the principle of this method when operated in the simplest case where no amplitude modulation is present. Application is provided notably in the context of digital holography. Then we consider the case where there is an additional amplitude modulation in the signal field. The first case which is discussed is related to unbalanced interferometry where the phase modulation is achieved via a power modulation of the laser source. Finally, we also discuss the case of phase-sensitive near-field imaging.

2. Theory: introduction to G-LIA

In order to introduce the G-LIA technique, we need to provide an expression for the detected signal. We consider the simplest configuration of a 2-arm interferometer comprising a reference arm and a signal arm (cf. **Figure 1**). The system is illuminated by a monochromatic radiation. The detected signal intensity $I(t)$ can be expressed as:

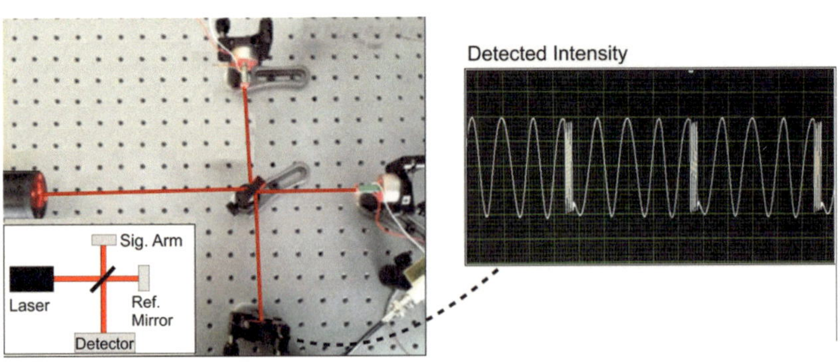

Figure 1. Pseudo-heterodyne approach. A quasi-linear phase modulation is achieved by a sawtooth modulation of the optical path using a piezo-actuated optical mirror in the reference arm. The detected intensity exhibits a sine modulation except during the flyback time of the mirror. Any phase change in the reference arm will produce a detectable phase shift of the observed quasi-sinusoidal pattern.

http://dx.doi.org/10.5772/66657

$$I(t) \propto E_r^2 + E_s^2 + 2mE_rE_s \cos\left(\Delta\phi(t)\right) \quad (1)$$

where E_s and E_r are, respectively, the amplitude of the field of interest and the reference field impinging on the detector. The phase difference between the two fields is $\Delta\phi(t)$ while the factor $m \leq 1$ in the interferometric term accounts for the interference contrast. Alternately, it is common to express the detected intensity as a function of the laser power P:

$$I(t) \propto P[1 + s\cos\left(\Delta\phi(t)\right)] \quad (2)$$

where s is proportional to the unknown signal amplitude E_s, considered as constant during a measurement. The unknown spatial phase ϕ_s of the signal field is also supposed to be constant during the measurement although a phase modulator can be included inside the signal arm. Any time dependence in $\Delta\phi(t)$ is therefore arbitrarily considered as coming from the reference field:

$$\Delta\phi(t) = \phi_R(t) - \phi_S \quad (3)$$

From the expression of I given by Eq. (2), different strategies can be proposed in order to recover amplitude and phase information. In order to extract the two unknowns (s, ϕ_S) from the signal $I(t)$, the phase of one of the two beams can be modulated in time by a frequency shifter or another phase modulator. When the time dependence is linear $\phi_r(t) \propto t$ as in the former case, (s, ϕ_S) are precisely determined by a Lock-in Amplifier (LIA) locked at the single frequency component present in $I(t)$. For other functions of time such as a sine waveforme (i.e. $\phi_R = \alpha \sin \Omega t$, where a is the modulation depth), the use of a conventional LIA is less trivial as the signal information is typically spread over a number of frequency components ($n\Omega/2\pi$) having different weights. The G-LIA method was introduced to handle such cases, while keeping an extraction procedure very similar to a LIA. Hereafter, the two approaches are provided to highlight the similitudes and differences.

2.1. Amplitude and phase determination using a standard LIA

2.1.1. Case of linear phase modulation (LIA)

Considering the general Eq. (2), the use of LIA is direct when $\Delta\phi(t)$ is linearly modulated by $\phi_R(t) = \Omega t = 2\pi\Delta F t$, where the phase modulation rate Ω can be induced in different ways. These ways include notably the use of a frequency shifter in one of the two arms (where ΔF is the frequency shift), a linear translation of one of the mirror (where ΔF is the associated Doppler shift), or a linear variation of the laser frequency[1] if the two arms are unbalanced. Amongst the cited methods, heterodyne measurement based on the use of frequency shifters is often considered as more favorable as a purely linear displacement of the mirror is hardly achievable in practise without alignment or coherence issues, while unbalanced

[1] In this case ΔF is function of the unbalance, that is, the optical path difference.

interferometry is subject to noise [8] induced by small wavelength fluctuation typically related to temperature drifts.

The two unknowns (s, ϕ_S) are simply the amplitude and phase of the sine waveform present in Eq. (1) which is modulated at the frequency shift ΔF. This determination is optimally carried out by a dual-output (X, Y) LIA locked at the frequency ΔF that is precisely provided by the modulator driver. Depending on the output, the detected signal I is multiplied by an in-phase or a quadrature sinusoids modulated at the same angular frequency ΔF and it is averaged over a time t_{int}:

$$X(I) = \frac{1}{t_{int}} \int_0^{t_{int}} I \cos(\Omega t) dt \propto s \cos(\phi_S) \tag{4}$$

$$Y(I) = \frac{1}{t_{int}} \int_0^{t_{int}} I \sin(\Omega t) dt \propto s \sin(\phi_S) \tag{5}$$

From these two outputs, the quantities $s \propto \sqrt{X^2 + Y^2}$ and $\phi_S = atan2(X, Y)$ are obtained with a signal to noise ratio that can be increased using longer integration time. We note that the two in-phase and quadrature sine waveforms are the *LIA reference signals* built from the frequency shift precisely provided to the LIA. If the LIA is not locked exactly at the angular frequency Ω, the measured values of ϕ_s will drift in time, while the extracted signal amplitude will decay for long integration time.

2.1.2. Case of a non-linear phase modulation (LIA)

Achieving ϕ_r (t) $\propto t$ is not possible for a number of phase modulators, notably because they have a finite range of phase modulation. Pseudo-heterodyne approaches were proposed long ago to circumvent this problem [9] by using a sawtooth modulation of the optical path, where the peak to peak amplitude of the sawtooth corresponds to an integer number of times 2π in term of phase. The approach is illustrated by **Figure 1**, in the case where the phase ramp is achieved by a piezo-actuated mirror in a balanced interferometer.

As can be seen, the detected intensity mimics the sinusoidal beating observed in heterodyne setups. Such approach is not widely used since errors are induced during the flyback time on the sawtooth edges, especially if the modulation is fast.

As mentioned, the use of sine modulation $\phi_R = \alpha \sin \Omega t$ is regarded as much more desirable as most of the modulator can operate better and faster when they are sinusoidal excited. It was early highlighted in this context [10] that for such modulation, the Fourier spectrum of the signal has harmonic sidebands coming from the interferometric term $I_{mod} \propto P\, s \cos(\Delta \phi (t))$ in Eq. (2), as shown in **Figure 2**.

The amplitudes of these frequency components are obtained by developing the term in $I_{mod} \propto P$ [$\cos(\phi_s) \cos(\phi_R) + \sin(\phi_s) \sin(\phi_R)$] and using the Jacobi-Anger expansion of $\cos(\phi_R)$ (even harmonics) and sin (ϕ_R) (odd harmonics) [11]. From this expansion, we see that a LIA locked at an angular frequency harmonics mΩ, with $m \neq 0$, gives:

$$X_m(I) \propto s\cos(\phi_S)\left| \begin{matrix} J_m(a) & for\ m\ even \\ 0 & for\ m\ odd \end{matrix} \right. \quad (6)$$

$$Y_m(I) \propto s\ \sin(\phi_S)\left| \begin{matrix} 0 & for\ m\ even \\ J_m(a) & for\ m\ odd \end{matrix} \right., \quad (7)$$

where J_m (a) is the m-th Bessel function. The amplitude and phase can then be extracted using both odd and even harmonics, for example, using $s \propto \sqrt{X_1^2/J_1^2(a) + Y_2^2/J_2^2(a)}$ and $\phi_S = atan2\left(X_1/J_1(a),\ Y_2/J_2(a)\right)$.

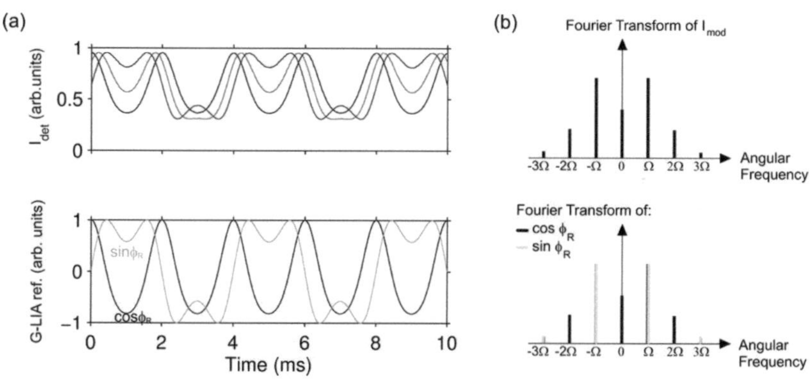

Figure 2. Signals in the case of a sinusoidal phase modulation. (a) Top: example of detected intensity for different signal phases. Bottom: corresponding reference functions. (b) Schematic example of Fourier transform of and associated references and in the case of an arbitrary sine phase modulation and $\phi_s = \pi/4$.

When only two harmonics m = (1, 2) are used, caution must be exercised in the choice of the modulation depth a in order to maximize the power density on the selected harmonics. The optimum value of a in this case is a = 2.19 rad (maximum of $J_1(a)^2 + J_2(a)^2$).

2.2. G-LIA method

The main benefit of the G-LIA method is that all the weighted harmonics are used to retrieve phase and amplitude with an operation similar to that of a LIA. To introduce this method, we also remark that the interferometric term $I_{mod} \propto P\ s\ \cos(\Delta\phi(t))$ in Eq. (2) can be expressed as $I_{mod} \propto P\ s\ [\cos(\phi_s)\ C(t) + \sin(\phi_s)\ S(t)]$, with $C(t) = cos(\phi_R)$ and $S(t) = sin(\phi_R)$. From this expression, it appears that $C(t)$ and $S(t)$ can be used as relevant reference signals within a modified LIA having the two following outputs:

$$X_{\phi R}(I_{mod}) = \frac{1}{t_{int}} \int_0^{t_{int}} I_{mod} C(t)\, dt = < I_{mod} C(t) > , \tag{8}$$

$$Y_{\phi R}(I_{mod}) = \frac{1}{t_{int}} \int_0^{t_{int}} I_{mod} S(t) dt = < I_{mod} S(t) > , \tag{9}$$

These references contain the same frequency components than the interferometric term since $I_{mod}(t)$ is a function of $C(t)$ and $S(t)$. These frequency components are naturally weighted so that the contribution of stronger harmonics will be favored. **Figure 2(b)** exemplifies the case where the phase modulation function is a sine function.

We note that in the particular case where $\phi_r(t) = \Omega t$, $C(t) = \cos(\Omega t)$ and $S(t)=\sin(\Omega t)$ and the G-LIA operation degenerates to that of a standard LIA. More generally, by replacing the expression of I_{mod} in Eqs. (8) and (9), we see that for any phase modulation we have:

$$X_{\phi R}(I_{mod}) \propto S[k_x * \cos(\phi_s) + k_x^{'} * \sin(\phi_s)], \tag{10}$$

$$Y_{\phi R}(I_{mod}) \propto s[k_y^{'} * \cos(\phi_s) + k_y * \sin(\phi_s)], \tag{11}$$

where $k_x = < C^2(t) >$, $k^{'}_x = k^{'}_y = \; < C(t)S(t) >$, and $k_y = \; < S^2(t) >$ are constants that can be calculated numerically or analytically for the considered phase modulation. But for most of the phase modulation functions that can be used $C(t)$ and $S(t)$ are orthogonal, that is, $k^{'}_x = k^{'}_y = \; < S(t)C(t) >= 0$, so that the $X_{\phi R}$ and $Y_{\phi R}$ outputs are:

$$X_{\phi R}(I_{mod}) \propto s k_x \cos(\phi_s), \tag{12}$$

$$Y_{\phi R}(I_{mod}) \propto s k_y \sin(\phi_s), \tag{13}$$

The G-LIA outputs are then similar to that of the LIA in the linear case (cf. Eqs. (4)–(5)). The difference is the presence of the additional proportionality constants k_x and k_y which need to be evaluated by calculating the average values $<C^2(t)>$ and $<S^2(t)>$, respectively. In the case of sine phase modulation $\phi_R = a \sin \Omega t$, the constants have analytical expressions obtained from the integral representations of Bessel functions and simple trigonometric developments:

$$k_x = 1 + J_0(2a) \tag{14}$$

$$k_y = 1 + J_0(2a) \tag{15}$$

where J_0 is the Bessel function of first kind. However, it is difficult to extract I_{mod} from $I(t)$ to feed the G-LIA input as it requires to precisely remove the signal and reference field intensities from the detected intensity $I(t)$. In general, it is not possible either to use directly $I(t)$ to perform the G-LIA operation described above. The reason is that for phase modulation function such as sine or triangle, the useful term I_{mod} also contains a DC component. In consequence the references $C(t)$ and/or $S(t)$ also contain a DC term so that the constant, non-interferometric term P in Eq. (2) is also detected by the G-LIA if $I(t)$ is used directly. To avoid this problem, specific phase modulation depths for which both $C(t)$ and $S(t)$ do not have a DC component can be used.[2] For a sine phase modulation, this specific depth of modulation a corresponds to the zeros of the $J_0(a)$, for example, $a = 2.405$ rad. We note that there is no prejudice in term of signal to noise ratio using this a since all the harmonic contents is detected by the G-LIA operation.

Alternately, a satisfactory solution is to filter the detected intensity to remove all DC component from the signal. In fact, such operation is easy to do and is often highly desirable to directly remove the ambient light contribution in normal conditions [12]. In this case where the signal is filtered, the G-LIA operation is:

$$X_{\phi R}(\tilde{I}) = < \tilde{I}\, C(t) > \propto s\, K_x \cos(\phi_s), \tag{16}$$

$$Y_{\phi R}(\tilde{I}) = < \tilde{I} S(t) > \propto s\, K_y \sin(\phi_s), \tag{17}$$

where the $\tilde{I}$ denotes a DC-filtered quantity,[3] that is, $\tilde{I}(t) = I(t) - < I(t) >$. The new proportionality constants are $K_x = < \tilde{C}^2(t) >$ and $K_y = < \tilde{S}^2(t) >$, where $\tilde{C}(t) = C(t) - < C(t) >$ and $\tilde{S}(t) = C(t) - < C(t) >$ are used to evaluate these proportionality factors numerically or analytically.[4] Amplitude and phase are then provided by:

$$s \propto \sqrt{X^2_{\phi R}(\tilde{I})/K_x^2 + y^2_{\phi R}(\tilde{I})/K_y^2} \tag{18}$$

$$\phi_s = a\ tan\ 2\left(X_{\phi R}(\tilde{I})/K_x, Y_{\phi R}(\tilde{I})/K_y\right) \tag{19}$$

In the useful case of a sine modulation of the form $\phi_R = a \sin \Omega t$, the constants K_x and K_y have the following analytical expressions:

$$K_X = k_x - J_0^2(a) = 1 + J_0(2a) - J_0^2(a) \tag{20}$$

$$K_Y = k_y = 1 - J_0(2a) \tag{21}$$

---

[2] Amplitude and phase are then determined using: $s \propto \sqrt{X^2_{\phi R}(I)/k_x^2 + Y^2_{\phi_R}(I)/k_y^2}$ and $\phi_s = a\ tan2(X_{\phi_R}(I)/k_x^2, Y_{\phi_R}(I)/k_y^2)$.

[3] An analog filter can be used. Alternately, it is possible to filter the DC component of the reference functions C(t) and S(t) only, or to filter both I and the references, with the same result. The operations $< \tilde{I}C(t) >$, $< I\tilde{C}(t) >$ and $< \tilde{I}\tilde{C}(t) >$ are theoretically equivalent. The interest of filtering both the signal and the references is that if the system operates at small modulation frequencies some filters may create a distortion of the modulated signal by changing the amplitudes of peaks and by creating phase shifts for the lowest frequency components. By filtering both the references and the signal, the distortion is similar for both the signal and references so that the distortion effect is cancelled out.

[4] A comment should also be made regarding the references C(t) and S(t). Building these references require the knowledge of $\phi = a \sin \Omega t$. In a number of setup ϕR can be monitored with sensors and it is then possible to take the sine and cosine of this quantity. The references can also be built numerically from the knowledge of the modulation depth a and frequency Ω, but C(t) and S(t) must be synchronized with $\phi_R = a \sin \Omega t$. In other word, we should not use an ersatz $\phi'_R = a \sin(\Omega t + \phi_o)$ as an argument for $C(t) = cos(\phi_R)$ and $S(t) = sin(\phi_R)$. If a phase shit ϕ0 exists, a phase adjustment of the references or the modulation drive signal can be made. This phase shift can be measured by the phase output of a standard LIA locked at the frequency Ω.

where the negative extra term in K_x comes from the filtering of the DC component which is not zero for $C(t)$. As shown in **Figure 3**, for certain phase modulation amplitude the two constants are identical and approximately equal to unity. This is obtained for $a \approx 2.814$ rad, but any phase modulation can be used. A better signal to noise ratio (SNR) is naturally achieved when there is no DC component in $C(t)$ and $S(t)$, since this part is filtered. These cases correspond to the zeros of $J_0(a)$ as previously mentioned (e.g. $a = 2.405$ rad), however the SNR is nearly optimum for a continuous range of values above $a = 2$ rad. Other analytical expressions can be given, for example, for a triangular modulation [7], however the constants estimation can be made numerically without difficulty for a variety of phase modulation functions.

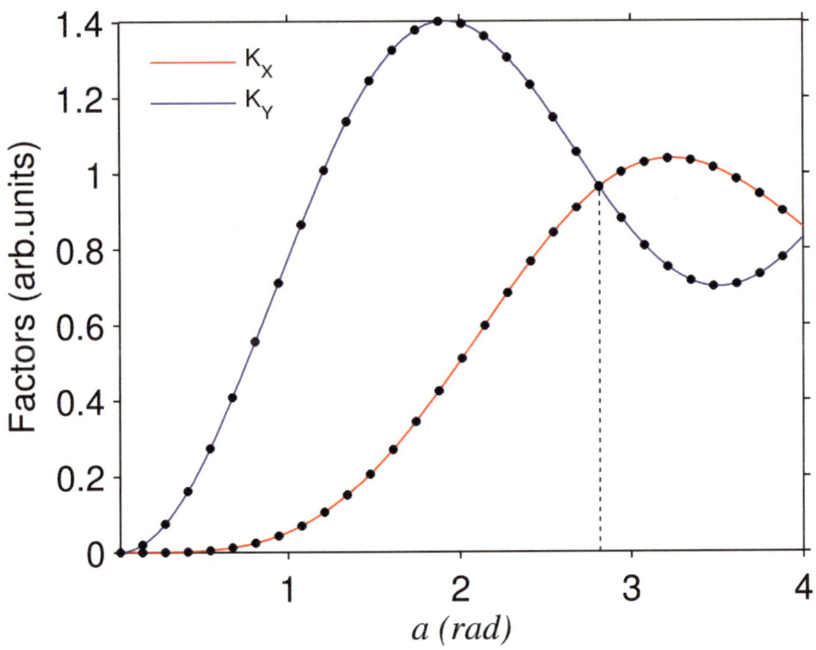

Figure 3. Proportionality factors and used in a G-LIA working with a sine phase modulation as a function of the phase modulation depth. The analytical evaluations are plotted in solid lines; the markers correspond to the numerically calculated values.

3. Application examples

In this section we review and present several results of interferometric measurements performed with the G-LIA approach described in the previous section. Results include measurement with a point detector reported elsewhere and interferometric measurement with 2D detector in the framework of holographic measurement.

3.1. Measurement with a point detector

Figure 4 shows measurement results adapted from the Ref. [7], where the G-LIA can be used with or without filtering to monitor an arbitrary displacement (here a triangle-shaped displacement).

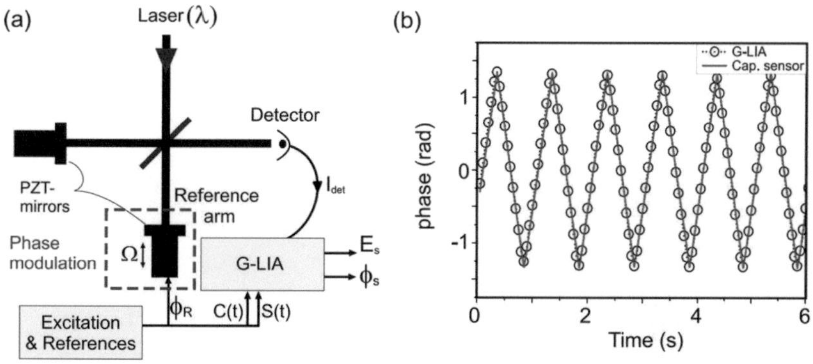

Figure 4. Interferometric measurement with a single detector. (a) Setup for displacement measurement including capacitive sensors for comparison. (b) Displacement measurement obtained with G-LIA and the capacitive sensor.

3.1.1. Displacement measurement

The setup is shown on **Figure 4(a)**. In this example, the phase modulation is a sine function $\phi_R = a \sin \Omega t$, with a modulation depth approximately equals to $a = 2.405$ rad. As explained before, with this value it is not necessary to filter the detected intensity to extract amplitude and phase from the G-LIA operations $X = \langle I\ C(t) \rangle$ and $Y = \langle I\ S(t) \rangle$. However, it is a common practise to filter the DC signal directly after the detector to get rid of the environmental light condition before acquisition in order to optimize the analog to digital conversion. In the Michelson configuration shown here, the value of a corresponds to a peak to valley oscillation of the reference mirror of about 38.3% of the wavelength. If no position sensor is present on the reference mirrors, several methods can be used to achieve the desired phase modulation depth a used in the references $C(t)$ and $S(t)$, notably by inspecting the signal $I(t)$ whose shape changes continuously with increasing value of a. Alternately the modulation depth can be precisely adjusted to recover precisely a controlled displacement of the signal mirror without affecting the amplitude output.

3.1.2. Sensing

Determining the phase rather than the amplitude is known to offer potential advantage in term of sensitivity in optical sensing systems [13]. More precisely, the phase detection coupled with surface plasmon resonance (SPR) is known to improve the measurement sensitive by one to several order of magnitude depending on the exact system geometry. Many different designs on combining interferometry or heterodyne detection on Kretschmann configuration-based SPR sensor have been done [14, 15].

Figure 5 shows the demonstration setup used in [7] to demonstrate the applicability of the G-LIA for phase sensitive sensing application. The setup is similar to that of **Figure 4(a)**, except that an I_2 gas cell was added in the reference arm. The phase modulation is still achieved with a piezo-actuated mirror and the wavelength of a laser diode emitting near 660 nm is ramped over 4pm across an absorption line of the gas. To avoid phase drift induced by the wavelength ramp, the length of the arms must be precisely balanced, since even minute wavelength fluctuation can create phase fluctuation in unbalanced interferometer. This balanced setting can be achieved by ramping the laser diode wavelength outside an absorption band, and adjusting the position the reference mirror until the phase output of the G-LIA remains constant.

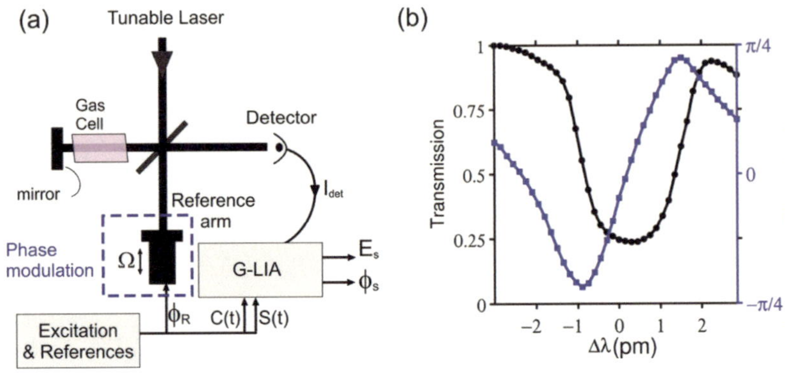

Figure 5. (a) Interferometric measurement with a single detector, applied to gas sensing. (b) Phase-sensitive detection of an absorption line. The obtained spectrum is adapted from Ref [7].

As can be seen, the phase varies more abruptly at the absorption peak center. However, the benefit of measuring the phase for monitoring a gas concentration is not clear since the amplitude has similar variation on the two sides of the absorption peak which indicates a similar sensitivity than the phase if the detection is made where the slope is maximum on the amplitude.

The interest of phase sensitive detection in SPR-based measurement is more obvious. In fact, strong plasmonic resonances can be reached by carefully adjusting the opto-geometrical parameters of the plasmonic layer in order to obtain very sharp phase variation across a resonance. One possible combination of phase sensitivity SPR bio-sensor using G-LIA for phase extraction is proposed in **Figure 6**, where a cuvette is put on a plasmonic chip to convey a fluid on the surface of a plasmonic chip. A coupling prism makes it possible to satisfy the Kretschmann condition for which the reflectivity of a p-polarized incident beam reaches a minimum corresponding the excitation of the plasmon-polariton surface mode. In order to have a stable phase, immune to wavelength fluctuations, the length of the two arms are made equal. **Figure 6(b)** presents the numerically calculated complex reflectivity as a function of the incident beam angle in the case of a glass coupling prism coated by a gold layer of thickness h covered by water and excited by a red laser. The best coupling angle is close to $\alpha = 70°$ in the provided examples.

As can be seen, the phase variation across the resonance can be made very sharp by adjusting the metal thickness h. It should be noted however that this fast variation is associated with a strong attenuation of the reflected beam and a compromise between signal level and sensitivity can be made depending on the available laser power. For example, on **Figure 6(b)** the reflection is about 1.6% for $h = 48$ nm at Kretschmann angle, but it drops to 2.5‰ for $h = 50$ nm.

If we consider a reasonable phase resolution of 10^{-3} rad, simple calculation show that the case h = 50 nm shown on the figure leads to sensitivity slightly better than 8×10^{-7} RIU (refractive index unit). On the other hand, if the thickness is slightly inferior, the RIU sensitivity drops rapidly (about 2×10^{-6} for h = 48 nm). To obtain a similar sensitivity with the only amplitude signal, the noise level on the amplitude measured at the maximum slope should be smaller than 10^{-3}%, which is hardly achievable.

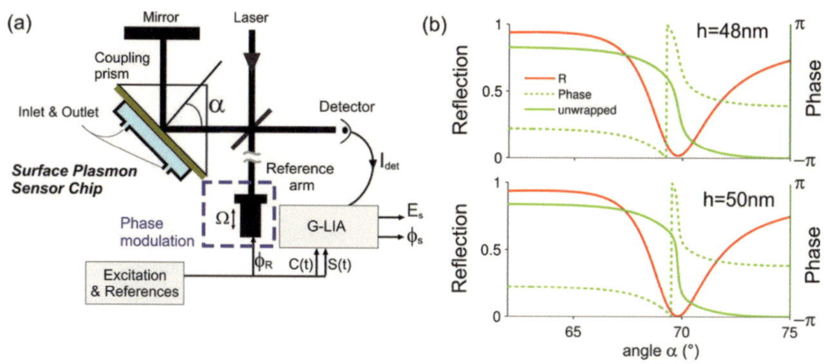

Figure 6. Example of possible experimental setup for phase sensing based on an SPR chip. (b) Simulation of the complex reflectivity (magnitude and phase) for a light beam impinging a gold layer with a thickness h as a function of the incident angle. The wavelength is 670 nm.

3.2. Digital holography

In digital holography, the holograms of a sample object are recorded on a 2D detector such as a Charge-coupled Device (CCD) or a Complementary Metal-Oxide-Semiconductor (CMOS) camera. Such system can notably be used as an optical profilometer, or for sensing applications [1, 6, 16]. **Figure 7(a)** presents the experimental setup of a lensless, compact, digital microscope working with the G-LIA extraction method.

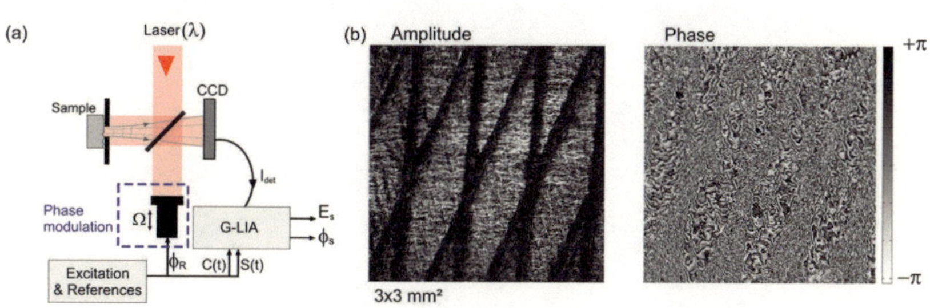

Figure 7. (a) Lensless digital holography setup. A diaphgram may be to select the central zone of interest in the sample. (b) Amplitude and phase of a grating of straight and tilted slits made in a steel surface.

In the provided example, a metallic grid of slit is imaged in amplitude and phase. The Lead Zirconate Titanate (PZT) oscillates in the reference arm at 10 Hz to generate the phase modulation function $\phi_R = a \sin \Omega t$ at a wavelength λ = 640 nm in the reference arm. During the piezo oscillation, a video is recorded at a frame rate of 120 Hz. Then, amplitude and phase of the detected field is obtained on the camera by performing a G-LIA operation on each pixel using a program code. In this 2D case, the operation was not made in real-time because of the non-negligible processing time (In the order of 1 min or less depending on the computing resource). Once the detected complex field is retrieved, the associated plane wave spectrum can be obtained by Fourier transform. Then each plane wave can be back-propagated numerically to any position before the CCD [17], typically up to the sample plane or surface.

In this example, the raw signal $I(X,Y)$ is processed by the numeric, software-based, G-LIA, without filtering. For a correct operation it is then mandatory to use an amplitude modulation a of about 2.405 rad. **Figure 7(b)** shows the amplitude and phase of the complex field which is back-propagated up to the sample plane. As we are dealing with a rough surface, the recorded phase has a speckle-like distribution. Because the illumination direction is normal to the sample surface, the system is strongly sensitive to any out-of-plane displacement of the structure. To give an idea of the system sensitivity, the sample is slightly rotated. By subtracting the phase image before rotation to the phase image after rotation, we obtain the phase-shift associated with the out-of-plane displacements, as shown in **Figure 8**.

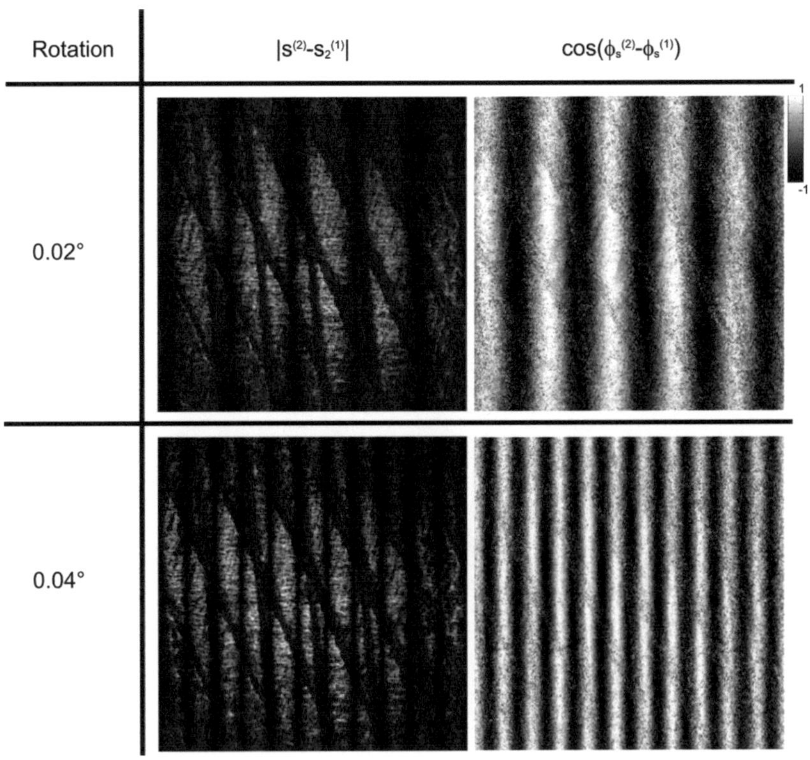

Figure 8. Effect of slight rotation on the holographic images. By subtracting the complex field after rotation $s^{(2)}(X,Y)$ from the complex field before rotation $s^{(1)}(X,Y)$, the out of plane displacement is revealed.

4. G-LIA in the case of amplitude and phase-modulated signal

In some important cases, the signal field is modulated both in phase and amplitude. The modulated term $I_{mod} \propto P\, s \cos(\Delta\phi(t))$ in the detected intensity can then be expressed as:

$$I_{mod} \propto f(t) \cos\left(\Delta\phi(t)\right) \tag{22}$$

where $f(t)$ accounts for the amplitude modulation function. We can mention two relevant examples in the context of *phase-sensitive nanoscopy* and *unbalanced interferometry.*

4.1. Unbalanced interferometry

In interferometers having a path unbalance, a phase modulation can be efficiently induced by a wavelength modulation of the emission wavelength. For this purpose a spectrally single-mode laser diode working at a central wavelength λ_0 can be used. The wavelength modulation is typically obtained by a current modulation which is associated with a power modulation. In air, the phase modulation is related to the small wavelength modulation $\delta\lambda(t)$ by: $\phi_R(t) = \frac{4\pi\Delta l\ \delta\lambda(t)}{\lambda_0^2}$, where Δl is the length difference between the two arms.

For modulation frequency below the MHz range, the change in wavelength is considered to be primarily due to a change of temperature that increases with the current. Therefore, using a sawtooth function to create a quasi-linear phase change is usually not an excellent choice, as the thermal inertia of the system prevents the wavelength to precisely follow the driving excitation. On the other hand, a sine power modulation will typically induce the desired sine wavelength and phase modulation. In this case, the detected intensity within an unbalanced interferometer is:

$$I \propto P_0\left(1 + \mu \sin(\Omega t)\right)\left(1 + s\cos(\phi_r - \phi_s)\right) \tag{23}$$

with $\phi_R(t) = a \sin(\phi_R)$, where $a = \frac{4\pi\Delta l a_\lambda}{\lambda_0^2}$ with a_λ corresponding to the depth of wavelength modulation.

It is clear that in the case where the amplitude modulation is small ($\mu \ll 1$ and $\mu \ll s$), the amplitude modulation disappears and the G-LIA method can be applied directly using the signal $\tilde{I}(t)$ where the DC component is filtered, with the references $C(t) = cos(\phi_R)$ and $S(t) = sin(\phi_R)$. The operations are identical to those given by Eqs. (16)–(21). However, while μ can be small compared to 1, the signal s of interest can also be very small in some experiments and setups so that $\mu \ll s$ cannot be satisfied in general. To see how to handle this problem, we can express the filtered intensity $\tilde{I}(t)$:

$$\tilde{I} \propto s[\cos(\varphi_r - \varphi_s)\left(\mu \sin(\Omega t) + 1\right)]_{filtered} + \mu \sin(\Omega t) \tag{24}$$

where we have normalized the detected intensity by the constant laser power factor. The brackets indicate the quantity is filtered from its DC component. We see that the main issue comes from the modulated term outside the bracket which is independent from the signal s. A direct solution to get rid of this unwanted term is to use references without harmonic component at Ω. This is achieved by choosing $J_1(a) = 0$ (e.g. $a = 3.832$ rad). With this choice of a, the G-LIA operation given by Eqs. (16)–(24) gives excellent results provided that μ is kept reasonably small in comparison with 1 (e.g. for $\mu = 0.1$ the maximum error on the G-LIA X and Y output is about 2%).[5] This condition is reasonably achieved in many cases. For example, if we consider a standard single mode laser diode such as a Vertical Cavity Surface Emitting Laser (VCSEL) with a tunability of 0.5 nm/mA and a bias current of 1 mA. We see that a current modulation of 4.4% is sufficient to achieve $a = 3.8$ rad for an unbalance of 10 mm at a wavelength of 850 nm. The corresponding power modulation μ depends on its $P(I)$ but will be typically smaller than 0.1.

4.1.1. *Improved unbalanced configuration*

Despite its advantages in term of cost, unbalanced interferometry is not currently widely used. The main reason is also related to the extreme sensitivity of the system to minute wavelengths changes. **Figure 9(a)** represents a compensation scheme to solve this issue. The idea is to illuminate the interferometer with a linear polarization at 45° with respect to the horizontal and vertical axis and to discriminate the two s and p polarization using polarization beam splitters. An additional signal arm is equipped with a fixed mirror in order to measure the phase fluctuation induced by any wavelength drifts in time. The light impinging on this mirror is s-polarized and is selectively detected by the photodiode PD1, using a polarization beam splitter in reflection. On the other hand, the p-polarized light impinging on the piezo-actuated mirror is reflected back onto the second photodiode (PD2). Both amplitude and phases are recorded with the above described G-LIA operation.

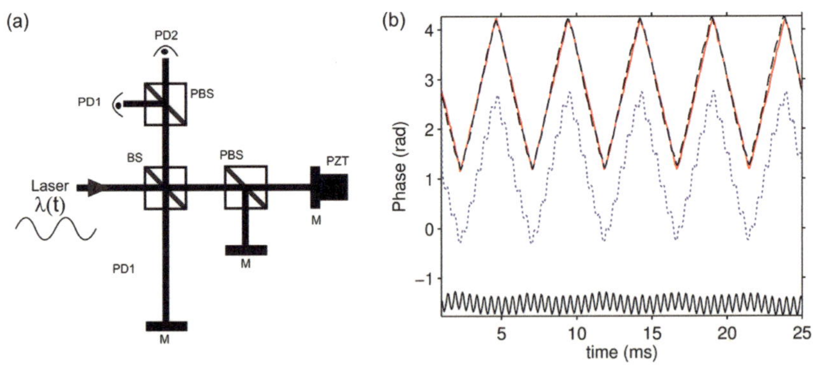

Figure 9. (a) Unbalanced interferometer with an extra arm for wavelength drifts compensation. The sine phase modulation is induced by a power modulation of the VCSEL laser source. BS: Beam splitter. (b) Actual displacement of the piezo actuated mirror (red); measured displacement without drift compensation (dotted blue); phase fluctuation induced by the intentional wavelength fluctuation (black line), and final measurement (dashed black line) obtained by subtracting the black line to the blue dotted line.

[5] We note that the cases where μ is too large to be neglected can be handled exactly without approximation but it requires to know μ in order to determine analytically or numerically all the coefficients of the G-LIA outputs (4 in this case). In general, the percentage of power modulation μ can be measured without difficulty. The condition J1(a) = 0 is still required.

http://dx.doi.org/10.5772/66657

Figure 9(b) shows a controlled triangular displacement which is correctly determined despite the presence of intentional wavelength drifts. In this experiment, the wavelength of the VCSEL is driven sinusoidally at about 10 kHz to create the phase modulation. The important wavelength drifts are artificially created by adding a low frequency sine to this excitation signal. The compensation is obtained by plotting the phase of the p-polarized light minus the phase of the s-polarized light which is coming from the fixed mirror. Both signal phases are obtained by the G-LIA method with a approximately equal to 3.83 rad.

Such system is really interesting in term of performance since VCSELs are very affordable laser sources that can be driven at very sinusoidally at very high speed. In the described experiment the phase modulation frequency was only limited by the acquisition card used to perform the G-LIA measurement.

4.2. Phase-sensitive nanoscopy

A modulation of the amplitude at an angular frequency Ω_A can be introduced in the signal arm to discriminate the amplitude-modulated signal from other unwanted contributions able to interfere with the reference field. This is the case in near-field nanoscopy where the signal light is coming from a near-field probe in interaction with a surface, oscillating at an angular frequency Ω_{probe}. The situation is depicted in **Figure 10**.

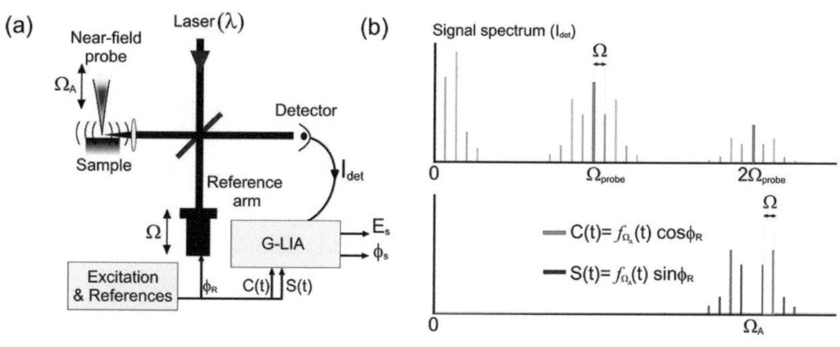

Figure 10. Phase-sensitive nanoscopy experiment based on G-LIA, where the phase modulation is $\varphi_R = a \sin \Omega t$. (a) Basic setup. (b) Illustration of the frequency spectra of (top) the detected intensity and (bottom) the references signal used in the extraction process. In this example $\Omega_A = 2\Omega_{probe}$. The condition (e.g. a = 2.405 rad) removes the unwanted peaks from the references.

In **Figure 10**, the near-field head is included in the signal arm of a Michelson interferometer. Alternately the near-field microscope can be used in the signal arm of a Mach-Zhender which is well adapted to the characterization of waveguiding photonic devices as in [18–20]. Here, the sample is scanned under a nano-tip which is precisely positioned in the focus spot of an objective lens. The light backscattered by the oscillating probe operating in tapping mode contains information on the local optical properties of the sample. This backscattered light can have a rich harmonic content due to its near-field interaction with a sample. The amplitude modulation function appearing in I_{mod} is therefore $f(t) = cte + s_1 \cos(\Omega_{probe}t + \Phi_1) + s_2 \cos(2\Omega_{probe}t + \Phi_2) + \ldots$

We note that I_{mod} in this case refers to the interference term between the *near-field signal* from the probe and the phase-modulated *reference signal*. However other parasitic fields can be backscattered by the probe-sample system. When using a standard, cantilevered, AFM probe, the oscillation amplitude is typically larger than the probe radius. In this case it is often interesting to detect the contribution from a higher harmonics $\Omega_A = k\Omega_{probe}$ as the unwanted part of the light which is modulated by the shaft of the probe (rather than the apex) only contributes to the first harmonic(s). This unwanted contribution modulated by the probe shaft can be referred as *modulated background contribution* (MBC) in contrast with the *unmodulated background contribution* (UBC) coming, for example, from the sample backscattering (nano-dusts, roughness, …) which is also unwanted. In less stringent configurations, especially when the oscillation is small compared to the tip radius (the near field varies almost linearly on the excursion of the probe), the MBC is not perceived[6] while keeping k=*1*.

As shown in **Figure 10(b)**, because of the interference between the probe signal and the reference field, the signal is split into sidebands at $k\Omega_{probe} \pm p\Omega$ where Ω characterize the phase modulation frequency. In order to collect the information spread throughout all these side-bands, the G-LIA can use the following references:

$$C(t) = f_{\Omega A}(t) cos(\phi_R t) \tag{25}$$

$$S(t) = f_{\Omega_A}(t) sin(\phi_R t) \tag{26}$$

where $f\Omega_A(t) = cos(\Omega_A t + \psi)$ is the amplitude carrier at the frequency of interest Ω_A. But even in this favorable case where the MBC is easily excluded, the UBC can be especially detrimental and has been described by several authors. UBC however can be efficiently removed with interferometric detection. As noted elsewhere, the UBC is perceived because the unmodulated background interferes with the modulated signal from the tip on the detector creating signal peaks at the harmonics frequencies $k\Omega_{probe}$. Therefore a direct solution consists in excluding the frequencies $\Omega_A = k\Omega_{probe}$ from the references $C(t)$ and $S(t)$. In the case of a sine modulation, the condition $a = 2.405$ rad ($J_0(a) = 0$) is sufficient to exclude the unwanted frequencies component by removing the possible DC contribution in $\cos(\phi_R t)$.

First examples of phase-sensitive near-field imaging based on G-LIA can be found in Ref. [7]. In **Figure 11**, a simple demonstration experiment is made by using a bare tuning oscillating fork to modulate part of the signal at an angular frequency Ω_A. The trace of the experimentally detected signal intensity exhibits clearly a slow modulation due the phase modulation ($\phi_R = a \sin \Omega t$ with a frequency of 1 kHz) as well as a sine amplitude oscillation at the oscillation frequency of the tuning fork Ω_A (at about 32 kHz). An additional coverglass can be added to

[6]Such case occurs when using elongated probes like tungsten probes, mounted on tuning fork working in tapping mode. The elongated shape minimizes the possible modulation of the background light, while an oscillation amplitude of few nanometers can also prevent a detectable modulation of the background light. In some other case where a Mach-Zehnder interferometer is used, only the apex of the probe can be illuminated (e.g.when imaging waveguiding structures). In general, reducing the amplitude of modulation of the probe reduces the background contributions more efficiently than the near-field contribution.

http://dx.doi.org/10.5772/66657

check the system immunity to UBC. In this example, the references are those given by Eqs. (25)–(26) with $f\Omega_A(t) = cos(\Omega_A t + \psi)$ where ψ is adjusted to be in phase with the fork oscillation, resulting in a maximized amplitude signal (not shown here). With a = 2.405 rad to exclude the UBC, the two G-LIA outputs provide:

$$X =< I(det) * C(t) > \propto k_x s_1 \cos(\phi_s), with\ k_x = \langle C^2(t) \rangle \tag{27}$$

$$Y =< I(det) * C(t) > \propto k_y s_1 \cos(\phi_s)\ with\ k_y = \langle S^2(t) \rangle \tag{28}$$

where s_1 corresponds to the amplitude of signal field modulated at Ω_A. For a sine phase modulation, an analytical expression can be derived for the proportionality constants k_x and k_y.These expressions can be found in the summary table given in Appendix A.

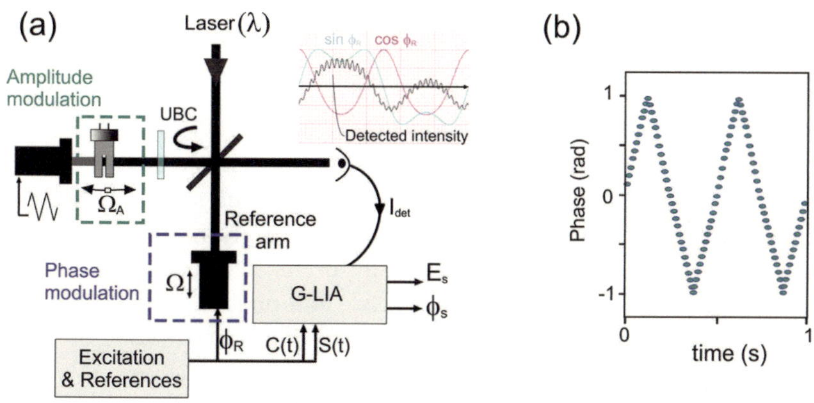

Figure 11. (a) Demonstration setup with $\phi_R = a \sin \Omega t$. An UBC can be added via a glass coverslip in the signal arm. (b) Recovered signal phase for a triangular displacement of the signal mirror.

Figure 11(b) shows the phase determined with this method when a triangular phase modulation having a peak to peak phase modulation depth of about 2.0 rad is induced by the signal mirror. The signal phase is precisely retrieved.

The value of ψ includes the mechanical phase shift existing between the driving signal and the actual motion of the fork. In fact, in a near-field experiment, this shift can vary from one position to another on the sample depending on the material in interaction with the probe. Depending on the system, the value of ψ in $f\Omega_A(t) = cos(\Omega_A t + \psi)$ is not necessarily known if only the driving signal is accessible. We note that the retrieved optical phase value is not affected by the value of ψ, which only affects the amplitude, but the SNR can be strongly decreased if we omit ψ.

If ψ is unknown, the G-LIA can be applied twice to solve this issue with two quadrature amplitude modulation functions. In other word, we can calculate for outputs signals $(X, Y, X', Y') =< I{*}reference(t) >$, using, respectively, the following references:

$$C(t) = cos\left(\Omega_A(t)\right)cos(\phi_R t) \tag{29}$$

$$S(t) = cos\left(\Omega_A(t)\right)sin(\phi_R t) \tag{30}$$

$$C'(t) = sin\left(\Omega_A(t)\right)cos(\phi_R t) \tag{31}$$

$$S'(t) = sin\left(\Omega_A(t)\right)sin(\phi_R t) \tag{32}$$

The four outputs (X, Y, X', Y') can be evaluated numerically or analytically, they provide, respectively, $< C(t)^2 > cos(\phi_s)$, $< S(t)^2 > sin(\phi_s)$, $< C'(t)^2 > cos(\phi_s)$, $< S'(t)^2 > sin(\phi_s)$ whose analytical expressions are given in Appendix A for the sine phase modulation $\phi_R(t)$.

5. Conclusion

We have detailed the principle of the G-LIA method, first in the case of pure phase modulations then in the case where the amplitude of the signal is also modulated. For pure phase modulations, the interest of the approach was illustrated in different contexts: position monitoring, sensing, and digital holography. In these experiments, the non-linear phase modulation was achieved by mirrors mounted on sinusoidally driven piezo-actuators. In this case, the main advantage of the G-LIA is to extract amplitude and phase information directly from all the harmonic contents created by the phase modulation function. While the examples only considered sine phase modulation functions which is often the most desirable one, the G-LIA also provides a unified treatment to handle arbitrary phase modulation function.

We have also detailed the case where an amplitude modulation can be present. This is notably the case in unbalanced interferometry where a non-negligible amplitude modulation can be perceived at the same frequency than the phase modulation. Experimentally, we considered the case of unbalanced interferometers where a fast sine phase modulation is provided by a current-driven single mode laser diode. A simple yet efficient setup was described to neutralize the impact of wavelength fluctuation on the system. Such approach offers the opportunity to develop simple and cost-efficient system without sacrificing precision. Finally, we discussed the case where the signal of interest is modulated in amplitude at a frequency different from that of the phase modulation. This case was detailed in the context of phase sensitive SNOM, where the low available signal requires to exploit all the available sidebands induced by the phase modulation. Notably, the condition to cancel the effect of the unmodulated background light was presented and attention was paid to the impact of the mechanical phase of the oscillating probe.

Appendix A

The table (**Figure 12**) provides a summary of case handled by the G-LIA method.

http://dx.doi.org/10.5772/66657

Modulation Functions		References	G-LIA on I_{det}		G-LIA on $\tilde{I}_{det}$
Phase $\phi_R(t)$	Amplitude $f(t)$	(X, Y)	(k_x, k_y)	Condition	(K_x, K_y)
Ωt	-	$C(t) = \cos\phi_R$, $S(t) = \sin\phi_R$	1,1	-	1,1
$a\sin(\Omega t)$	-		$1+J_0(2a)$, $1-J_0(2a)$	$J_0^2(a)\approx 0$	$k_x-2J_0^2(a)$ k_y
$a\,Tr(\Omega t)$	-		$1+2\,\mathrm{sinc}(2a)$, $1-2\mathrm{sinc}(2a)$	$\mathrm{sinc}^2(a)\approx 0$	$k_x-\mathrm{sinc}^2(a)$ k_y
other	-		analytic or numeric evaluation of $\langle C^2(t)\rangle$ and $\langle S^2(t)\rangle$	No DC term in the references	analytic or numeric evaluation of $\langle \tilde{C}^2(t)\rangle$ and $\langle \tilde{S}^2(t)\rangle$
Ωt	$\cos(\Omega_A t+cte)$ cte: known	$C(t) = 2f(t)\cos\phi_R$, $S(t) = 2f(t)\sin\phi_R$	1,1		
$a\sin(\Omega t)$			$1+g(a)$, $1-g(a)$		
other	other		analytic or numeric evaluation of $\langle C^2(t)\rangle$ and $\langle S^2(t)\rangle$		

Modulation Functions		References	G-LIA on I_{det}
Phase $\phi_R(t)$	Amplitude	(X, Y, X', Y')	(k_x, k_y, k_x', k_y')
$a\sin(\Omega t)$	$\cos(\Omega_A t+\psi)$ ψ unknown	$\cos(n\Omega t)\cos\phi_R$, $\cos(n\Omega t)\sin\phi_R$, $\sin(n\Omega t)\cos\phi_R$, $\sin(n\Omega t)\sin\phi_R$	$(1+g(a))\cos\psi$, $(1-g(a))\cos\psi$, $-(1+g(a))\sin\psi$, $-(1-g(a))\sin\psi$

$g(a)=J_{2n}(2a)+J_0(2a)$

$n=\Omega_A/\Omega$ is preferentially large

The case n=1 is treated separately

For a sine phase modulation, an eventual UBC is canceled if $J_0(a)=0$ (e.g. a=2.405 rad)

The 4 unknowns (E_S, ϕ_S, $\cos\psi$, $\sin\psi$) are retrieved from the 4 outputs X, Y, X', Y'

Figure 12. Summary table.

Author details

Aurélien Bruyant*, Julien Vaillant, Tzu-Heng Wu, Yunlong Zhu, Yi Huang and Abeer Al Mohtar

*Address all correspondence to: aurelien.bruyant@utt.fr

ICD-LNIO, Université de Technologie de Troyes, Troyes, France

References

[1] O. Sasaki, and H. Okazaki. Sinusoidal phase modulating interferometry for surface profile measurement. Appl. Opt. 1986;**25**(18):3137–3140. DOI: 10.1364/AO.25.003137

[2] O. Sasaki, and H. Okazaki. Analysis of measurement accuracy in sinusoidal phase modulating interferometry. Appl. Opt. 1986;**25**(18): 3152–3158. DOI: 10.1364/AO.25.003152

[3] O. Sasaki, T. Suzuki, and K. Takahashi. Sinusoidal phase modulating laser diode interferometer with feedback control system to eliminate external disturbance. Opt. Eng. 1990;**29** (12):1511–1515. DOI: 10.1117/12.55754

[4] G. He, and X. Wang. Real-time micro-vibration measurement in sinusoidal phase-modulating interferometry. Optik. 2009;**120**(3):101–105. DOI: 10.1016/j.ijleo.2007.05.010.

[5] Z. Li, X. Wang, P. Bu, B. Huang, and D. Zheng. Sinusoidal phase-modulating laser diode interferometer insensitive to the intensity modulation of the light source. Optik. 2009;**120** (16):799–803. DOI: 10.1016/j.ijleo.2008.03.015

[6] E. Bo, F. Duan, C. Lv, F. Zhang, and F. Feng. Sinusoidal phase modulating interferometry system for 3D profile measurement. Opt. Laser Technol. 2014;**59**:137–142.

[7] A. Al Mohtar, J. Vaillant, Z. Sedaghat, M. Kazan, L. Joly, C. Stoeffler, J. Cousin, A. Khoury, and A. Bruyant. Generalized lock-in detection for interferometry: application to phase sensitive spectroscopy and near-field nanoscopy. Opt. Express 2014;**22**(18):22232–22245. DOI: 10.1364/OE.22.022232

[8] G. Economou, R. Youngquist, and D. Davies. Limitations and noise in interferometric systems using frequency ramped single-mode diode lasers. J. Lightwave Technol. 1986;**4** (11):1601–1608. DOI: 10.1109/JLT.1986.1074672

[9] D.A. Jackson, A.D Kersey, M. Corke, and J.D.C. Jones. Pseudoheterodyne detection scheme for optical interferometers. Electron Lett. 1982;**18**(25–26):1081. DOI: 10.1049/el:19820740

[10] A. Dandridge, A. Tveten, and T. Giallorenzi. Homodyne demodulation scheme for fiber optic sensors using phase generated carrier. IEEE Trans. Microwave Theory Tech. 1982;**30** (10): 1635–1641. DOI: 10.1109/JQE.1982.1071416

[11] G.B. Arfken, H.J. Weber, and F.E. Harris. Mathematical methods for physicists: a comprehensive guide. Academic Press, Waltham, USA & Kidlington, UK; 2011.

[12] L. Shen, J.A. Hagen, and I. Papautsky. Point-of-care colorimetric detection with a smartphone. Lab Chip. 2012;**12**(21):4240–4243. DOI: 10.1039/C2LC40741H

[13] Y.H. Huang, H.P. Ho, S.Y. Wu, S.K. Kong. Detecting phase shifts in surface plasmon resonance: a review. Adv. Opt. Technol. 2011. DOI: 10.1155/2012/471957

[14] C.M. Wu, Z.C. Jian, S.F. Joe, and L.B. Chang. High-sensitivity sensor based on surface plasmon resonance and heterodyne interferometry. Sens. Actuators B. 2003;**92**(1):133–136. DOI: 10.1016/S0925-4005(03)00157-6

[15] W. Yuan, H.P. Ho, C.L. Wong, S.K. Kong, and C. Lin. Surface plasmon resonance biosensor incorporated in a Michelson interferometer with enhanced sensitivity. IEEE Sensors J. 2007;**7**(1):70–73. DOI: 10.1109/JSEN.2006.884982

[16] B. Javidi, and E. Tajahuerce. Three-dimensional object recognition by use of digital holography. Opt. Lett. 2000;**25**(9):610–612. DOI: 10.1364/OL.25.000610

[17] K.A. Stetson. Mathematical refocusing of images in electronic holography. Appl. Optics. 2009;**48**(19):3565–3569. DOI: 10.1364/AO.48.003565

[18] B. Dagens, M. Février, P. Gogol, S. Blaize, A. Apuzzo, G. Magno, R. Mégy, and G. Lerondel. Direct observation of optical field phase carving in the vicinity of plasmonic metasurfaces. Nano Lett. 2016. DOI: 10.1021/acs.nanolett.6b00435

[19] X. Wu, L. Sun, J. Wang, and Q. Tan. Real-time phase error compensation in phase sensitive scanning near-field optical microscopy. Appl. Optics. 2015;**54**(19):6128–6133. DOI: 10.1364/AO.54.006128

[20] N. Rotenberg, and L. Kuipers. Mapping nanoscale light fields. Nature Photon. 2014;**8**(12):919–926. DOI: 10.1038/nphoton.2014.285

Chapter 11

120° Phase Difference Interference Technology Based on 3 × 3 Coupler and its Application in Laser Noise Measurement

Yang Fei, Xu Dan, Cai Haiwen and Qu Ronghui

Additional information is available at the end of the chapter

http://dx.doi.org/10.5772/66129

Abstract

A 120° phase difference interferometer technology based on an unbalanced Michelson interferometer composed of a 3 × 3 optical fiber coupler is proposed, and based on this technology, the differential phase of the input laser can be obtained. This technology has many applications. This paper introduced its application in laser phase and frequency noise measurement in detail. The relations and differences of the power spectral density (PSD) of differential phase and frequency fluctuation, instantaneous phase and frequency fluctuation, phase noise, and linewidth are derived strictly and discussed carefully. The noise features of some narrow-linewidth lasers are also obtained conveniently without any specific assumptions or noise models. Finally, the application of this technology in the phase-sensitive optical time domain reflectometer (φ-OTDR) is also introduced briefly.

Keywords: interferometer, 120° phase difference, 3 × 3 optical fiber coupler, laser noise measurement, phase-sensitive optical time domain reflectometer

1. Introduction

Single-frequency narrow-linewidth lasers are fundamental to a vast array of applications in fields including metrology, optical frequency transfer, coherent optical communications, high-resolution sensing, and light detection and ranging (LIDAR) [1–9]. In these applications, the phase and frequency noise is one of the key factors to affect the system performance. The characterization and measurement of the phase and frequency noise are very important for the applications, and thus have been one of the most attractive subjects of researches in laser and photonics field. The phase and frequency noise of such lasers can be conveniently described either in terms of linewidth or in terms of the power spectral density (PSD) of their phase or frequency noise. The linewidth gives a basic and concise parameter for characterizing laser coherence but lacks detailed information on frequency noise and its Fourier frequency spectrum, which is needed for understanding the noise origins and improving laser performances. Therefore, the measurement of frequency noise PSD is a focus of attention in the field, especially for lasers of very high coherence, whose linewidth is not easy to be measured.

To measure the phase and frequency noise, many methods have been proposed, such as beat note method [10], recirculating delayed self-heterodyne (DSH) method [11], DSH technique based on Mach-Zehnder interferometer with 2 × 2 coupler [12, 13], or Michelson interferometer (MI) with 2 × 2 coupler [14]. These methods can obtain good measurement results but need some strict conditions. The beat note method needs a high coherent source as a reference. The recirculating DSH method needs very long fiber delay lines. The DSH interferometers with 2 × 2 coupler need to control the quadrature point by some active feedback methods and accurate calibration.

To overcome these difficulties, we introduce a robust technique that can demodulate directly the laser differential phase accumulated in a delay time and then derive strict mathematical relations between the laser differential phase and the laser phase noise or frequency noise that can describe the complete information on laser phase and frequency noise. Because 3 × 3 optical fiber coupler acts as a 120° optical hybrid, it can demodulate the differential phase of the input light and has been used for DxPSK signal demodulation [15], optical sensors [16], optical field reconstruction, and dynamical spectrum measurement [17]. In this chapter, 120° phase difference interference technology based on an unbalanced Michelson interferometer, which is composed of a 3 × 3 optical fiber coupler and two Faraday rotator mirrors, is utilized to demodulate the differential phase of a laser. The structure has the advantage of being polarization insensitive and adjust-free. Especially, it does not need any active controlling operation that is used in the DSH methods with 2 × 2 coupler. Furthermore, based on the differential phase and strict physical and mathematical derivation, the PSD of the differential phase fluctuation and frequency fluctuation, the PSD of the instantaneous phase fluctuation and frequency fluctuation, laser phase noise, and linewidth are completely calculated and discussed.

2. 120° phase difference interference technology

2.1. Symmetric 3 × 3 optical fiber coupler

Figure 1 shows the schematic diagram of the proposed 120° phase difference technology. As we know, the interferometric signals are 180° out of phase because of 2 × 2 optical fiber coupler. In an ideal 3 × 3 optical fiber coupler, there is a 120° phase difference between any two of the three output ports. A symmetric 3 × 3 optical fiber coupler is described by the matrix.

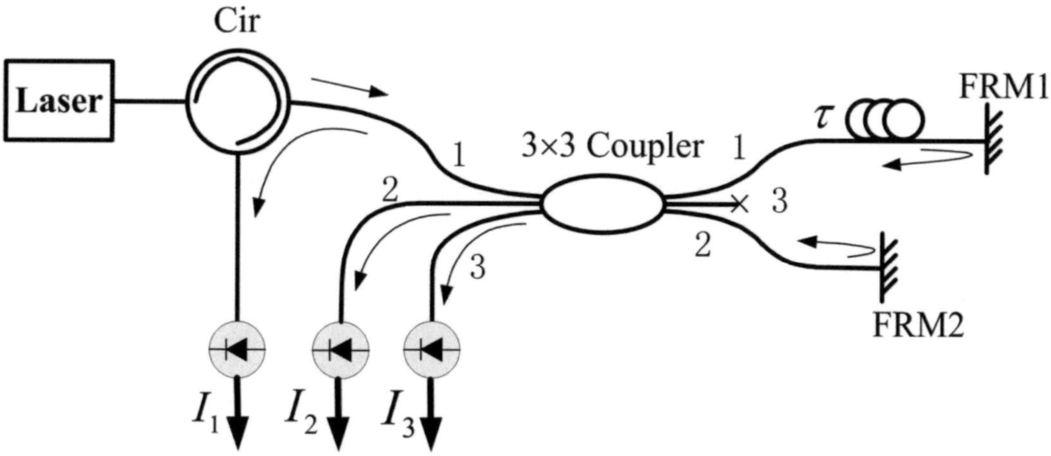

Figure 1. Schematic diagram of the proposed 120° phase difference technology.

$$T_3 = \frac{1}{\sqrt{3}}\begin{pmatrix} 1 & \exp\frac{i2\pi}{3} & \exp\frac{i2\pi}{3} \\ \exp\frac{i2\pi}{3} & 1 & \exp\frac{i2\pi}{3} \\ \exp\frac{i2\pi}{3} & \exp\frac{i2\pi}{3} & 1 \end{pmatrix} \tag{1}$$

The matrix of the two arms of the Michelson interferometer is given by,

$$P = \begin{pmatrix} \exp(i\varphi_1) & 0 & 0 \\ 0 & \exp(i\varphi_2) & 0 \\ 0 & 0 & 0 \end{pmatrix} = \begin{pmatrix} 1 & 0 & 0 \\ 0 & \exp(i\Delta\varphi) & 0 \\ 0 & 0 & 0 \end{pmatrix} \exp(i\varphi_1) \tag{2}$$

Where $\Delta\varphi = \varphi_2 - \varphi_1 = 2\pi n\Delta L/\lambda$ is the phase difference of the two arms of the Michelson interferometer, n is the refractive index of fiber, λ is the wavelength of transmit light, ΔL is the length difference of the two arms. Setting $2\pi/3 = \theta$, the operation of the whole interferometer is then described by the system matrix.

$$M = T_3 P T_3 = \frac{e^{i\varphi_1}}{3}\begin{pmatrix} 1 + e^{i(\Delta\varphi+2\theta)} & (1+e^{i\Delta\varphi})e^{i\theta} & e^{i\theta}(1+e^{i(\Delta\varphi+\theta)}) \\ e^{i\theta}(1+e^{i\Delta\varphi}) & e^{i2\theta} + e^{i\Delta\varphi} & e^{i\theta}(e^{i\theta}+e^{i\Delta\varphi}) \\ e^{i\theta}(1+e^{i(\Delta\varphi+\theta)}) & e^{i\theta}(e^{i\theta}+e^{i\Delta\varphi}) & e^{i2\theta}(1+e^{i\Delta\varphi}) \end{pmatrix} \tag{3}$$

The electric field of input laser is expressed by,

$$E(t) = |E(t)| \exp[i\omega_0 t + i\varphi(t)] \tag{4}$$

with amplitude $|E(t)|$, center frequency ω_0, and instantaneous phase fluctuation $\varphi(t)$. The input field $K^{in} = [E(t), 0, 0]^T$ is a vector whose elements are the amplitudes of the input modes to the interferometer. K^{in} is transformed by the interferometer into an output vector.

$$\left[E_1^{out}, E_2^{out}, E_3^{out}\right]^T = MK^{in} = \frac{e^{i\varphi_1}}{3} \cdot \begin{pmatrix} 1 + e^{i(\Delta\varphi + 2\theta)} \\ e^{i\theta}(1 + e^{i\Delta\varphi}) \\ e^{i\theta}(1 + e^{i(\Delta\varphi + \theta)}) \end{pmatrix} \cdot E(t) \tag{5}$$

The measured intensities $I_n = |E_n^{out}|^2$ (n=1, 2, 3) of the three outputs are,

$$\begin{pmatrix} I_1 \\ I_2 \\ I_3 \end{pmatrix} = \frac{2I_0}{9} \begin{pmatrix} 1 + \cos(\Delta\varphi - 2\pi/3) \\ 1 + \cos(\Delta\varphi) \\ 1 + \cos(\Delta\varphi + 2\pi/3) \end{pmatrix} \tag{6}$$

where $I_0 = |E(t)|^2$ is the input intensity. Then the differential phase can be obtained.

$$\Delta\varphi = \arctan\left[\frac{\sqrt{3}(I_3 - I_1)}{I_3 + I_1 - 2I_2}\right] \tag{7}$$

2.2. Asymmetric 3 × 3 optical fiber coupler

However, the commercially available 3 × 3 optical fiber coupler is usually asymmetric and lossy. So the transmissivity of 3 × 3 coupler from port m to port n (m, n = 1, 2, 3) is $b_{mn}\exp(i\theta_{mn})$, where b_{mn} and θ_{mn} are the splitting ratio and phase delay of coupler, respectively. The forward transmission matrix of 3 × 3 coupler is then given by.

$$T_{3nm} = b_{nm} \exp(i\theta_{nm}) \tag{8}$$

Similarly, the backward transmission matrix of 3 × 3 coupler is then given by,

$$T'_{3nm} = b'_{nm} \exp(i\theta'_{nm}) \tag{9}$$

where b'_{mn} is the splitting ratio and θ'_{mn} is the phase delay of 3 × 3 coupler from port n to port m. The matrix of the two arms of the Michelson interferometer is given by,

$$P = \begin{pmatrix} p_{11} & 0 & 0 \\ 0 & p_{22}\exp(-i\omega\tau - i\delta) & 0 \\ 0 & 0 & 0 \end{pmatrix} \tag{10}$$

where p_{11} and p_{22} are the transmissivity of the two arms, respectively; τ and δ are the differential delay time and the differential phase delay between the signals in two arms of the Michelson interferometer. The operation of the whole interferometer is then described by the system matrix.

$$\begin{aligned} M_{nm} &= \left(T_3' P T_3\right)_{nm} \\ &= b'_{n1} p_{11} b_{1m} \exp\left[i\left(\theta'_{n1} + \theta_{1m}\right)\right] + b'_{n2} p_{22} b_{2m} \exp\left[i\left(\theta'_{n1} + \theta_{2m}\right)\right] \exp(-i\omega\tau - i\delta) \end{aligned} \tag{11}$$

The output light from the output port n is detected by a photodetector having a responsivity of r_n. The detected intensities from the output port n, I_n, can be expressed as.

$$\begin{aligned} I_n(t) &= r_n \left|E_n^{\text{out}}\right|^2 \\ &= r_n c_{n1}^2 \left|E(t)\right|^2 + r_n c_{n2}^2 \left|E(t-\tau)\right|^2 + \eta_n \cos\left(\delta + \Delta\varphi\right)\left|E(t)E(t-\tau)\right| + \\ &\quad \zeta_n \sin\left(\delta + \Delta\varphi\right)\left|E(t)E(t-\tau)\right| \end{aligned} \tag{12}$$

Define an intermediate matrix,

$$X' = \left(X_1'(t) \quad X_2'(t) \quad X_3'(t)\right)^{\mathrm{T}} = \begin{pmatrix} \cos\left(\delta + \Delta\varphi\right)\left|E(t)E(t-\tau)\right| \\ \sin\left(\delta + \Delta\varphi\right)\left|E(t)E(t-\tau)\right| \\ \left(\left|E(t)\right|^2 + \left|E(t-\tau)\right|^2\right)/2 \end{pmatrix} \tag{13}$$

Then,

$$\begin{pmatrix} I_1(t) \\ I_2(t) \\ I_3(t) \end{pmatrix} = \begin{pmatrix} \eta_1 & \zeta_1 & \xi_1 \\ \eta_2 & \zeta_2 & \xi_2 \\ \eta_3 & \zeta_3 & \xi_3 \end{pmatrix} \begin{pmatrix} X_1'(t) \\ X_2'(t) \\ X_3'(t) \end{pmatrix} + \begin{pmatrix} r_1\left(c_{11}^2 - c_{12}^2\right) \\ r_2\left(c_{21}^2 - c_{22}^2\right) \\ r_3\left(c_{31}^2 - c_{32}^2\right) \end{pmatrix} \frac{\left|E(t)\right|^2 - \left|E(t-\tau)\right|^2}{2} \tag{14}$$

Where

$$\begin{aligned} \eta_n &= 2r_n c_{n1} c_{n2} \cos\theta_n \\ \varsigma_n &= 2r_n c_{n1} c_{n2} \sin\theta_n \\ \xi_n &= r_n\left(c_{n1}^2 + c_{n2}^2\right) \\ c_{nm} &= p_{mm} b'_{nm} b_{m1} \\ \theta_n &= \left(\theta'_{n2} + \theta_{21}\right) - \left(\theta'_{n1} + \theta_{11}\right) \\ \Delta\varphi &= \varphi(t) - \varphi(t-\tau) \end{aligned} \quad , \quad \begin{pmatrix} n = 1,2,3 \\ m = 1,2 \end{pmatrix} \tag{15}$$

The parameters η_n, ς_n, and ξ_n are constant for the setup, once the devices and structure are determined. They can be obtained by a broadband light source without measuring each parameter one by one as said in Ref. [15]. Eq. (13) can be transformed as.

$$X' = \begin{pmatrix} \eta_1 & \varsigma_1 & \xi_1 \\ \eta_2 & \varsigma_2 & \xi_2 \\ \eta_3 & \varsigma_3 & \xi_3 \end{pmatrix}^{-1} \begin{pmatrix} I_1(t) \\ I_2(t) \\ I_3(t) \end{pmatrix} - \begin{pmatrix} \eta_1 & \varsigma_1 & \xi_1 \\ \eta_2 & \varsigma_2 & \xi_2 \\ \eta_3 & \varsigma_3 & \xi_3 \end{pmatrix}^{-1} \begin{pmatrix} r_1\left(c_{11}^2 - c_{12}^2\right) \\ r_2\left(c_{21}^2 - c_{22}^2\right) \\ r_3\left(c_{31}^2 - c_{32}^2\right) \end{pmatrix} \frac{\left|E(t)\right|^2 - \left|E(t-\tau)\right|^2}{2} \tag{16}$$

Eq. (16) shows the relation between the differential phase $\Delta\varphi$ and the detectors outputs (I_1, I_2, I_3). We note that the second term on the right-hind side of Eq. (16) becomes zero or can be omitted under the following conditions: (a) the splitting ratios of the 3 × 3 coupler are uniform (i.e., $c_{n1} = c_{n2}$ for all n), or (b) the intensity of the laser under test is constant or periodical (i.e., $|E(t)|^2 - |E(t-\tau)|^2 = 0$), or (c) the extinction ratio of the Michelson interferometer. As a result, the differential phase $\Delta\varphi(t)$ accumulated in delay time τ can be obtained in the following simple form

$$\Delta\varphi(t) = \varphi(t) - \varphi(t-\tau) = \arctan\left(\frac{X'_2(t)}{X'_1(t)}\right) - \overline{\arctan\left(\frac{X'_2(t)}{X'_1(t)}\right)} \tag{17}$$

3. The application in laser noise measurement

3.1. Experimental setup

Based on the 120° phase difference technology, the experimental setup of laser noise measurement is shown in **Figure 2** [18]. It consists of a commercially available 3 × 3 optical fiber coupler (OC), a circulator (C), two Faraday rotator mirrors (FRMs), three photodetectors (PDs), a data acquisition board (DAC) or a digital oscilloscope, and a computer. The 120° phase difference Michelson interferometer is composed of the 3 × 3 coupler and the FRMs. On one hand, the FRM will remove the polarization fading caused by external disturbance on the two beam fibers of the interferometer. On the other hand, the length or index of the fiber configuring the interferometric arms would change randomly because of temperature fluctuations, vibration, and other types of environmental disturbances; thus it induces low-frequency random-phase drifts in the interferometric signal. So in the proposed experimental setup, the complete interferometer is housed in an aluminum box enclosed in a polyurethane foam box for thermal and acoustic isolation. Meantime, the two fiber arms of the Michelson interforometer are placed closely in parallel to improve the stability against the perturbation.

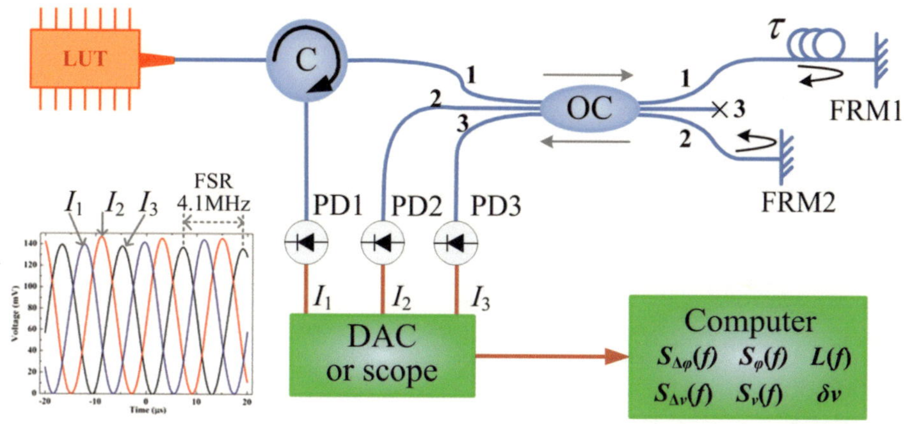

Figure 2. Experimental setup used to measure the laser phase and frequency noise, and the output interference fringe of the PD1, PD2, PD3 (inset). LUT: laser under test, C: circulator, OC: optical fiber coupler, FRM: Faraday rotation mirror, PD: photodetector, DAC: data acquisition board [18].

The laser under test (LUT) injects the left port 1 of the 3 × 3 optical fiber coupler through a circulator and then splits into three parts by the coupler. Two of them interfere mutually in the coupler after being reflected by Faraday mirrors and with different delay times, and the third part of them is made reflection-free. Then the interference fringes are obtained from the left port 1, 2, and 3 of the coupler and read by a DAC or a digital oscilloscope.

In the experimental setup, a swept laser source with linewidth of about 2.5 kHz [19] is used as the broadband light source to show clearly the small free spectral range (FSR) of the MI. The measured interference fringe is shown in the inset figure of **Figure 1**. On the other hand, all parameters of the devices are considered in the differential phase fluctuation calculation process, so the possible errors from device defects such as imperfect splitting ratio or phase difference are removed, and the requirements for the device performance parameters are also relaxed. In our experimental setup, the final setup parameters are $\tau = 244$ ns (corresponding FSR of the MI is 4.1 MHz), and

$$\begin{pmatrix} \eta_1 & \varsigma_1 & \xi_1 \\ \eta_2 & \varsigma_2 & \xi_2 \\ \eta_3 & \varsigma_3 & \xi_3 \end{pmatrix} = \begin{pmatrix} 0.1198 & 0.0002 & 0.1223 \\ -0.0519 & 0.1366 & 0.1491 \\ -0.0634 & -0.1094 & 0.1290 \end{pmatrix} \tag{18}$$

When the setup parameters are calibrated, the whole setup is in a state of plug-and-play.

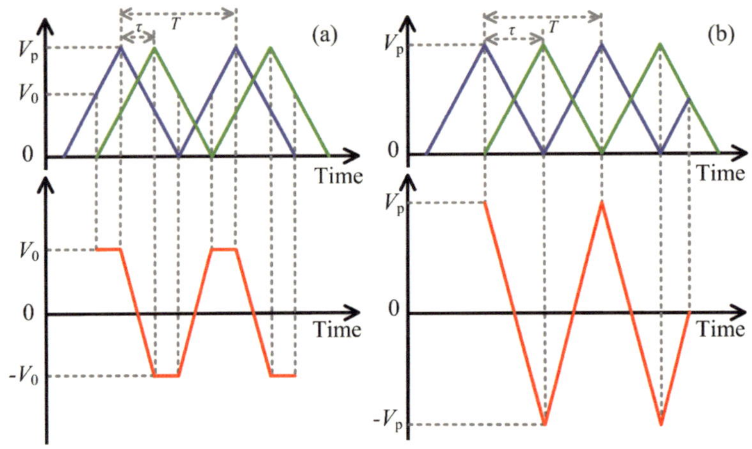

Figure 3. Demodulation principle of the setup for triangle waveforms with different modulation periods (a) $T \neq 2\tau$ and (b) $T = 2\tau$ [18].

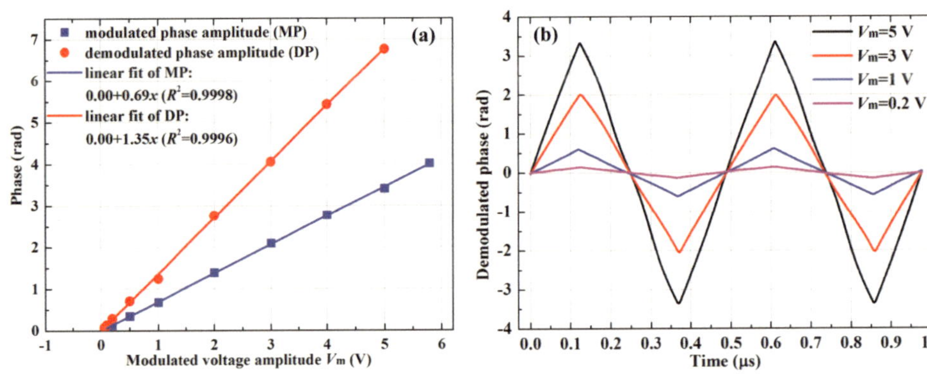

Figure 4. Modulated and demodulated (a) triangle phase amplitude and (b) waveform at different modulated voltage amplitudes [18].

To verify the correction of the phase demodulation of the setup, a demodulated test for a preset modulated phase by a LiNbO3 phase modulator is demonstrated. A narrow linewidth laser that is phase-modulated with the LiNbO3 phase modulator is used as LUT. The triangle waveform is selected to the modulation waveform for holding the frequency components as such and much more. Hence, the fact of the interference at the coupler is a subtraction between two triangle waveforms with delay time τ as shown in **Figure 3**. The modulation period T needs to be twice of the delay time difference τ. Otherwise, there are some constants in the interference fringe as shown in Figure 3(a), accordingly the triangle wave cannot be demodulated. Once the triangle waveform is demodulated correctly as shown in Figure 3(b), the demodulated amplitude would be twice the input modulated amplitude. **Figure 4**. shows the demodulated triangle phase amplitude and their waveforms at different modulated voltage amplitudes. The results confirmed the correction of the differential phase demodulation both in terms of waveform and amplitude. **Figure 5** shows the time-domain interference fringes of the MI and the corresponding demodulated phase waveform at a fixed modulated voltage. The red curve and the black curve represent the results of two independent tests at different time, respectively. Despite the interference fringe changes at different time due to the environment variation, the demodulated phase would not change and hold the consistence with the input phase modulation, so the consistency and robustness of the proposed setup is verified.

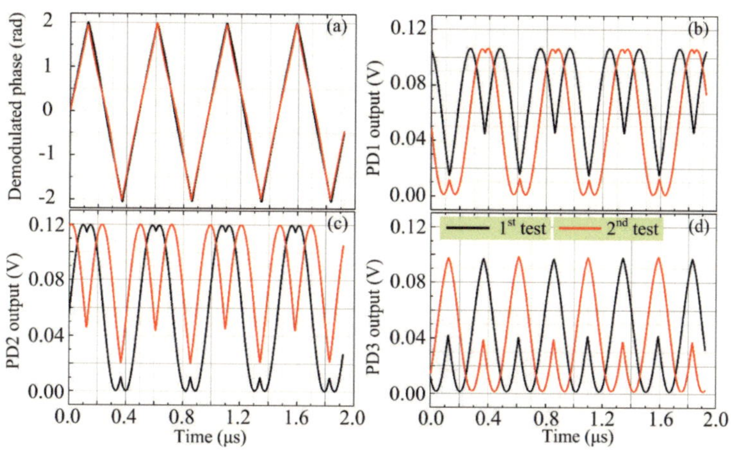

Figure 5. Output voltages of (b) PD1, (c) PD2, (d) PD3, and (a) corresponding demodulated phase waveforms at a fixed modulated voltage $V_m = 3$ V. The red and black lines represent the first and second test, respectively [18].

3.2. Laser noise theory

Considering the relation between the delay phase φ and frequency ν;

$$\varphi = 2\pi\frac{nl\nu}{c} = 2\pi\tau\nu \tag{19}$$

the differential phase variation is from the laser frequency variation in the time interval τ, because the delay time difference τ of the MI is fixed and the random variation of the fiber is eliminated carefully by some techniques as described above. So from the differential phase fluctuation $\Delta\varphi(t)$, the laser frequency fluctuation in time interval τ defined as differential frequency fluctuation $\Delta\nu(t)$ can be expressed as

$$\Delta\nu(t) = \Delta\varphi(t)/(2\pi\tau). \tag{20}$$

Hence, the PSD of differential phase fluctuation $\Delta\varphi(t)$ and differential frequency fluctuation $\Delta\nu(t)$ can be calculated, respectively, in the computer by PSD estimation method [20] and denoted as $S_{\Delta\varphi}(f)$ and $S_{\Delta\nu}(f)$, respectively, where f is the Fourier frequency. Meantime, from the linearity of the Fourier transform, they have a fixed relation

$$S_{\Delta\nu}(f) = \left(\frac{1}{2\pi\tau}\right)^2 S_{\Delta\varphi}(f). \tag{21}$$

So far, the differential phase fluctuation $\Delta\varphi(t)$ accumulated in the delay time difference τ of the MI, corresponding differential frequency fluctuation $\Delta\nu(t)$, and their PSD is calculated. But these values are not the instantaneous information of the LUT. Furthermore, considering the relation of differential phase fluctuation $\Delta\varphi(t)$ and instantaneous phase fluctuation $\varphi(t)$ expressed in Eq. (17), the definition of the PSD [21], linearity, and time-shifting properties of Fourier transform, we derived strictly the PSD of laser instantaneous phase fluctuation and frequency fluctuation, which can be expressed as,

$$S_{\varphi}(f) = \frac{1}{4\left[\sin(\pi f\tau)\right]^2} S_{\Delta\varphi}(f) \tag{22}$$

.

$$S_{\nu}(f) = \frac{f^2}{4\left[\sin(\pi f\tau)\right]^2} S_{\Delta\varphi}(f) = \frac{1}{\left[\text{sinc}(\pi f\tau)\right]^2} S_{\Delta\nu}(f) \tag{23}$$

From Eq. (22), the single-side-band phase noise can also be obtained with [22].

$$L(f) = \frac{1}{2} S_{\varphi}(f) = \frac{1}{8\left[\sin(\pi f\tau)\right]^2} S_{\Delta\varphi}(f) \tag{24}$$

Eq. (22) means that, at the low Fourier frequency domain, the PSD of the laser instantaneous phase fluctuation $S_{\varphi}(f)$ would be larger than the PSD of the differential phase fluctuation $S_{\Delta\varphi}(f)$, but at the high Fourier frequency domain, the former is smaller than the latter. Eq. (23) means that, however, the PSD of the laser instantaneous frequency fluctuation $S_{\nu}(f)$ is larger than the PSD of the differential frequency fluctuation $S_{\Delta\nu}(f)$ at any positive Fourier frequency. On the other hand, it is observed that if the differential phase and frequency fluctuation are normalized in 1 m delay fiber (τ ~ 5 ns), then

$$S_{\nu}(f) \approx S_{\Delta\nu}(f), \quad \text{for } f < 5\,\text{MHz}. \tag{25}$$

http://dx.doi.org/10.5772/66129

Eq. (25) means that the PSD of the laser instantaneous frequency fluctuation $S_\nu(f)$ would approximately equal to the PSD of the differential frequency fluctuation $S_{\Delta\nu}(f)$ at the Fourier frequency less than MHz level. In physics, the results can be so explained that the frequency is the differential of the phase and the delay of the MI is equivalent to the differential operation for the phase. The conclusions are very important that the characterization of differential phase and instantaneous phase (laser phase) should be carefully distinguished, but the instantaneous frequency can be replaced by the differential frequency sometimes in the practical engineering applications.

3.3. Laser noise measurement results

First, the phase and frequency noise of an external-cavity semiconductor laser (RIO ORIONTM) [23] with a wavelength of 1551.7 nm and a linewidth of about 2 kHz are measured. The PSD of the differential phase fluctuation is normalized to 1 m delay fiber ($S_{\Delta\varphi}(f)$ @ 1 m), the PSD of the differential frequency fluctuation is normalized to 1 m delay fiber ($S_{\Delta\nu}(f)$ @ 1 m), and the PSD of instantaneous frequency fluctuation $S_\nu(f)$, the PSD of the instantaneous phase fluctuation $S_\varphi(f)$, and the laser phase noise $L(f)$ are shown in **Figure 6**. The data are very close to that given in the product datasheets or the typical data given in the website [23]. The curves clearly demonstrate the relations between these physical quantities as described above. At the focused frequency range (<1 MHz), $S_{\Delta\nu}(f)$ @ 1 m approximately equals to $S_\nu(f)$, $S_{\Delta\varphi}(f)$ @ 1 m is much less than $S_\varphi(f)$, and laser phase noise $L(f) = S_\varphi(f)/2$. So the usage of PSD of differential phase as the laser phase noise is not strictly correct and needs more careful definition and consideration.

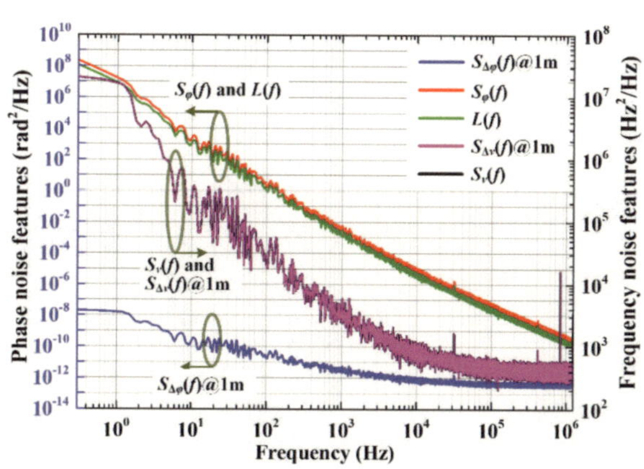

Figure 6. PSD of the differential phase fluctuation ($S_{\Delta\varphi}(f)$ @ 1 m), differential frequency fluctuation ($S_{\Delta\nu}(f)$ @ 1 m), instantaneous phase fluctuation $S_\varphi(f)$, instantaneous frequency fluctuation $S_\nu(f)$ and phase noise $L(f)$ of the RIO laser [18].

From the PSD of the frequency fluctuation $S_\nu(f)$, the linewidth at different observation time can be calculated. **Figure 7** shows the linewidth calculated with the approximated model presented in Ref. [24] for different values of the integration bandwidth. The results indicate that the linewidth is very dependent on the integration bandwidth. Linewidth would increase with the increase of observation time (in other words, linewidth increases with the decrease of the lower limit of the integration bandwidth). It mainly results from the presence of the $1/f^{\alpha}$ type noise in the PSD of frequency fluctuation. In a high-frequency domain (>100 kHz), there is only white noise, and the minimum linewidth of about $\delta v_1 = 2$ kHz is calculated. Meantime, the inset figure shows the linewidth of the same laser measured by the self-delay heterodyne (SDH) method [14] with heterodyne frequency of 80 MHz and optical fiber delay length of 45 km. The Lorentz fitted linewidth at −20 dB from the spectrum measured by the SDH method is about 51.7 kHz, so the laser linewidth is about $\delta v_2 = 2.6$ kHz, and the fitted linewidth would not vary with the observation time. It indicates that the nonwhite noise components are not revealed in the Lorentz fitted results of SDH method, but the values are also conservative for white noise components and about 30% larger than the values calculated by the PSD of frequency fluctuation. This is because the tail of the spectrum measured by SDH method is not taken into consideration in the Lorentz fitting process, resulting in the fitted value being larger than the real value. Therefore, PSD of frequency fluctuation is recommended to completely describe the frequency noise behavior, and a specified linewidth value should be reported with the corresponding integration bandwidth or observation time.

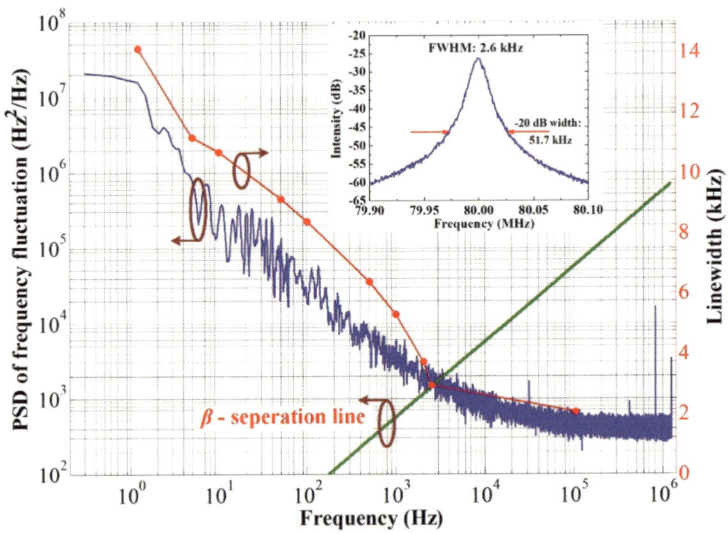

Figure 7. PSD of instantaneous frequency fluctuation of the RIO laser and the β – separation line given by $S_v(f) = 8\ln 2f/\pi^2$ [24] (left axis), laser linewidth (FWHM) obtained by the method in Ref. [24] (right axis) and by the SDH method (inset) [18].

Second, the noise features of commercial distributed feedback (DFB) semiconductor laser are measured under different operating temperatures (24 and 22.7°C) and different operating currents (71, 91, 111, and 136 mA), shown in Figures 8 and 9. **Figure 8** shows that this DFB laser is more suitable for working at 24°C than at 22.7°C. **Figure 9** shows that the PSDs of frequency fluctuation $S_v(f)$ decrease with the increase of operating currents from 71 to 136 mA. According to above results, if we measure the PSDs of frequency fluctuation $S_v(f)$ of the laser under a wide range of operating temperatures and currents, the optimum operating temperature and current can be found out.

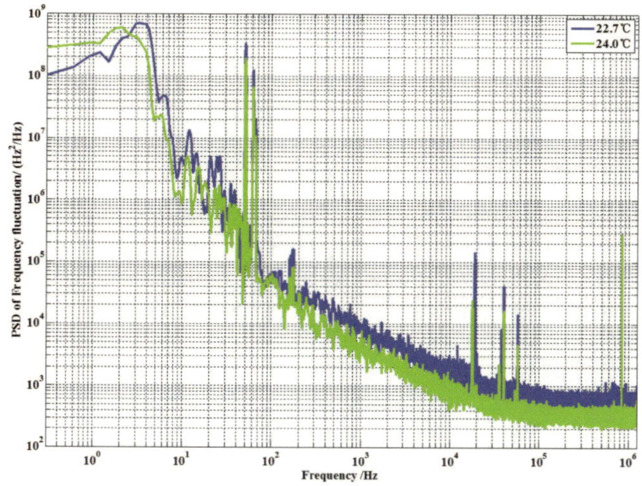

Figure 8. $S_\nu(f)$ of the DFB laser at operating temperature of 22.7 and 24.0°C.

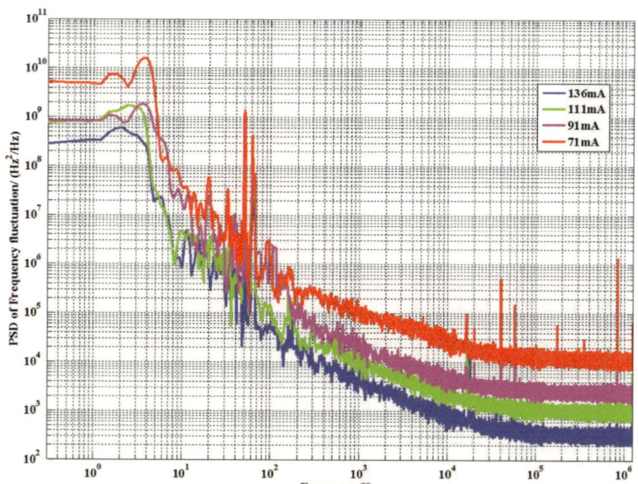

Figure 9. $S_\nu(f)$ of the DFB laser at operating current of 71, 91, 111, 136 mA.

4. The application in phase-sensitive OTDR

The experimental setup is shown in **Figure 10** [25]. The light source is a DFB laser with an output power of 20 mW and a wavelength of 1551.72 nm, which is injected into the acoustic-optic modulator (AOM) to generate the pulses with a width of 40 ns and a repetition rate of 100 kHz. Before being injected into the fibre under test (FUT), the pulses would be amplified by the Erbium-doped fiber amplifier 1 (EDFA1). The backward Rayleigh scattering is amplified by the EDFA2 and then launched into the circulator2, and ASE noise of EDFA2 has been filtered by the fibre Bragg grating (FBG). Then the amplified scattering with better signal-to-noise ratio (SNR) is injected into one port of the 3 × 3 coupler through circulator3. There are two ports on the other side of the 3 × 3 coupler connected to FRMs and the other one port made reflection-free. The interferometer with 4 m delay is housed in a sealed box for thermal and acoustic isolation, avoiding disturbance from the environment. The fiber length between PDs and each coupler port is set as equal to guarantee the same optical path. A trichannel digital acquisition (DAQ) card is used to acquire the voltage signal, and a radio frequency driver (RFD) is used to trigger the AOM and DAQ card simultaneously for synchronization. The collected trichannel signals are processed by the software program to demodulate phase information by using Eq. (16).

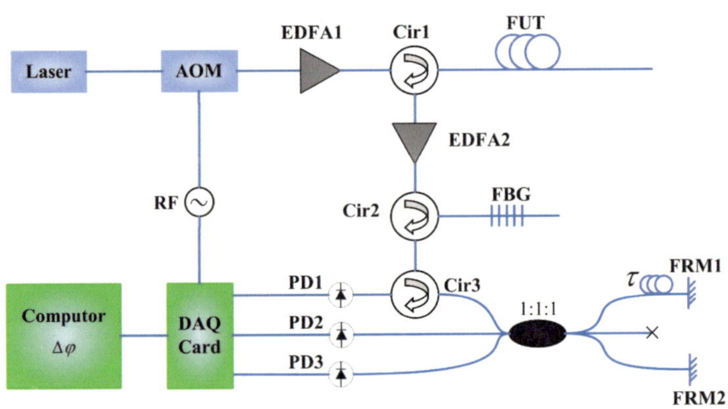

Figure 10. Experimental setup used to demodulate the distributed phase in φ-OTDR [25].

Phase sensitive optical time domain reflectometer (φ-OTDR) based on 120° phase difference Michelson interferometer is a new demodulation scheme used to demodulate the distributed phase. The 120° phase difference interferometer φ-OTDR can detect the phase along a 3 km fiber and the acoustic signal within the whole human hearing range from 20 Hz to 20 kHz is reproduced accurately and fast. The results show that the scheme has high accuracy and real-time response. The acoustic vibration system can be used in audio monitoring, and the health and state monitoring of railway or other structures [25].

5. Conclusion

A laser phase and frequency noise measurement technique based on a 120° phase difference unbalanced Michelson interferometer composed of a 3 × 3 optical fiber coupler and two Faraday rotator mirrors is proposed. In the method, the laser differential phase fluctuation accumulated by the interferometer delay time is demodulated directly at first and then the phase and frequency noise is calculated by the PSD estimation for the differential phase. Also the concepts and differences of differential phase and frequency fluctuation PSDs, instantaneous phase and frequency fluctuation PSDs, and phase noise are defined strictly and discussed carefully. The method can obtain the noise features of a narrow linewidth laser without any specific assumptions or noise models. Meantime, the technique is used to characterize a narrow linewidth external-cavity semiconductor laser, which confirmed the correction of the method and revealed the fact that the linewidth would increase with the increase of observation time, and the Lorentz fitted linewidth measured by the SDH method only includes the contribution of the white noise components and would be larger than the real value. Moreover, the technique can monitor the state change of commercial DFB semiconductor lasers in different processes of designing, installation, debugging, routine test, and final check test, and it offers suggestions to optimize design and improve its performance. The 120° phase difference technology is also applied to test the phase difference between Rayleigh scattering in phase sensitive optical time domain reflectometer (φ-OTDR).

Author details

Yang Fei[1*], Xu Dan[1,2], Cai Haiwen[1] and Qu Ronghui[1]

*Address all correspondence to: fyang@siom.ac.cn

1 Shanghai Institute of Optics and Fine Mechanics, Chinese Academy of Sciences, Shanghai, China

2 LNE-SYRTE, Observatoire de Paris, Paris, France

References

[1] M Trobs, L D'arcio, G Heinzel, et al. Frequency stabilization and actuator characterization of an ytterbium-doped distributed-feedback fiber laser for LISA. J. Opt. Soc. Am. B, 2009, 26(5): 1137–1140.

[2] O Lopez, A Haboucha, F Kéfélian, et al. Cascaded multiplexed optical link on a telecommunication network for frequency dissemination. Opt. Express, 2010, 18(16): 16849–16857.

[3] K Predehl, G Grosche, S M F Raupach, et al. A 920-kilometer optical fiber link for frequency metrology at the 19th decimal place. Science, 2012, 336(6080): 441–444.

[4] M Fujieda, M Kumagai, S Nagano, et al. All-optical link for direct comparison of distant optical clocks. Opt. Express, 2011, 19(17): 16498–16507.

[5] N Chiodo, K Djerroud, O Acef, et al. Lasers for coherent optical satellite links with large dynamics. Appl. Opt., 2013, 52(30): 7342–7351.

[6] F Lienhart, S Boussen, O Carraz, et al. Compact and robust laser system for rubidium laser cooling based on the frequency doubling of a fiber bench at 1560 nm. Appl. Phys. B-Lasers O., 2007, 89(2–3): 177–180.

[7] G A Cranch, G M H Flockhart, C K Kirkendall. Distributed feedback fiber laser strain sensors. IEEE Sens. J., 2008, 8(7–8): 1161–1172.

[8] J P Cariou, B Augere, M Valla. Laser source requirements for coherent lidars based on fiber technology. C. R. Phys., 2006, 7(2): 213–223.

[9] F Yang, Q Ye, Z Q Pan, et al. 100-mW linear polarization single-frequency all-fiber seed laser for coherent Doppler lidar application. Opt. Commun., 2012, 285(2): 149–152.

[10] H Lee, M G Suh, T Chen, et al. Spiral resonators for on-chip laser frequency stabilization. Nat. Commun., 2013, 4: 1–6.

[11] H Tsuchida. Laser frequency modulation noise measurement by recirculating delayed self-heterodyne method. Opt. Lett., 2011, 36(5): 681–683.

[12] T Okoshi, K Kikuchi, A Nakayama. Novel method for high resolution measurement of laser output spectrum. Electron. Lett., 1980, 16(16): 630–631.

[13] O Llopis, P H Merrer, H Brahimi, et al. Phase noise measurement of a narrow linewidth CW laser using delay line approaches. Opt. Lett., 2011, 36(14): 2713–2715.

[14] S Piazzolla, P Spano, M Tamburrini. Characterization of phase noise in semiconductor lasers. Appl. Phys. Lett., 1982, 41(8): 695–696.

[15] Y Takushima, H Y Choi, Y C Chung. Measurement of differential phasor diagram of multilevel DPSK signals by using an adjustment-free delay interferometer composed of a 3 × 3 optical coupler. J. Lightwave Technol., 2009, 27(6): 718–730.

[16] S K Sheem. Optical fiber interferometers with [3 × 3] directional couplers: Analysis. J. Appl. Phys., 1981, 52(6): 3865–3872.

[17] T Butler, S Slepneva, B O'shaughnessy, et al. Single shot, time-resolved measurement of the coherence properties of OCT swept source lasers. Opt. Lett., 2015, 40(10): 2277–2280.

[18] D Xu, F Yang, D J Chen, et al. Laser phase and frequency noise measurement by Michelson interferometer composed of a 3 × 3 optical fiber coupler. Opt. Express, 2015, 23(17): 22386–22393.

[19] F Wei, B Lu, J Wang, et al. Precision and broadband frequency swept laser source based on high-order modulation-sideband injection-locking. Opt. Express, 2015, 23(4): 4970–4980.

[20] J Rutman. Characterization of phase and frequency instabilities in precision frequency sources: fifteen years of progress. Proc. IEEE, 1978, 66(9): 1048–1075.

[21] P D Welch. The use of fast Fourier transform for the estimation of power spectra: a method based on time averaging over short, modified periodograms. IEEE Trans. Audio Electro., 1967, 15(2): 70–73.

[22] R P Scott, C Langrock, B H Kolner. High-dynamic-range laser amplitude and phase noise measurement techniques. IEEE J. Sel. Top. Quant., 2001, 7(4): 641–655.

[23] Rio Orion™ Laser Module. http://www.rio-inc.com/_products/orion.html.

[24] G D Domenico, S Schilt, P Thomann. Simple approach to the relation between laser frequency noise and laser line shape. Appl. Opt., 2010, 49(25): 4801–4807.

[25] Y L Cao, Y Fei, D Xu, et al. Phase sensitive OTDR based on 120° phase difference Michelson interferometer. Chin. Phy. Lett., 2016, 33(05): 050701.